Falling Short: The Limits of Traditional Conservation and the Need for New Approaches

Jahinson

First Printing, 2024

Table of Contents

Chapter 1

Introduction

1.1. Introduction

In 2011, the 2020 Aichi Targets set out to conserve 17% of terrestrial and inland water areas, and 10% of coastal and marine areas (CBD, 2011). These targets emerge from ongoing efforts within "the global conservation regime", which originate with *The Convention on Biological Diversity of 1992* (see Adams, 2004a). Recent estimates of 15% of terrestrial and 7% of coastal and marine areas under global protected area (PA) coverage attest to much progress, though the effectiveness of these PAs is questioned (UNEP-WCMC et al., 2020). Furthermore, a substantial proportion of the world's wild flora and fauna remain completely outside conventional PA boundaries, with the conservation importance of Indigenous-lands especially and increasingly recognized (Butchart et al., 2015; Garnett et al., 2018; Jones et al., 2018b; Schuster et al., 2019). Additionally, inadequate management of many conventional 'people-free' terrestrial and marine PAs, means insufficient protection of many species endures (Watson et al., 2014; Gill et al., 2017; Jones et al., 2018a&b; Coad et al., 2019). Therefore, despite widespread and concerted efforts to conserve biodiversity, human-triggered global biodiversity loss persists at an increasing and concerning rate, and requires change (Jones et al., 2018a&b; Diaz et al., 2019; Habel et al., 2019).

Game et al. (2014: p271) suggest, "Conservation is not rocket science; it is far more complex." This is because while natural and social systems are themselves complex, contemporary 'wicked' conservation problems involve greater complexity due to interactions within these interconnected social-ecological systems (SESs) (Game et al., 2014; Berkes et al., 2016; Colding & Barthel, 2019). Accordingly, contemporary wickedly complex conservation is forced to grapple with the diverse objectives of biodiversity protection and human development

(Thiault et al., 2018; Sarkki & Acosta García, 2019). Furthermore, increasing recognition of this complexity has led some to suggest conventional 'people-free' conservation approaches and institutional structures may better suit simpler systems (e.g. Game et al., 2014). Therefore, scholars note that while "ecological perspectives are vital, nature protection is a complex social enterprise" (Brechin et al., 2003: px), and propose that conservation might better be considered "primarily not about biology but about people and the choices they make" (Balmford & Cowling, 2006: p692). Consequently, as Shove and Walker (2007: p763) note:

> "Across the board there is growing recognition of the holistic, unavoidably inter-related nature of contemporary environmental problems and of the need for fresh approaches and forms of governance capable of engaging with complex challenges of this kind."

Therefore, conservation requires flexible, effective and equitable conservation governance arrangements (Gavin et al., 2018; Jones et al., 2019a). More specifically, calls exist for more nuanced, holistic and people-centred conservation governance. *Community-Based Conservation* (CBC) – which seeks to include local communities and their knowledge and priorities in conservation governance to promote 'pro-conservation' behaviour (Nilsson et al., 2016) – is one potentially viable, though context-specific approach (introduced further in *section 1.2.2.*).

Returning to the Aichi Targets, *Target 11* specifically promotes increased conservation coverage, "through effectively and equitably managed, ecologically representative and well-connected systems of protected areas and other effective area-based conservation measures" (CBD, 2011). More specifically, the importance of "other effective area-based measures" is increasingly recognized (e.g. Diz et al., 2018; Dudley et al., 2018; Schuster et al., 2019; Donald et al., 2019). *Community-Conserved Areas* (CCAs), which I use here to include any *community-owned, -managed* or *-conserved* PA, represent an "other effective area-based measure". Nevertheless, the extent to which these 'measures' may contribute to biodiversity conservation remains disputed (see Lemieux et al., 2019). More specifically, the question of whether conservation governance can be entrusted to communities, and achieve both ecological and socio-economic objectives, remains 'hotly' contested (Holt, 2005; Ostrom & Nagendra, 2006; Terborgh & Peres, 2017).

Undoubtedly, CBC implementation and governance has proven problematic, with mixed results emerging from global CBC reviews (e.g. Brooks et al., 2013), and mirrored in their African-specific counterparts (e.g. Galvin et al., 2018). However, as Murphree (2000: p12) stated at the time, specifically in reference to southern African CBC, "[CBC] has to date not been tried and found wanting; it has been found difficult and rarely tried!" Unfortunately, this remains symptomatic of South African CCA implementation efforts to date. Whilst some progress exists with wildlife CCAs, to date no forestry or coastal CCAs exist. This is particularly noteworthy considering the country's progressive and enabling legislation for CCA implementation and governance (introduced in *Chapter 4*). Accordingly, South African CCA implementation, most notably within the coastal realm, is experiencing a 'policy-praxis disjuncture'. Therefore, greater understanding of *how to change* toward a CBC mode of governance when contextually appropriate is required. Consequently, this book explores:

> *What factors, conditions and processes are required to facilitate a shift toward CBC initiation, implementation and governance in South Africa, when contextually appropriate, so as to realize desired social and ecological outcomes?*

Commons Theory is a field dedicated to the study of enabling factors and conditions for *Common Property Resource Management* (CPRM) (Dietz et al., 2003; van Laerhoven & Ostrom, 2007), and is considered highly applicable to addressing the above question. More specifically, the influence of the principles and factors for successful CPRM, as described by Ostrom (1990), Pomeroy et al. (2001), and Agrawal (2002) (introduced in *Chapter 3*), are widely recognized and used to assess the viability and success of CBC initiatives. However, whilst commons research has positively influenced CBC initiatives, this 'approach' may not be sufficient to facilitate a *shift* toward a community-based mode of governance, and greater understanding (and potentially success) may be obtained by exploring the aforementioned South African coastal CCA policy-praxis disjuncture as a change process.

Consequently, this chapter begins the 'journey' toward greater understanding of this change process, firstly by briefly introducing and better defining *Community-Based Conservation* (CBC) within the context of the prevailing "new conservation debate". Secondly, it discusses the *research rationale*, and thereafter stipulates the dissertation's specific *aim* and *objectives*. Lastly, the chapter culminates with a 'Research Roadmap' to orientate the reader through the subsequent chapters.

1.2. Background

1.2.1. The 'New Conservation' Debate

Whilst acknowledging the "conservation tree" consists of numerous "branches", Sandbrook (2015: p565) proposes a broad definition of conservation as "actions that are intended to establish, improve or maintain good relations with nature." Accordingly, he emphasizes a concept central to this book; that conservation is as an 'action-dependent' process (Sandbrook, 2015). Whilst acknowledging not all conservation actions successfully reach their desired outcomes, biodiversity loss and conservation are considered here, and widely within the literature, as a result of actions undertaken (e.g. Brooks et al., 2009; Leader-Williams et al., 2011; Game et al., 2014; Sandbrook, 2015; Montoya et al., 2018). Given the aforementioned recognition of human-triggered biodiversity loss (Jones et al., 2018; Coad et al., 2019), the importance of identifying deliberate and goal-centred actions that result in context-appropriate governance arrangements, better able to produce positive social and ecological conservation outcomes, is recognized (and discussed further in *Chapters 3 & 5*).

Plural views of conservation have led to an intense "new conservation debate" over the 'best' approach to conservation management (Miller, et al., 2011; Soulé, 2013; Marvier, 2014; Holmes et al., 2016). The two poles of this debate are notably "traditional conservation" (i.e. *biocentric* – focused on nature), and "new conservation" (i.e. *anthropocentric* – focused on human development) (Holmes et al., 2016). While the term "traditional conservation" is used by Holmes et al. (2016), and others within the 'new conservation' debate, it should be acknowledged that it is perhaps better to refer to "conventional conservation" or "top-down conservation" to avoid confusion with the concepts of 'traditions' or 'traditional communities' as used to describe customary aspects of specific ethnic groups. Therefore, hereafter I use the term "conventional conservation" to refer to this more biocentric approach to conservation.

Consequently, this new conservation debate – though perhaps limited in its representation of diverse and contentious conservation perspectives (Holmes et al., 2016; Sandbrook et al., 2019) – essentially comprises two sets of advocates. Firstly, the "natural protectionists" who strongly advocate a 'strict' and 'people-free' approach to PAs (e.g. Oates, 1995, 2006; Terborgh, 1999, 2004; Soulé, 2013; Terborgh & Peres, 2017). Secondly, the "social conservationists" who support various forms of sustainable use, and collective conservation- and welfare-oriented development approaches, with associated elements of poverty alleviation and social justice (e.g. Western & Wright, 1994; Brechin et al., 2003; West et al., 2006; Brockington et al., 2010). Therefore, in accordance with the 'new conservation' narrative, CBC emphasizes "the coexistence of people and nature, as distinct from protectionism and the segregation of people and nature" (Western & Wright, 1994: p8). The subsequent section introduces CBC more fully in.

1.2.2. Community-Based Conservation (CBC)
1.2.2.1. What is CBC?

Limited positive social and ecological conservation outcomes following decades of the aforementioned "conventional conservation" approach, has placed specific emphasis on the effects of human displacement *for* conservation, and the need for a greater understanding of our 'modern' engagement with nature (Neumann, 2004; Agrawal & Redford, 2009). Accordingly, the above limitations of protectionist conservation approaches have led policy makers and scholars to re-examine the role of 'community' in conservation (Agrawal & Gibson, 1999; Waylen et al., 2010; Brooks et al., 2012).

CBC operates under the premise of the *subsidiarity principle*, which dictates the lowest possible organizational level should possess governing responsibility (see Schäfer, 1991), since it assumes being closer to the issue increases one's ability to influence it (Pressman & Wildavsky, 1983). More specifically, CBC is grounded in the notion that community engagement in policy and management decisions, and community ownership over natural resources, can contribute to alleviation of poverty, improve social cohesion, increase 'pro-conservation' mindsets and behaviour, and subsequently reduce threats to biodiversity (Schultz, 2011; Clayton et al., 2013; Nilsson et al., 2016; Brooks, 2016). Consequently, CBC represents, "natural resources or biodiversity protection *by*, *for*, and *with* the local community" (Western & Wright, 1994: p7 – *emphasis* added).

CBC is multifaceted and used within this book as an overarching term collectively representing diverse conservation strategies involving communities to varying degrees in conservation management activities, including amongst others: planning, decision-making, monitoring and evaluation, and enforcement. Therefore, CBC may best be represented on a continuum comprising various governance arrangements with varying degrees of community powers and responsibilities and levels of external support in conservation management, including collaborative governance or co-management with partner organizations (Borrini-Feyerabend et al., 2013). The term partner organization is used here to refer to any organization outside of the local community involved in, and working together with other actors in a CBC initiative, and is inclusive of State departments,

NGOs, and other civil society and private sector partners. To clarify, NGOs refer to a broad category of organizations that operate neither for profit or a State department or organization. More specifically, I use the term NGO throughout to refer predominantly to non-governmental development organizations who are commonly involved in addressing poverty, human rights and environmental concerns (*Cf.* Fowler, 1997). This is inclusive of *Northern* (i.e. those from developed countries) and *Southern* NGOs (i.e. those from developing countries) (see Lewis, 2004). Moreover, at times I refer specifically to *Big International Non-Governmental Organizations* (BINGOs), which are international conservation NGOs largely based in developed countries of the western hemisphere, who work in various countries and regions, and possess substantial financial capital and public and policy influence (*Cf.* Davies et al., 2018).

Consequently, CBC is used in this book to incorporate various 'community-centred' modes of governance, and includes initiatives such as *Integrated Conservation & Development Projects* (ICDPs), *Indigenous Peoples' and Local Community Conserved Territories and Areas* (ICCAs), *Community-Conserved Areas* (CCAs), *Locally Managed Marine Areas* (LMMAs) and *Community-Based Natural Resource Management* (CBNRM - i.e. the term most often used in southern Africa). A brief history of the development of CBC follows.

1.2.2.2. A brief history of CBC

CBC developed largely in reaction to the aforementioned perceived 'failures' of "conventional conservation" to account for complex social dimensions. Protectionist PAs date back to the late nineteenth century (Runte 1987; 1990), and became the *de facto* model for many subsequent PAs (Nash, 1967; Igoe, 2005).

However, sacred groves and ancient royal forests, among many other examples of customary conservation governance, have long provided comparable protection for 'nature' (Mulder & Coppolillo, 2005; Dudley et al., 2009). Accordingly, 'commons research' has long advocated the potential of local communities to self-organise and under certain conditions sustainably use and manage their natural resources (e.g. Agrawal, 1999; Ostrom, 1992). However, as Berkes (2007) notes, critics of CBC often lack a foundational understanding of these local systems of governance. Nevertheless, not all resource-users protect their common property resources (CPRs) successfully, and outcomes of such systems are highly variable. All the same, the "new conservation" narrative promotes CBC as able to efficiently manage natural resources under certain conditions.

CBC experienced several distinctive development phases (Salafsky & Wollenberg, 2000; Redmore et al., 2018). In the 1980s attempts to integrate the role and interests of rural people in conservation, and its institutionalization, began to attract wider support (see Hutton et al., 2005). By the 1990s, the previously dominant 'fortress conservation' narrative – i.e. the forceful exclusion of local communities from PAs (see Brockington & Igoe, 2006) – no longer enjoyed supremacy either globally or in Africa (Murphree, 2002; Hutton et al., 2005). However, many preliminary 'community-involved' conservation projects were considered poorly conceived, and to have retained a protectionist PA foundation (Wells & Brandon, 1992, Kepe, 2009).

CBC initiatives aimed to connect biodiversity and livelihoods, and thus 'close the loop' and provide conservation impetus (Salafsky & Wollenberg, 2000; Hutton & Leader-Williams,. 2003). However, this conflicted with the "conventional conservation" narrative (Sanderson & Redford, 2003; Brechin et al., 2003). Resultant tensions strained relations between local communities, conservationists, donors, and governmental and non-governmental conservation organizations, resulting in what some described as the creation and institutionalization of "major political disjunctures in the intent and ideal of [CBC]" (Dressler et al., 2010: p7). Nonetheless, the emergence of the CBC narrative managed to produce a significant injection of funds enabling experimentation with 'community-involved' conservation approaches (Roe et al., 2000). Not surprisingly, the last two decades of the twentieth century observed CBC largely dominating the conservation debate, especially in the rural developing world (Hackel, 1999; Berkes, 2007). Therefore, like protectionist PAs before it, CBC threatened to become a *panacea*, (i.e. 'blueprint' approach) (Berkes, 2007). However, many scholars cited mixed results emerging from evaluations of CBC initiatives to promote a 'back-to-barriers' approach, reasserting people-free PAs as the main approach to biodiversity conservation (see Hutton et al., 2005). Consequently, whilst some scholars have long described a sound logic underlying CBC (Wells & Brandon, 1992; Agrawal & Gibson, 1999; Larson, 2003), many agree that early CBC-associated 'enthusiasm' has largely diminished. Therefore, greater understanding of the enabling factors, conditions and processes is required to facilitate CBC initiation, implementation and governance, and remains a key research topic. Accordingly, the *research rationale* of this book is discussed next.

1.3. Research Rationale

Like its global counterpart African CBC literature often focusses on reasons for failure and not success (e.g. Kellert et al., 2000; Blaikie, 2006; Zulu, 2008; Singleton, 2009). The focus of this research tends to be on reasons for policy and implementation failures (Child & Barnes, 2010; Nelson, 2012; Zulu, 2012; Child, 2019), including a failure to deliver tangible community-wide benefits (Child & Barnes, 2010; Zulu, 2012; Galvin et al., 2018; Child, 2019), and a high frequency of collapse (Balint & Mashinya, 2006; Child & Barnes, 2010). Yet, global CBC 'success' stories do exist and include, in the present coastal context, for example progress made by LMMAs in the Pacific Islands (Govan et al., 2009; Levine & Richmond, 2014). Furthermore, recent reviews of African CBC initiatives also include positive outcomes (Galvin et al., 2018). Notable 'positive' examples of regional CBC initiatives include ever-increasing networks of Namibian community wildlife conservancies (Boudreaux & Nelson, 2011; Mufune, 2015), and LMMAs in Madagascar (Harris, 2011; Oliver et al., 2015; Brenier & Vogel, 2017). Nevertheless, in all cases challenges remain, largely due to changes to CBC practices caused by market forces, recentralized control, and erosion of customary institutions (Cinner & Aswani, 2007; Jones et al., 2008; Brooks & Tshering, 2010; Fernández-Llamazares et al., 2018; Hebinck et al., 2020). Therefore, whilst a wide array of concerns provides 'fodder' for CBC's perceived 'crisis' of identity and purpose, optimism remains amongst many scholars regarding its potential effectiveness (e.g. Dressler et al., 2010; Mulrennan et al., 2012; Brooks, 2016; Galvin et al., 2018).

Therefore, scholars argue that CBC initiatives can succeed if implemented properly, and allowed enough time to work (Ribot, 2004; Lund & Treue, 2008).

However, various 'so-called' CBC initiatives have been described as "half-hearted, misdirected, and theory-ignorant" (Berkes, 2007: p15191). Accordingly, some authors suggest that a key obstacle to successful CBC implementation and governance is a lack of sufficient empirical evidence to inform improved practice (e.g. Geldman et al., 2013; Oldekop et al., 2016). Empirical evidence is particularly crucial for understanding the factors, conditions and processes that lead to successful context-specific conservation governance. Governance focuses on *who* has *power*, *responsibility*, and *accountability* in decision-making and implementation of actions (Jentoft, 2007a; Borrini-Feyerabend & Hill, 2015 – discussed further in *Chapter 3*). Consequently, governance is an important determinant for positive conservation outcomes, such as those summarized by Borrini-Feyerabend et al. (2013: pxii), which are depicted in *Table 1.1.* That said, attempts to produce a 'best system' for conservation governance should be considered both unattainable and unrealistic, and may even constrain conservation efforts (Stern et al., 2002; Game et al., 2014; Bennett & Satterfield, 2018).

Table 1.1.: The importance of governance to improved conservation outcomes.
Source: Borrini-Feyerabend et al. (2013: pxii)

Governance...
▪ is a main factor in determining the effectiveness and efficiency of management
▪ is the variable with greatest potential to affect coverage to meet Aichi Biodiversity Target 11 of the CBD Strategic Plan 2011-2020
▪ is a determinant of appropriateness and equity of decisions
▪ can help to maximise the ecological, social, economic and cultural benefits of protected areas without incurring unnecessary costs or causing harm, if improved
▪ can ensure that protected areas are better embedded in society
▪ can be improved and provide precious help in facing on-going challenges and global change

Whilst greater community involvement is arguably necessary for both positive social and ecological outcomes in contemporary conservation initiatives, in practice CBC initiatives involve 'nested' multi-actor collaborative governance arrangements comprising, in addition to local communities, State departments, non-governmental organizations (NGOs), the private sector and civil society partners (Ostrom et al., 2002; Seixas & Berkes, 2010; Baird et al., 2019a&b). The term *actors* is specifically used throughout this book to include any individuals, groups or organizations likely to *affect* or *be affected* by a conservation initiative, which better captures the complex reality of 'nested' multi-actor and multilevel conservation governance arrangements (Armitage et al., 2012; Schultz et al., 2015). Furthermore, the use of the concept of 'nesting' here seeks to emphasize the importance of combining higher- (e.g. State departments) and lower-level (e.g. local community) institutions for more 'robust' governance required to address 'wicked' conservation problems (Marshall, 2008).

Some scholars have noted that whilst borrowing concepts from other fields will not solve all wicked conservation problems, conservation policy and practice may benefit from this, as it "broadens our range of options" (Game et al., 2014: p275). Accordingly, this book draws on a diverse array of literature, most notably *Commons Theory, Governance Theory,* and *Theory of Change* (ToC), which are introduced in *Chapters 3* and *5*. Accordingly, I propose that greater understanding may be obtained by building upon the foundations of commons research to explore *how* CBC initiation (or planning), implementation and governance takes place as a change process. More specifically, this book seeks to contribute to addressing the aforementioned South African coastal CCA policy-praxis disjuncture, by drawing on the ToC approach to better understand *how* initiating CBC involves a process of change from one mode of governance to another. Consequently, this book strives to provide greater theoretical and practical empirical evidence on *how* to facilitate initiation, implementation and ongoing governance of coastal CCAs in South Africa, to address the current policy-praxis disjuncture, and realize desired social and ecological outcomes *when* contextually appropriate. Accordingly, the **aim** and **objectives** of the book are stipulated below.

1.4. Aim & Objectives

The **aim** of this book is:

To identify and explore the factors, conditions and processes that enable change to a CBC mode of governance in South Africa, with a view to better aligning law & praxis, so as to promote positive social and ecological conservation outcomes.

Furthermore, the **objectives** of the study are:

1. To conduct an extensive review of CBC literature with particular attention given to developing countries and focusing on the enabling factors, conditions and processes for shifting to a community-based mode of conservation governance;
2. To review progress with CBC in South Africa, within the context of national conservation legislation, and identify the current enabling and constraining factors, conditions and processes to its implementation;
3. To draw on theoretical ideas from Governance Theory, Commons Theory, and Theory of Change to develop a *Generic Theory of Change Pathway* that offers a theoretical understanding of factors, conditions and processes that enable change towards a community-based mode of conservation governance in the developing world;

4. To investigate the factors, conditions and processes that enabled and constrained CBC in two regional case studies to learn lessons for South African CBC initiation, implementation and governance;

5. To explore a South African coastal CBC *case-in-progress*, to better understand the factors, conditions and processes enabling and constraining its implementation;

6. To propose an *Empirical Theory of Change Pathway* for CBC, based on the empirical findings of this study, and thus provide recommendations for initiating, planning and implementing CBC governance in South Africa;

7. To contribute to the theory and practice of CBC

1.5. A Research Roadmap

In summation, this book strives to investigate and analyse the policy-praxis disjuncture in South Africa with respect to implementation and governance of CBC in coastal areas by: (a) developing and appraising a set of key factors, conditions and processes that enable its initiation, implementation, and governance, (b) proposing a *South African Empirical CBC ToC Pathway*, so as to make recommendations for future conservation planning and management from a more *people-centred* CBC perspective, and (c) thereby contribute to CBC theory and practice not only in South Africa, but regionally and globally. To achieve this, I organize this book according to the *Research Roadmap* presented in *Figure 1.1.* A more comprehensive discussion of the research phases as they pertain to subsequent chapters follows in *Chapter 2.*

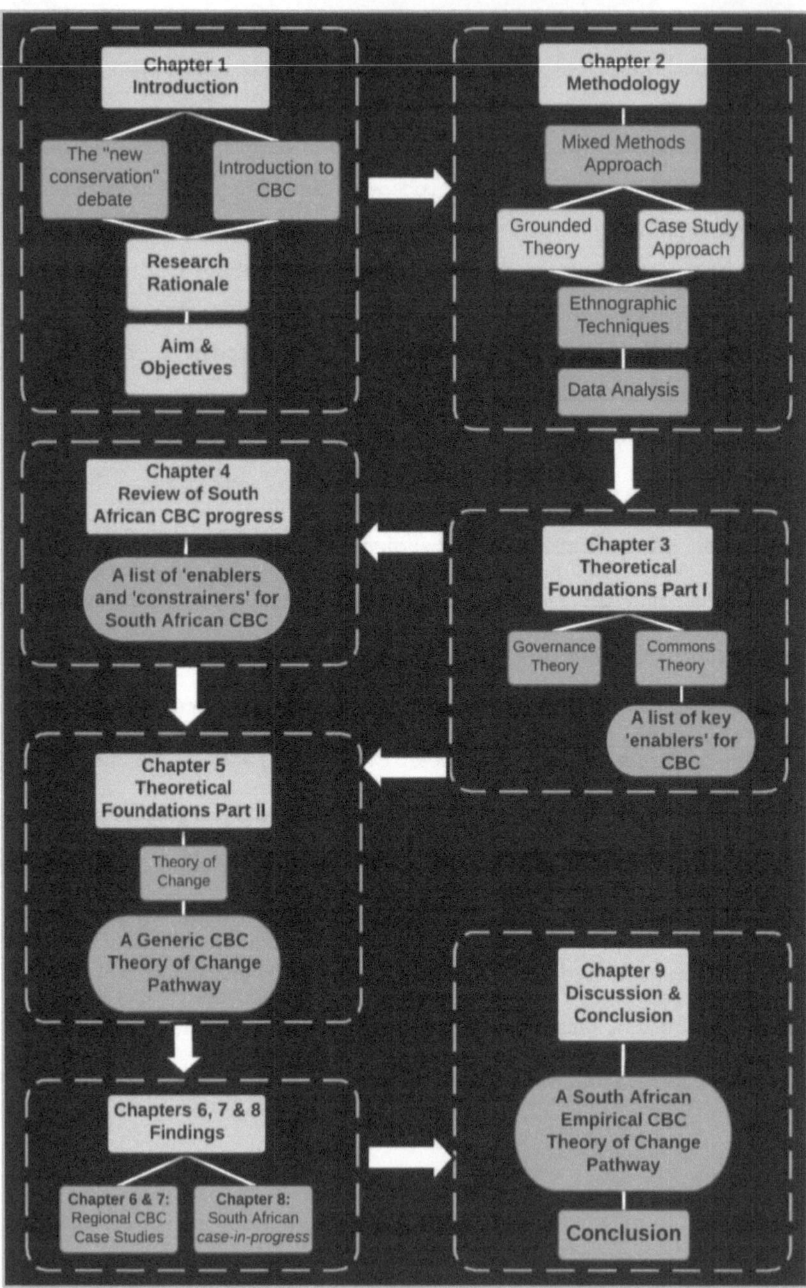

Figure 1.1.: A 'Research Roadmap'.

Methodology

2.1. Research Approach and Methods

This chapter introduces the dissertation's methodological foundation. Following a brief overview of the research approach, I introduce the five research phases and their associated methods. Thereafter, follows the case study selection and description. The chapter culminates with the dissertation's limitations and ethical procedures.

2.1.1. An Overview

In an effort to improve both the value and validity of results, this book makes

use of *mixed methods* (Johnson et al., 2007) and *triangulation* (Denzin, 1978; Jick, 1979). *Triangulation* reduces the risk of drawing false conclusions from unreliable data by cross-checking information from different sources (Jick, 1979). Furthermore, this approach enables greater inter- and trans-disciplinarity required to investigate complex environmental issues from different perspectives, specifically within small-scale fisheries contexts (Balmford & Cowling, 2006; Chuenpagdee & Jentoft, 2019). Such an approach is particularly useful since it promotes, "deepening and widening one's understanding" (Olsen, 2004: p1). However, promoting a shift toward increased 'interdisciplinarity' in conservation research is not easy (Balmford & Cowling, 2006). Nevertheless, it proved beneficial to this book by better capturing diverse sources of knowledge and practice, than a natural science approach could alone, and therefore, more robust insights into designing holistic, socially acceptable, and ecological appropriate conservation interventions.

A *classical grounded theory approach* formed the foundation of this 'interdisciplinarity', and specifically informed data collection and interpretation, and subsequently theory generation (Glaser & Strauss, 1967). This approach emphasizes both the emergence of theoretical categories from evidence, and an integrated and incremental approach to data gathering and case selection, to address *what*, *how* and *why* questions found within complex social settings (Glaser, 1992, 2002; Charmaz, 2008). Accordingly, this book strives for 'ground-truthing' of data through 'theoretical saturation', which is reached when further data collection no longer provides any substantial changes in findings. This specifically increases confidence in developing 'robust' *Theory of Change* (ToC) *Pathways* facilitating CBC initiatives.

A case study approach was employed to better understand the dynamics within a focused setting (Flyvbjerg, 2006). In accordance with Flyvbjerg (2006: p223) this book strongly argues that "the closeness of the case study to real-life situations and its multiple wealth of details", are important "for the development of a nuanced view of reality." Furthermore, understanding historical institutional development processes is highly relevant to all case studies presented (*Cf.* Tool, 1979), and provides greater insights required to develop 'robust' ToC Pathways.

Whilst quantitative methods focus more on the effect, they are limited in their exploration of the underlying mechanisms leading to an effect (Salazar et al., 2018). In contrast, qualitative approaches provide greater insight into underlying mechanisms, and of specific relevance here, are considered to provide, "a more nuanced understanding of complex conservation issues" (Salazar et al., 2018: p635). More specifically, qualitative social science methods have been shown crucial to greater understanding of conservation decision-making (Young et al., 2018; Moon et al., 2019a). This includes the rationale *for,* and the way conservation decisions *should be/are* made (Moon et al., 2019a). Moreover, these methods can lead to greater understanding of different conservation perspectives by exploring *why* conservation policies are not always implemented as intended; and *how* and *why* conservation actor's behaviours are expected to change (Moon et al., 2019a: p300). While qualitative methods and data offer much to the research of wicked complex conservation problems, challenges to their use are acknowledged (see Alexander et al., 2019a). Nevertheless, I consider qualitative social science methods especially valuable to understanding CBC implementation and governance as a change process.

This book employs methods thought best to address the specific research aim and objectives. Accordingly, ethnographic techniques, including semi-structured interviews, informal focus groups, and participant observation, form the foundation of data collection and analysis. Ethnography comprises two core activities, firstly, "first-hand participation in some initially unfamiliar social world," and secondly, "production of written accounts of that world" based on such participation (Emerson et al., 1995: p1). In doing so this book endeavours to incorporate diverse knowledge production systems – inclusive of *Western/ Scientific* and *Local/ Indigenous/ Traditional/ Ecological knowledge* (LEK) – which are increasingly considered beneficial to navigating environmental problems (Tengö et al., 2014; Alexander et al., 2019b). Data collection and analysis from the above ethnographic techniques benefitted specifically from, and incorporated modified versions of other research tools namely: *Social Network Analysis* (SNA - Wasserman & Faust, 1994), the *Most Significant Change* approach (MSC - Davies & Dart, 2005), and the *Strength, Weakness, Opportunities, Threats* analysis (SWOT - Weihrich, 1982). Lastly, reviewed secondary data such as technical country reports, government documents, and archive material contributed important information to case study investigations. Data collection and analysis is described in detail in the subsequent section.

2.1.2. The Five Research Phases

The research process comprised five phases (*Table 2.1.*). The *first phase* involved conducting an extensive review of published and unpublished literature on global developing nation CBC initiatives to identify a set of key *enablers* (i.e. *enabling* factors and conditions for CBC implementation and governance – presented in *Chapter 3*), and common *change elements* in the *CBC change process* (introduced in *Chapter 5*). This review of extant CBC literature also led to the development of a 'prior' *Generic CBC ToC Pathway* presented in *Chapter 5*.

Table 2.1.: An overview of the various research phases and associated methods.

Research Phase	Methods	Time Period
1) Developing of a list of CBC 'enablers', a *ToC design framework* & a *Generic CBC ToC Pathway*	1. Extensive review of published and unpublished literature 2. Theory of Change	• Ongoing between July 2015 and July 2019
2) Investigating South African CBC implementation and governance progress	1. Review of South African CBC Enabling Legislation 2. Extensive review of published and unpublished literature 3. Semi-Structured Interviews	• Ongoing between July 2016 and July 2019
3) Investigating two African coastal CCA case studies in Madagascar and Guinea-Bissau respectively	1. Semi-Structured Interviews 2. Informal focus groups 3. Social Relational Network Appraisal 4. Participant Observation 5. Secondary socio-economic and ecological data (when available) 6. Extensive review of published and unpublished literature	• Literature and secondary data review ongoing between July 2016 and July 2019 • Madagascar fieldwork: October-November 2016 • Guinea-Bissau fieldwork: February-March 2018
4) Investigating a South African CCA *'case-in-progress'*	1. Semi-Structured Interviews 2. Informal focus groups 3. Social Relational Network Appraisal 4. Participant Observation 5. Secondary socio-economic and ecological data (when available) 6. Extensive review of published and unpublished literature	• Literature and secondary data review ongoing between January 2018 and July 2019 • Fieldwork: August-September 2018
5) Developing an *Empirical CBC ToC Pathway*	1. Theory of Change 2. Case study analysis	• Ongoing between July 2015 and July 2019

An extensive review of global developing nation CBC literature was conducted focusing on the common enabling and constraining factors, conditions and processes for shifting to a community-based mode of conservation governance. Peer-reviewed journal articles, as well as grey literature, were obtained using a carefully constructed search string commencing with the year '1990' (see *Appendix 1*). This year was chosen since the 1990's represented a period of revised narratives promoting and funding CBC initiatives (refer *Chapter 1*). This allowed for the review of emerging trends from existing CBC initiatives already implemented and functioning. The search was performed on the *EBSCOHost* and *Thomas Reuters Web of Knowledge* platforms for peer-reviewed journal articles, and repeated in *Google Scholar* to incorporate further published but also notably relevant grey literature (especially practitioner reports) into the review. Further literature was consulted based on "snowballing" of literature emerging from the above search. A secondary goal of the search was to identify potential appropriate regional case studies. The literature review also informed the development of a *Generic CBC ToC Pathway* (presented in *Chapter 5*), which itself provided the platform for *phase five*. Lastly, since this review of literature was limited to English-language journals and grey literature it is acknowledged that this inhibits the potential insights gained from other sources.

2.1.2.2. Phase Two: Review of South African CBC Progress

Primary Data Collection and Analysis:

Semi-Structured Interviews:

Phase two involved semi-structured interviews conducted with 28 respondents from various South African conservation groups, namely: State officials and parastatal conservation organizations; academics and researchers; NGOs; the private sector; and civil society. Parastatal conservation organizations refer to conservation-orientated organizations that operate separately from, but perform service delivery on behalf of the State.

Semi-structured interviews allow for the flexibility required for complex issues, such as those encountered in conservation decision-making (Young et al., 2018). These interviews were conducted predominantly in English, but with some in Afrikaans. No translator was required due to my proficiency in both languages. Questions focused on the identification of enabling and constraining factors, conditions and processes for CBC implementation and governance in the country (see *Appendix 2*). Questions specifically aimed to elicit reasons for the current CCA *policy-praxis disjuncture* in South Africa, especially within coastal contexts, given the country's CBC enabling legislation. As with all semi-structured interviews conducted in this book, a "conversational technique" was employed in so far as questions were explained and/ or rephrased to promote conversational flexibility required for increased respondent understanding and response accuracy (Schober & Conrad, 1997; 2015). Furthermore, as with all respondent-based *phases* sample sizes were not determined *in priori* but attempted to achieve a perceived theoretical saturation (see Sim et al., 2018). Data analysis of interviews throughout this book involved coding of responses in order to cluster both key enabling and constraining factors, conditions and processes, as well as emerging themes regarding a CBC initiatives change process.

Secondary Data Collection:
All relevant national environmental legislative documentation pertaining to provisions for CBC initiatives were reviewed. In addition, relevant both published

and grey literature was consulted and consolidated to appraise the status of CBC in South Africa.

2.1.2.3. Phase Three: Regional Case Study Research
Primary Data Collection and Analysis:
Semi-Structured Interviews:

Case study selection and respondent information is introduced in detail in *section 2.2.* In all three case studies semi-structured interviews were conducted in a similar format, and targeted all key conservation actors, including State and local area managers, NGOs and other partner organizations, and local community members, community-based management organizations (CBOs) and traditional authorities. Once again, a conversational technique was employed and questions were structured used to obtain focused data on the enabling and constraining factors, conditions and processes for community-based governance, as well as elicit actor perceptions related to the CBC change process in each case (see *Appendices 4, 5 & 6* for partner organization, local representative, and community member example interview questions respectively).

Regional case study interviews were conducted with a translator, which though potentially limiting – due to potential personal bias and the accuracy of translations (Temple & Young, 2004) – were also deemed advantageous, as community respondents observably felt more at ease and could express themselves more fully in their local dialect. All translators were university graduates, and thus represented educated individuals. In Madagascar, my translator was born and raised in the south-west, and has worked extensively as a research assistant and interpreter. He is fluent in multiple *Malagasy* dialects, including *Vezo* and *Masikoro*. His proficiency in *Masikoro* proved especially beneficial when conversing with these respondents, who make up a minority in the predominantly *Vezo* local population. Furthermore, whilst many partner respondents (especially international NGO staff) were competent English speakers, some State representatives and NGO members were also interviewed with a translator in *Malagasy* for their convenience. In Guinea Bissau, my translator was fluent in both *Portuguese* (i.e. the official national language), and the more commonly spoken *Kriolu (i.e. Portuguese Creole)*. Furthermore, most partner respondents were also interviewed in *Kriolu* or *Portuguese* with a translator. Notwithstanding the proficiency of my translators, I acknowledge my limited control over the conversation and or accuracy of the translations, notably the strong possibility of finer details and cultural nuances being lost in translation. Nevertheless, the use of the aforementioned conservational technique (*section 2.1.2.2.*), and constant clarification with my translators enabled greater response accuracy. Consequently, whilst acknowledging the above shortcomings of translation, I believe that the research focus was not substantially impeded.

In both cases community respondents were approached largely at random (i.e. as encountered) within their villages/ homes, and/ or along the beach, but at times a "snowballing" technique was employed, especially to identify past and present local representatives (i.e. on CBOs and traditional authorities), and village elders. I refer here to elders simply as those viewed within the community as senior and respected individuals based on age and experience, and therefore, considered knowledgeable on past and present conservation management activities employed in the area.

<u>Informal Focus Groups:</u>

Focus group discussions offer a flexible qualitative method able to capture attitudes and perceptions of CBC actors, most notably shown useful concerning social relations and conservation of natural resources (Nyumba et al., 2017). Several informal focus groups were conducted when the opportunity arose, and frequently involved commonly marginalized demographics (e.g. women and youth). These focus groups proved particularly useful in capturing insights from female community members, who often spoke more freely away from their male counterparts. Detailed notes were taken as these discussions provided valuable knowledge, notably on actor perceptions and attitudes toward CBC initiatives, issues relating to local governance, and relations with partners. These notes were later analysed for common themes/ patterns emerging concerning the CBC change process that took place in each case. They also aided in identifying key community-perceived enabling and constraining factors, conditions and processes regarding the implementation and governance of their specific CCAs. Consequently, they proved especially effective in supplementing and confirming findings emerging from individual semi-structured interviews.

The basic structure of the informal focus groups included a brief introduction to the research, followed by conversations that were 'guided' by modified and very abbreviated versions of the *Most Significant Change* (MSC) and *Strength, Weakness, Opportunities, Threats* (i.e. SWOT) approaches. MSC is a 'story-based' technique used within diverse community initiatives involving collecting stories of community-perceived significant change (Davies & Dart, 2005). The objective of its use was to collect stories on the *rationale* behind *why* an event was important to the respondent, and *who* did *what, when* and *why* (*Cf.* Dart, 1999). This was deemed useful, since as Moon et al. (2019b: p427) state, "social science is not just answers, but stories." MSC possesses numerous advantages within the context of community-based research, as it provides a culturally appropriate means of identifying and communicating unexpected changes within initiatives across diverse cultures that actively involves participants; allows for holistic, meaningful and practical discussions of what is of *greatest importance* at a local level; and is inclusive of 'non-quantifiable' factors (Davies & Dart, 2005). Consequently, data analysis of themes/ patterns emerging from modified 'MSC-guided' informal focus group conversations were deemed useful in appraising the CBC change process from a community perspective.

A modified SWOT approach also informed informal focus groups. SWOT is a diagnostic approach used to identify key *factors* leading to the success or failure of an approach or strategy, and is considered useful and adaptable to a diverse array of contexts (Weihrich, 1982; Hill & Westbrook, 1997; Vonk et al., 2007). Consequently, modified versions of both MSC and SWOT informed informal focus group discussions, and data analysis of these discussions proved especially useful to identifying key community-perceived *issues* and *opportunities,* as well as *enabling* and *constraining* factors and conditions in the CBC change process.

<u>Participant Observation:</u>
In addition to semi-structured interviews and informal focus groups, participant observation shed further light on CCA implementation and management, and insights from LEK. This involved accompanying fishers and others on harvesting activities, and attending local CBO, village and other relevant multi-actor meetings. Detailed noted were taken during observation, which when analysed

29

complimented other methods in identifying enabling and constraining factors, conditions and processes, and appraising the CBC change process in each case. This method proved especially useful to obtaining further data on community-perceptions of both resource users and their local representatives. However, various practical constraints limited time spent in each of the regional cases to a total of six weeks.

<u>Social Relations & Network Appraisal (SRNA):</u>
Social networks comprise sets of actors linked by socially meaningful relations or ties. *Social Network Analysis* (SNA) inspired this research as it facilitates improved understanding of the influential nature of network structure and actor's positions to promote or hinder collaborative conservation governance (Crona & Hubacek, 2010). More specifically, SNA in conservation social networks can provide greater understanding ranging from formal policy and governance (e.g. Sandstrom & Carlsson 2008) to informal modes of governance contained within CBC initiatives (e.g. Lauber et al., 2008; Vance-Borland & Holley, 2011).

The *Social Relations & Network Appraisal* (SRNA) designed for this research was only conducted with community-members to emphasize the *community-perspective*. Therefore, it does not represent a complete SNA (i.e. does not include partners), but an adapted version. Three SRNA themes were selected to address the research's aim and objectives, and more specifically appraise the CBC change process in each case. These themes were *Interactional Support, Knowledge Acquisition and Diffusion, and Power and Politics*. *Interactional Support* refers to those actors community-members deemed most approachable to interact with regarding their concerns related to natural resource access and use. *Knowledge*

Acquisition and Diffusion identified those actors community-members deemed influential as *sources of knowledge* (i.e. possess the knowledge) and *knowledge providers* (i.e. share the knowledge), over three topics: legal rights to natural resources; obtaining and managing both monetary and non-monetary resources (these resources were questioned separately); and ecological aspects of governance. *Power and Politics* refers to actors with whom the community perceives decision-making power resides on the following topics: legal rights of natural resource access and use; local/customary conservation practices associated with natural resource access and use; the election of CBO representatives; obtaining and managing financial resources; distribution of any tangible benefits; and changing institutional and governance arrangements (see *Appendix 7* for SRNA questions). In addition, community respondents were specifically asked who they perceived had *Ultimate Decision-Making Power* regarding the management of their specific CCA.

SRNA responses were limited to a maximum of three actors for each question, and could include both organizations (e.g. State department or NGO) and individuals. Furthermore, other 'specifications of emphasis' (e.g. *No Trust* or *Don't Know*) emerged from responses, and are discussed within findings chapters. Data was recorded in two-dimensional matrices in *Microsoft Excel* for each community respondent. An entry of "1" or "0" in a cell indicated response/ no response for that actor respectively. Matrices were then developed into datasets and subsequently social network maps using *UCINET 6 Social Network Analysis* and *Netdraw* software packages (Borgatti et al., 2002). *Degree of centrality* – which reflects actor centrality based simply on the number of ties to other actors – was selected as the centrality measure to emphasize the (potentially) influential role of an actor for the CBC change process (Borgatti et al., 2009).

Whilst the SRNA provided an accurate and deeper understanding of community-perceived social relations – and thus present and potential issues and opportunities regarding collaboration among diverse actors from a community-perspective – if not corroborated with other data could lead to simplistic conclusions about actor relations in a conservation management setting (Prell et al., 2009). Consequently, the SRNA complemented qualitative data gathered during semi-structured interviews, informal focus groups and participant observation, as well as understanding obtained from analysing secondary data.

Secondary Data Collection:
Every effort was made to obtain all relevant socio-economic, ecological and legal information and secondary data sources on each of the regional case studies at both national- and local-levels.

2.1.2.4. Phase Four: Review of CBC case-in-progress in South Africa
Primary Data Collection
Semi-Structured Interviews:
Like *phase three*, semi-structured interviews were conducted with all key actors, targeting State and local area management, NGOs and other partner organizations, and local community, CBO representatives and local leaders. Once again, a conversational technique promoted greater understanding and therefore greater accuracy in responses to obtain focused data addressing the research aim and objectives. As with regional cases, questions were structured to provide greater information on the CBC change process which is in progress to establish a CCA at the mouth of the Olifants estuary. In particular, questions focused on gaining

information on actor perceptions and attitudes to the proposed CCA, and respondent's views on enabling and constraining factors, conditions and processes required for CCA implementation and governance (see once again *Appendices 4, 5 & 6* for example interview questions).

Data collection of community respondents was conducted exclusively in Afrikaans, with partners interviewed in both Afrikaans and English. Once again, no translator was required due to my proficiency in both languages. As with regional case studies, community respondents were approached within their settlements largely at random, but a snowballing technique was employed to identify the past and present local representatives, and village elders.

<u>Informal Focus Groups:</u>
As with regional cases, several opportunities arose in the South African *case-in-progress* for informal focus groups. This was facilitated by my established relations in the community due to substantial previous research that I have conducted in the community. These informal focus groups provided key demographic-specific insights into community perceptions, and progress with the CBC change process.

<u>Participant Observation:</u>
Participant observation was conducted to shed further light on implementation and management issues regarding the proposed CCA. This once again involved attending local multi-actor and CBO meetings, which included meetings pertaining to the proposed CCA boundary and follow-up meetings on implementation progress, as well as several fishing expeditions with a variety of local fishers to obtain LEK insights. However, as with the regional cases, various practical constraints limited time spent in the community to a total of six weeks. Furthermore,

as with regional cases described above, data was analysed to derive patterns emerging, and factors and conditions enabling or constraining the CBC change process underway.

<u>Social Relations & Network Appraisal (SRNA):</u>
The SNRA was conducted in the South African case-in-progress, in the same way as that within regional cases, to provide an accurate and deeper understanding of social relations among diverse actors from a community-perspective. Consequently, once again the SRNA endeavoured to complement qualitative data gathered during the semi-structured interviews, informal focus groups and participant observation, and in particular emphasize the (potentially) influential role of an actor within the CBC change process of the proposed CCA.

<u>Secondary Data Collection:</u>
Once again, every effort was made to obtain all relevant socio-economic, ecological and legal information, and secondary data sources, on the South African case study site to supplement analysis of primary data.

<u>2.1.2.5. Phase Five: Developing a South African Empirical CBC ToC Pathway</u>
Phase five entailed consolidating findings from *phases one* to *four* to develop a *South African Empirical CBC ToC Pathway* (presented in *Chapter 9*). The process of developing a ToC pathway is described in detail in *Chapter 5*.

2.2. Case Studies
2.2.1. Case Study Site Selection
Phase one's extensive review of extant CBC literature identified potential regional CCA case studies. Thereafter, logistical feasibility, and monetary and time constraints were considered. Furthermore, as an 'outsider' regional case selection proved highly contingent upon the support of a local organization, in both cases an

NGO. Case studies were selected from different areas of the continent to be more regionally representative and holistic (i.e. *East* and *West Africa*). No appropriate coastal CCA could be identified within the literature in North Africa at the time. Furthermore, case study selection was based on the presence of a local CBO (i.e. local community-led CCA management committee), and a post-implementation time frame of preferably 10 years at the time of the fieldwork visit. This time frame is considered to offer greater insight on local governance capacity (Capistrano et al., 2005). Moreover, while the three cases selected represent distinct natural systems (i.e. a coastal lagoon, an island, and an estuary) the rationale for this was to provide a more holistic approach to the topic at hand. The natural systems, and the resources commonly harvested by each community, in each case is discussed in more detail within each case study's respective chapter (i.e. *Chapters 6, 7* and *8*)

In East Africa, *Madagascar* was selected as a potential area of interest due to its well-established and much-publicized 'success' in implementing coastal CBC initiatives (e.g. Harris, 2011; Brenier & Vogel, 2017). The basis of this 'success' is the first LMMA established in the south-west of the country (i.e. the *Velondriake LMMA* - Harris, 2011; Oliver et al., 2015). Unfortunately, this site was unavailable due to extensive research already being conducted there, however, its progress heavily factored into the interpretation of national findings. Nonetheless, the *Bay of Ranobe* – also located in the south-west (*Figure 2.1.*) – provided an equally compelling and appropriate case, and met the above selection criteria, with the first of two LMMAs established in May 2007 (and fieldwork was conducted in October/ November 2016) (Belle et al., 2009).

An extensive search for West African coastal CCAs meeting the above criteria revealed two appropriate and relevant cases. The first was the *Kawawana ICCA* in central Casamance, Senegal. Unfortunately, the NGOs contacted were not supportive of further research being conducted in the area. Nonetheless, the second site in the *Urok Islands* of the Bijagós Archipelago, in Guinea-Bissau met the above criteria (*Figure 2.1.*). The *Urok Islands Community Marine Protected Area* (CMPA) was established in 2005 (and visited in February/ March 2018), and has received a *Ramsar Award for Management* (RAMSAR, 2012). Furthermore, the Bijagós Archipelago is internationally recognized as a *UNESCO Biosphere Reserve*, a *RAMSAR* site, and is currently pursuing *UNESCO World Cultural and Natural Heritage* status (Brenier et al., 2009; Tiniguena, 2019). Moreover, since the fieldwork visit this CBC initiative received recognition as one of the *2019 Equator Initiative* prize-winners (Equator Initiative, 2019).

In addition to the two regional case studies a South African 'case-in-progress' was selected. The site is the Olifants Estuary situated on the west coast (*Figure 2.1.*). This site is in the process of declaring a CCA located at the estuary mouth, which if successful will be the first coastal-marine CCA in the country. Not only does this site represent the costal-marine CCA example closest to being declared but it was also the site of my masters research and I therefore have a strong working relationship with this community. This established relationship was deemed advantageous as the time spent in the field at this site would be limited. It also enabled numerous informal focus group discussions with community members and in particular local leaders which proved highly informative.

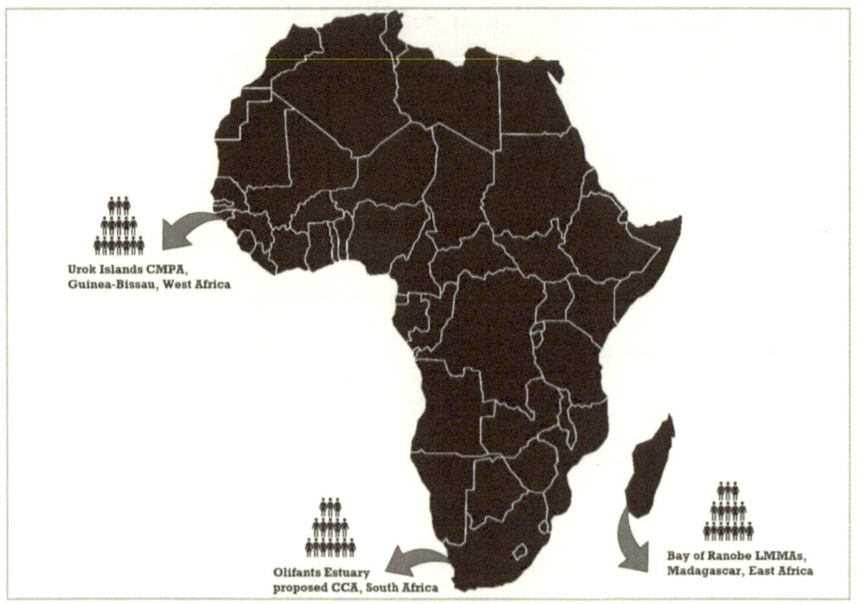

Figure 2.1.: Map indicating the location of the three case studies of the Bay of Ranobe, south-west Madagascar, East Africa; the Urok Islands, Guinea-Bissau, West Africa; and the Olifants Estuary, South Africa. **Source:** Designed by author.

2.2.2. Regional Case Studies

2.2.2.1. Regional Case Study One: The Bay of Ranobe, Madagascar

A total of 107 semi-structured interviews were conducted with partner organization and community respondents (*Table 2.2.*). Partner organization respondents totaled 25, inclusive of Bay of Ranobe specific and national State and NGO partners. Furthermore, 82 Bay of Ranobe community respondents were interviewed inclusive of past and present *FIMIHARA* village representatives (i.e. the CBO managing the CCAs), the current *FIMIHARA* president and three of the four village presidents (i.e. local traditional authorities). Community respondents encompassed four of the 13 fishing villages, with a minimum of 20 community respondents per fishing village. The four focus villages were selected due to their proximity to the two LMMAs, and limited time spent on site. These included two southern villages, *Beravy* and *Ifaty*, located near the *Massif des Roses* LMMA, and two northern villages, *Ambolomailaka* and *Andrevo*, located near the *Ankaranjelita* LMMA (refer to *Figure 2.2.*). Whilst the village of *Mangily* is located near, and its villagers were primary participants in the establishment of FIMIHARA and the subsequent implementation of the *Massif des Roses* LMMA, it was omitted from community interviews at the advice of Reef Doctor representatives due to most community members being involved in non-fishing related industries. Community respondent demographic composition attempted to be gender inclusive (i.e. 56% male & 44% female), though this was not explicitly accounted for in data analysis (refer to *Figure 2.3.*).

Table 2.2.: An overview of Madagascar and Bay of Ranobe partner organization and community respondents.

Respondent Group	Number of Respondents	Respondent Affiliations
Partner Organizations	25	
State		• Ministry of Fisheries and Marine Resources (MFMR – Bay of Ranobe partner), • Ministry of Ecology, Environment and Forests (MEEF) • Service d'Appui à la Gestion de l'Environnement (i.e. Support Service for Environmental Management – SAGE – Bay of Ranobe partner)
NGOs		• Blue Ventures • Conservation International (CI) • MIHARI LMMA Network • ReefDoctor (Bay of Ranobe partner NGO) • Worldwide Fund for Nature (WWF) • Wildlife Conservation Society (WCS)
Other		• Institut Halieutique et des Sciences Marines at the University of Toliara (i.e. Institute of Fisheries and Marine Sciences - IHSM) • Managed Resources for Protected Areas (MRPA - Malagasy UNDP-funded project) • Bay of Ranobe Hoteliers & Tour Operators
Community Members	82	• Local community members • FIMIHARA village representatives • Village presidents

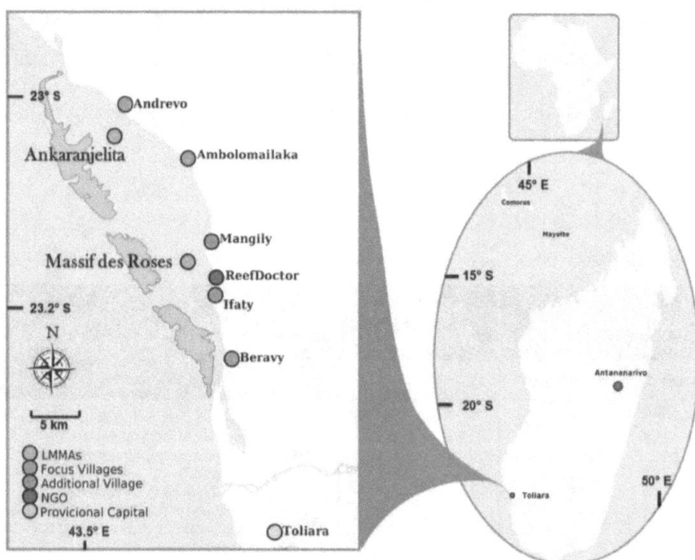

Figure 2.2.: Map indicating the location of the Bay of Ranobe in Toliara province, south-west Madagascar. In addition, the location of Madagascar off the east coast of Africa, and the location of Toliara City relative to the national capital of Antananarivo, are depicted. **Note:** See enclosed legend for the location of the four focus villages; two Locally-Managed Marine Areas (LMMAs), Reef Doctor's local headquarters and other sites mentioned in the text. Icons are not indicative of actual sizes of locations. See also the location of the barrier reef with north and south passes. **Source:** Designed by author.

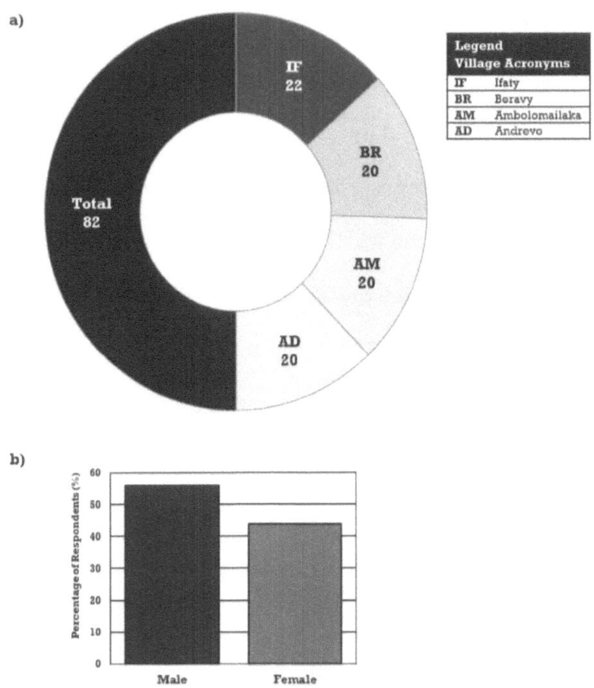

Figure 2.3.: Bay of Ranobe community respondent demographic composition reflecting a) the total and per village number of respondents, and b) the percentage of male and female respondents. Note: see legend for village acronyms.

2.2.2.1. Regional Case Study Two: The Urok Islands, Guinea-Bissau

A total of 94 semi-structured interviews were conducted with both partner organization and community respondents (*Table 2.3.*). Partner organization respondents totaled 14, inclusive of State and NGO partners, both those specific to the *Urok Islands CMPA* and additional national partners. In addition, 80 community respondents were interviewed, encompassing 11 of the 12 villages on the main Urok Island of *Formosa* (*Table 2.3.* & refer to *Figure 2.4.* for locations). Although the CMPA includes the three islands of *Formosa, Chediã* and *Nago*, due to the limited scope of the study, most of the Urok Islands population residing on *Formosa*, and

40

logistical challenges preventing reaching the other two islands, fieldwork focused on villages on *Formosa Island* (*Figure 2.4.*). However, one women from *Nago* was interviewed (whilst visiting family and friends in *Formosa*) and two separate informal focus group discussions were completed with community respondents from *Chediã* and *Nago* (i.e. one with a group of 6 men, and one with a group of 8 women). Although limited, these did shed some light on the perceptions of these community respondents for the CMPA and the challenges to its implementation and governance. Community respondents included past and present local CBO representatives of *Village Management Committees* (VMCs) and *Urok Management Committee* (UMC) representatives. Six out of seven *UMC* local representatives were interviewed, unfortunately the seventh declined to be interviewed.

Table 2.3.: An overview of Guinea-Bissau and Urok Islands partner organization and community respondents.

Respondent Group	Number of Respondents	Respondent Affiliations
Partner Organizations	14	
State		• Institute for Biodiversity and Protected Areas (IBAP – the parastatal national conservation agency) • National Centre for Applied Fisheries Research (CIPA) • National Institute of Studies and Research (INEP) t • National Coastal Planning Office (CPO)
NGOs		• Action for Development (AD) • Associação de Desenvolvimento Integrado das Mulheres (i.e. Association for the Integrated Development of Women [ADIM] - working country-wide inclusive of the Bijagós Islands) • BirdLife • Manitese (a locally-based branch of an Italian NGO working country-wide inclusive of the Bijagós Islands) • Nantinyan (local NGO working exclusively in the Bijagós Islands) • Palmeirinha (local NGO) • SWISSAID • Tiniguena (Urok Islands CMPA partner NGO) • Wetlands International
Other		• IUCN
Community Members	80	• Local community members • Urok Management Committee (UMC) local representatives • Village Management Committee (VMC) representatives • Village chiefs

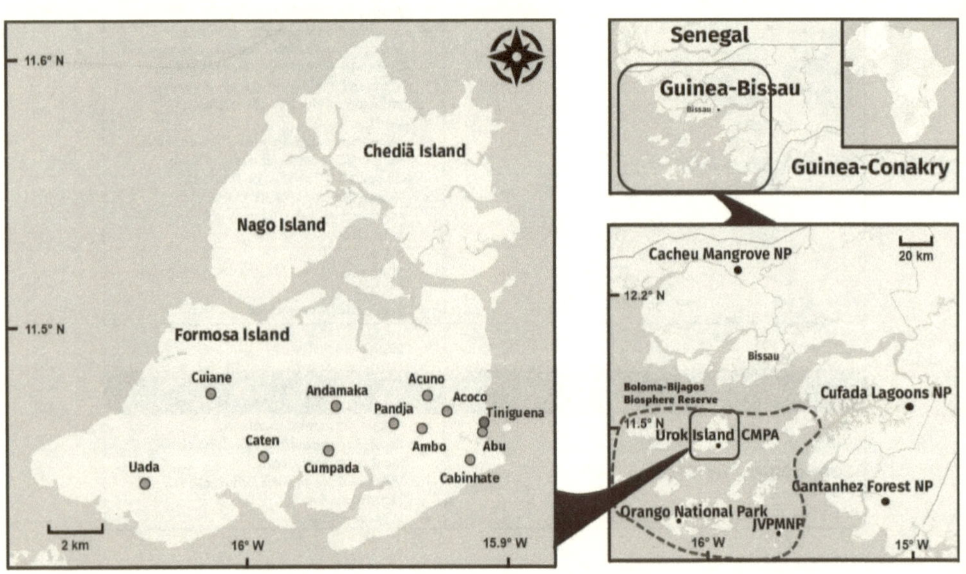

Figure 2.4.: Map indicating the location of the Urok Islands in the Bijagós Archipelago, Guinea-Bissau. Note the location of the eleven focus villages on Formosa Island (in green) and the local offices of Tiniguena (in red). Icons are not indicative of actual sizes of locations. Inset depicts the location of Guinea-Bissau in West Africa. The location of the Urok Islands within the *UNESCO Boloma-Bijagós Biosphere Reserve*, along with two State Marine National Parks, i.e. *Orango* and *João Vieira e Poilão* Marine National Park (JVPMNP), are also provided.

Unfortunately, recruiting community respondents on *Formosa Island* at times proved problematic. Locals were often closed off, perhaps due to the Island's socio-political history (discussed in *Chapter 7*), general reluctance, mistrust of foreigners or a (cultural) desire to retain *their* knowledge. The latter observed elsewhere in the country (see Davidson, 2010). This was most notable as one moved further from the 'project hub' of *Abu* village (i.e. the location of the *UMC* headquarters and Tiniguena's local offices) most notably in *Kuian* and *Caten* (refer to *Figure 2.4.* above for locations). Furthermore, women often declined requests to be interviewed, some citing the need for male family members to be present, or simply a reluctance, even when encouraged by male respondents who had completed interviews. Thus, accounting for gender was somewhat problematic. Nonetheless, approximately 36% of respondents interviewed were *female* (*Figure 2.5.*).

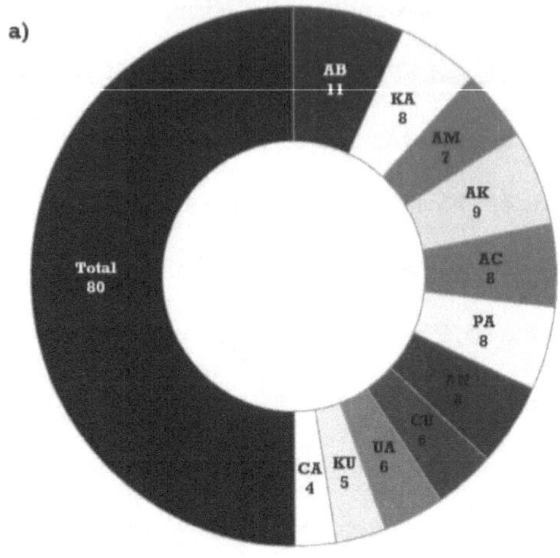

Legend
Village Code

AB	Abu
KA	Kabinhato
AM	Ambo
AK	Akoco
AC	Acuno
PA	Pandja
AN	Andamaka
CU	Cumpada
UA	Uada
KU	Kuian
CA	Caten

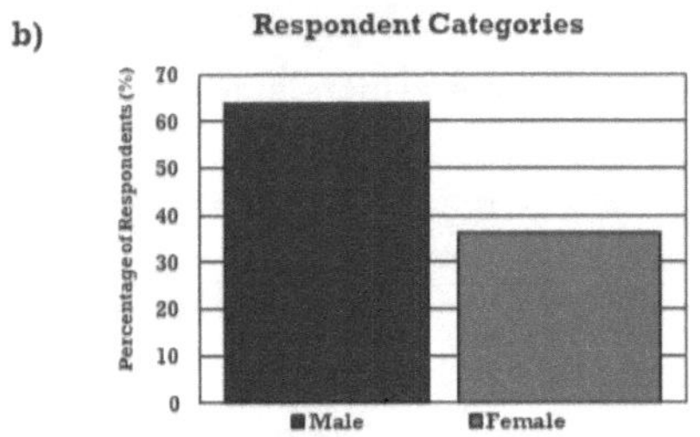

Figure 2.5.: Urok Island community respondent demographic composition reflecting a) the total and per village number of respondents, and b) the percentage of male and female respondents. Note: see legend for village acronyms.

2.2.3. South African Case Study: The Olifants Estuary, South Africa

A total of 56 semi-structured interviews were conducted with both partner organization and community respondents within the *Olifants Estuary* case study (*Table 2.4.*). More specifically, 10 partner organization, inclusive of State and other partners, and 46 community respondents were interviewed. Community respondents encompassed members of the four settlements of *Papendorp*, *Olifantsdrif*, *Nuwestasie*, and *Nuwepos*. The latter three settlements are all located close together, and although *Papendorp* is located further away, the four settlements together constitute the greater *Ebenhaeser* community (*Figure 2.6.*).

Table 2.4.: An overview of Olifants Estuary partner organization and community respondents.

Respondent Group	Number of Respondents	Respondent Affiliations
Partner Organizations	10	
State		• Matzikama Local Municipality • West Coast District Municipality (WCDM) • Western Cape Provincial Department of Environmental Affairs and Development Planning (DEADP)
Other		• CapeNature (parastatal provincial conservation agency) • Environmental Evaluation Unit (EEU) at the University of Cape Town (UCT) • Olifants Estuary Management Forum (OEMF) • Civil society partners
Community Members	46	• Local community members • Olifants Fishing Committee (OFC) • Ebenhaeser Communal Property Association (CPA)

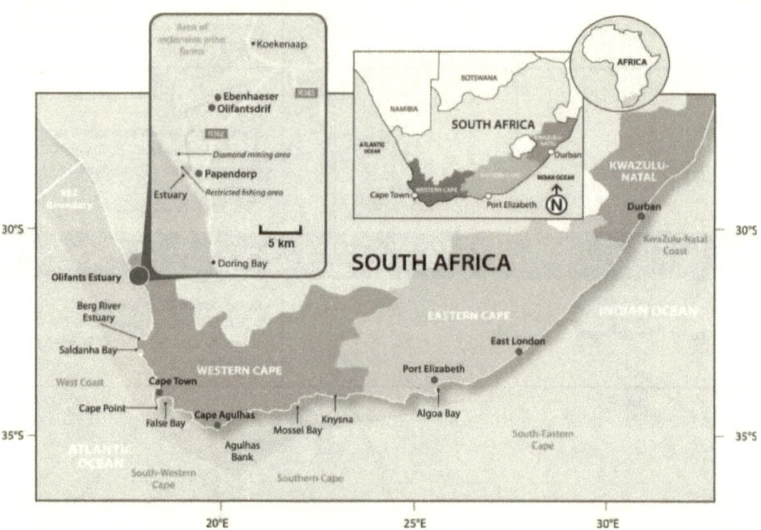

Figure 2.6.: Map indicating the location of the Olifants estuary in the Western Cape Province, South Africa. Insets indicate the location of South Africa on the southern tip of Africa, and the location of Cape Town (i.e. the provincial capital of the Western Cape Province). Map also indicates the four focus settlements, *Papendorp*, *Olifantsdrif*, *Nuwepos* and *Nuwestatsie* (i.e. the latter three collectively considered the settlement of *Ebenhaeser* and indicated as such on the map). The current restricted fishing area (i.e. no-take zone) and the proposed CCA are located near the estuarine mouth near *Papendorp*. Icons are not indicative of actual sizes of locations. **Source:** Rice et al. (2017).

47

Most interviews reflected fisher households in *Papendorp* and *Olifantsdrif* (*Figure 2.7.*). These settlements account for the vast majority of river (as well as marine) fishing permit holders. Nevertheless, an effort was made to include other community groups as well. However, these yielded no useful information regarding the research'aim and objectives as most of these respondents claimed they were completely unaware of the proposed CCA. This was not a surprising finding given diverse local livelihood strategies, with fishers making up the minority. This resulted in 10 interviews with residents from *Nuwestasie* being eliminated from data analysis. Nonetheless, theoretical saturation was deemed to have been reached by the remaining 36 community respondents. Community respondents included representatives on both the *Olifants Fishing Committee* (OFC) and the local *Ebenhaeser Communal Property Association* (CPA). This was inclusive of the current and previous presidents of both CBOs. Community respondent composition once again attempted to be gender inclusive, however, this did not materialize, as fishing is exclusively a male occupation within this community. Furthermore, numerous women approached, especially in Nuwestasie, were unaware of the proposed CCA. Accordingly, male and female respondents made up 82 and 18% respectively (*Figure 2.7.*).

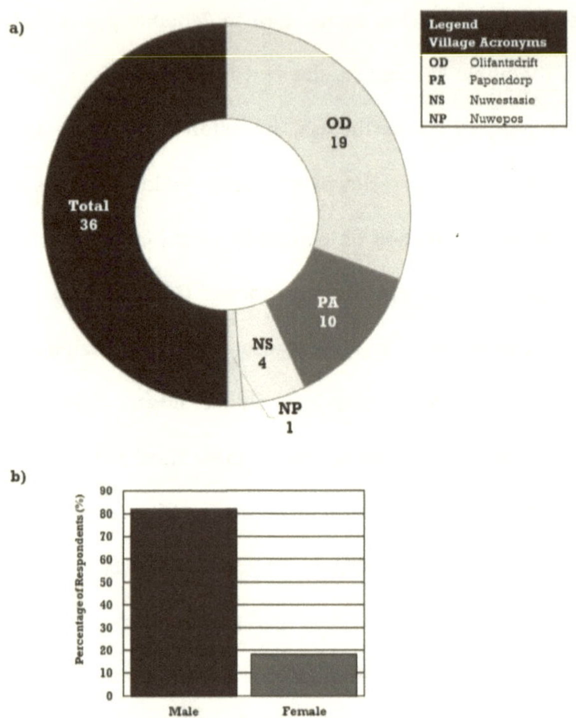

Figure 2.7.: Olifants Estuary community respondent demographic composition reflecting a) the total and per village number of respondents, and b) the percentage of male and female respondents. Note: see legend for village acronyms.

2.3. Limitations

In accordance with Johnson and Onwuegbuzie (2004: p14) this book specifically acknowledges:

"that multiple-constructed realities abound, that time- and context-free generalizations are neither desirable nor possible, that research is value bound, that it is impossible to differentiate fully causes and effects, that logic flows from specific to general and that knower and known cannot be separated because the subjective knower is the only source of reality."

Accordingly, I acknowledge that my own positionality may have affected the objectivity of my interpretations (Sidaway, 2000 – see also Koot et al., 2020). More specifically, limited prior knowledge of the context, especially in the regional cases, is acknowledged to have potentially affected data collection and interpretation. Moreover, as an 'outsider' and male researcher, a lack of cultural understanding and knowledge of the subtleties in language and customs, most notably again in the regional cases, may have affected my interpretations, and the ability of respondents to be completely open and honest in interviews and focus groups (see Sim et al., 2018; Moon et al., 2019b for further discussion on these research limitations). Furthermore, whilst regional case study data collection required the use of a translator, which may be limiting, this was not deemed to have substantially affected data collection (refer to *section 2.1.1.*). Accordingly, every effort was made with participants of informal focus groups, and especially semi-structured interviews, to avoid acquiescence and 'satisficing', i.e. the tendency for the respondent to provide answers perceived to be positive, polite or agreeable irrespective of the content of questions so as to appear in a positive light (Bentler et al., 1971; Krosnick, 1991). Nevertheless, limited control over this is acknowledged. In addition, whilst semi-structured approaches allow for greater flexibility, the possibility of pre-conceived interviewer-, question-, and interpretation-bias, is acknowledged. It is also specifically acknowledged that interview coding and analysis, and more specifically, the development of ToC pathways is subjective, especially since the latter process requires making assumptions.

Challenges to data collection were encountered on occasion, most notably at times the ability to recruit respondents. Unfortunately, this was particularly influential within *phase two* (i.e. concerning South African conservation actor interviews - *section 2.1.2.2.*). Many potential respondents relating to this phase were unresponsive to repeated attempts to make contact. Furthermore, some of these respondents even at times actively attempted to guide the conversation in a particular direction in order to paint themselves or their organizations in a favourable light regarding their community conservation efforts. In addition, occasionally issues were encountered regarding recruiting community respondents within some villages on *Formosa Island*, in the Guinea-Bissau case (refer to *section 2.2.1.3.*). Accordingly, it is acknowledged that these challenges may have impacted the data obtained, and therefore the findings and conclusions drawn. In particular, a greater number of interviews, especially with State respondents, in the aforementioned *phase two* may have proven beneficial to the appraisal of the status of progress and current constraints with CBC initiatives in South Africa. Nonetheless, it is believed that in all cases at the very least the full range of actors were consulted, and due to the mixed-method approach, each phase appeared to reach theoretical saturation, and meet the research's aim and objectives. That said greater research into this topic is strongly encouraged. Lastly, data collection, analysis, and interpretation of secondary data is subject to human error and bias, which need to be taken into consideration. Appropriate measures taken to address the aforementioned issues included: collecting, assessing and analysing data from a variety of sources; and comparing primary data with that obtained through secondary data collection when available (i.e. consolidating data through *triangulation*).

As Game et al. (2014: p272) note that, "There is no "right" solution to wicked problems in complex systems." Furthermore, they suggest that whilst it will not solve all our wicked conservation problems, borrowing concepts from other fields "broadens our range of options" (Game et al., 2014: p275). Accordingly, this book has considered and incorporated literature and methods from diverse social science fields, in addition to those conventionally associated with conservation, notably those from the behavioural and health sciences, psychology, sociology, and economics and management studies. More specifically, it makes use of the ToC approach – a well-established and proven tool in the development sector – to offer a 'change perspective' to the initiation, implementation and governance of CBC initiatives. However, since this change perspective focuses solely on the initial planning and implementation stages of a CBC initiative, and does not seek comprehensively monitor and evaluate a specific initiative's results, it is acknowledged that this inhibits the ability to conclude its effectiveness or accurately portray causality of the *ToC pathways* developed. However, as stated, establishing causality was never the objective of this research.

2.4. Ethical Considerations

Ethical considerations include the need to obtain prior informed consent and preserve respondent anonymity; not directly affect or endanger the lives of respondents or one's self; and ensure transparency toward respondents and the reporting of findings (AISSR, 2017). Following an introduction to the research's aim and objectives, verbal informed consent was obtained from both customary authorities in case study sites, and individual respondents. Furthermore, respondent anonymity was maintained, since nowhere was a respondent's name recorded, with each respondent merely allocated a code for data collection, analysis and reporting purposes. Moreover, security of raw data was maintained as this was either on my person or in a locked bag at all times. The aforementioned steps are considered to have minimized any direct harm or danger to the respondent, researcher or translators. Nonetheless, it is acknowledged that the possibility exists that respondents may have 'feared' the implications of participating in this research, perhaps most notably regarding questions related to possible illegal resource harvesting activities.

This study obtained the required ethical clearance prior to commencing any data collection activities from the *University of Cape Town's Faculty of Science Research Ethics Committee* (Approval code: FSREC 02 – 2016). Lastly, the aforementioned steps were taken to meet the ethical guidelines as stipulated by the *University of Amsterdam's Amsterdam Institute of Social Science Research* (AISSR) (AISSR, 2017).

Chapter 3

Theoretical Foundations Part 1:

Governance and Commons Theory

3.1. Introduction

This chapter firstly introduces the concept of *Governance*, and its important theoretical considerations. More specifically, given that the three cases selected are all CCAs, it provides context by briefly introducing *Protected Areas* (PAs), and notably different 'modes' or 'types' of *PA governance*. Thereafter, it introduces *Commons Theory*, a section which culminates with the identification of 14 proposed CBC enablers considered key to facilitating the initiation, implementation and governance of CBC initiatives. In doing so, this chapter addresses **objective 1** (*Box 3.1.*).

Box 3.1.:

> ***Objective 1:*** To conduct an extensive review of CBC literature with particular attention given to developing countries and focusing on the enabling factors, conditions and processes for shifting to a community-based mode of conservation governance

3.2. Governance

3.2.1. What is Governance?

Governance is one of the most important factors affecting effective environmental and conservation management (Lockwood et al., 2010; Armitage et al., 2012; Borrini-Feyerabend et al., 2013). Accordingly, an improved understanding of the attributes and processes of governance is necessary to better understand *how* a change from one 'mode' or 'type' of governance to another occurs – in the present case from *state-centred* or *top-down* to *community-based governance* (types of governance are introduced in relation to PAs in *section 3.2.4.* below). Whilst the concept of governance, and environmental governance in particular, possesses a long history and has evolved extensively especially from 2000 onwards, I focus primarily here on a 'more modern' understanding of governance as it relates to interactions among multiple actors and multiple scales which strive for an equilibrium between the powers of *government* and *other governance actors* (see Morin & Orsini, 2020).

Governance focuses on who has power, responsibility, and accountability in decision-making and the implementation of actions (Jentoft, 2007a; Borrini-Feyerabend & Hill, 2015). Furthermore, governance can take on an analytical or a normative perspective, that is, it is "both *what is* and *what should be*, reality as well as potential" (Kooiman & Bavinck, 2005: p16 – *emphasis* added). Governance essentially comprises a *governing system* and a *system-to-be-governed* (Jentoft, 2007a). Governing systems are institutional mechanisms or sets of rules for directing the system-to-be-governed, which can be viewed as partially natural (i.e. ecosystems and their resources), and partially social (i.e. the users and, or stakeholders) (Jentoft, 2007a). Therefore, governance requires cognisance of not only interactions between/ among actors but also their interactions with the natural system. That said, the subsequent sections focus primarily on providing a brief introduction to the theoretical foundation of interactions between/ among governance actors.

Therefore, a key governance consideration is the "quality of the totality of the interactions between those governing and those governed" (Kooiman & Bavinck, 2005: p19). Accordingly, Kooiman et al. (2005a) refer to the collaborative efforts between diverse actors to emphasize integrated communicative and politically informed approaches, which generate opportunities and provide solutions to societal problems. However, empowering some actors may disempower others, and therefore disturb the *status quo*, and create new governance challenges (Jentoft, 2007a; García-López, 2019). Consequently, a brief introduction to the topic of *governance interactions and power* follows.

3.2.2. Interactions and Power

Governance interactions comprise two key components, "actors" and "structures" (Kooiman & Bavinck, 2005). As introduced in *Chapter 1*, *actors* are those affected by or affecting conservation governance. More specifically, *actors* refer to those "possessing agency or power of action", while structures represent "the frameworks within which actors operate" (Kooiman & Bavinck, 2005: p18). However, actors may be *constrained* or *enabled* by these structures (Bavinck et al., 2013).

A key consideration is that diverse governance actors – as found in CBC governing systems – will contribute a diverse range of perspectives to governance interactions (Baird et al., 2019a&b; Armitage et al., 2020). Accordingly, numerous scholars and practitioners emphasize the importance of interactions amongst diverse governance actors (e.g. Graham et al., 2003; Kooiman & Bavinck, 2005; Jentoft, 2007a, 2017; Borrini-Feyerabend & Hill, 2015; Baird et al., 2019a&b). Therefore, governance needs to be motivated by collaborative, multi-actor participation, where governing activities should go 'beyond government', and thus represent more than just a conventional hierarchical and top-down institutional process (Kooiman & Bavinck, 2005; Jentoft, 2007b; Bavinck et al., 2013; Baird et al., 2019a).

A system's 'governability' relies upon the ability of the governing system to exercise power and resolve conflict within the system-to-be-governed (Graham et al., 2003; Armitage et al., 2012; Jentoft, 2017; Fisher et al., 2018; Baynham-Herd et al., 2018). Consequently, environmental governance research increasingly endeavours to integrate the role of power into institutional analysis (Jentoft, 2007a; Bennett, A. et al., 2018; Morrison et al., 2019). *Power* can represent both a capacity to act, and a method to mobilise *collective action* (Etzioni, 1968; Lukes, 1982). I use the term *collective action* here in accordance with Wright et al. (1990: p995) who suggest, "a group member engages in collective action any time that she or he is acting as a representative of the group and the action is directed at improving the conditions of the entire group." Nevertheless, the motivations of individuals for collective action are diverse, and include perceptions related to costs and benefits, the ability of the collective action to succeed, and perceived collective identity (Bamberg et al., 2015; Bodin, 2017; Bennett, N. et al., 2018). Furthermore, collective action of group members may also be subject to coercion by a member possessing power and motivated by their own ability to benefit from the collective action. This is discussed further below under the concept of *elite capture*.

Therefore, governance is "the conscious determination of action via the use of various forms of power" (Borrini-Feyerabend & Hill, 2015: p171). Furthermore, this use of power extends to encapsulate both *policy* and *practice* (Borrini-Feyerabend et al., 2013). Accordingly, environmental governance must account for diverse manifestations of power, which relate to among others: control over material resources, who possesses the decision- and rule-making authority, and how power is expressed through discourse (Clement, 2013; Epstein et al., 2014; Kashwan,

2016, 2019; Bennett, A. et al., 2018). Not surprisingly, scholars describe power as a "slippery concept" (Jentoft, 2007a: p434), since it is "difficult to capture the invisible workings of power" (Cleaver, 2017: p16). Nevertheless, acknowledging the significance and diverse representations and applications of power is crucial to better understanding a CBC change process. Of specific relevance here, Graham et al. (2003: p13) distinguish between five types of PA governance power, which include: planning powers, regulatory powers, spending powers, revenue-generating powers, and the power to enter into agreements. They specifically emphasize the importance of *regulatory power* (i.e. concerning the use of land and resources), since this power directly influences desired PA objectives (Graham et al., 2003).

The ability of individuals to exercise power over others due to their societal position and vested interests, may manifest to either coerce or constrain other's actions, can result in inequity and injustice, and challenges institutional design (Jentoft, 2007a; Raik et al., 2008; Boonstra, 2016; Calfucura, 2018). This is a result of the strength of ties amongst actors within a governance network, which determines the level of influence over one another, similarities of perspectives, knowledge-diffusion, and trust and support (Coleman, 1990; Crona & Bodin, 2006; Borgatti et al., 2009; Dressel et al., 2020). Accordingly, some scholars argue CBC initiatives often fail due to local power dynamics that may lead to injustices or incompetent leadership and elite-capture (Lane & Corbett, 2005; Balint & Mashinya, 2006; Warren & Visser, 2016; García-López, 2019). The term *elite-capture* is used here to describe "the capture of the distribution of resources, project implementation and decision making which negatively impacts non-elites or the target population or is deemed to be corrupt under the law" (Musgrave & Wong, 2016: p92). However, elite-capture is not limited to local elites, but can originate with other actors from the upper-echelons of governance regimes, i.e. within state departments (Williams & Le Billon, 2017) and/ or NGOs (e.g. Kamat, 2004; Brass, 2012). Accordingly, and of specific relevance here, Normann (2006) discusses in reference to South African fisheries reform, how power tends to corrupt at all governing levels.

Therefore, the risk exists of encouraging a community-based mode of governance without considering the inequalities and power relations defining *who* participates, the *type* of participation, and its *outcomes* (Baird et al., 2019a; Kashwan, 2019). Consequently, the need exists to account for power as a source of social change, and create greater distinction of power within both informal and formal social groups, both of which are considered central to change (Simon & Oakes, 2006; Cleaver, 2017). Thus, based on the above discussion, *governance* can perhaps better be considered as, "the interactions among structures, processes, and traditions that determine how power and responsibilities are exercised, how decisions are taken, and how citizens or other stakeholders have their say" (Graham et al., 2003: p2).

3.2.3. Principles for 'robust' governance
Governance principles can be defined as "codes of conduct, operating guidelines, or yardsticks to internally refer to when decisions and actions are made, evaluated, criticized and when changes are proposed" (Song et al., 2013: p168). A lack of basic principles means "no human relation or governing interaction can last" (Kooiman & Bavinck, 2005: p17). Accordingly, numerous scholars have proposed principles for 'good' governance (e.g. Kooiman & Bavinck, 2005; Jentoft, 2007a;

Lockwood et al., 2010; Charles, 2011). These principles include the need for governance to be *sensitive* to the *diversity* of diverse actors and inputs, *flexible* to *dynamic* actor-interactions, and provide for *context, coordination, learning* and *safeguarding* within systems-to-be-governed (Jentoft, 2007a). Furthermore, Lockwood et al. (2010), emphasize among others the need for *legitimacy, transparency,* and *accountability* for decision-making and actions, *inclusiveness* and *fairness* of actor participation, and the ability of the governing system to systematically reflect on performance and learn. More recently, and of specific relevance to a community-based/ collaborative mode of governance, Bennett & Satterfield (2018) emphasize the need for connected networks and 'nested' institutions able to assign tasks to appropriate levels, and empower and support devolved decision-making to the lowest-level possible.

Environmental governance has also increasingly grappled with the issue of *institutional fit,* which is the extent to which institutions and policy match the social-ecological context (Epstein et al., 2015). This builds on Jentoft's (2007a) emphasis of governing systems accounting for context within the system-to-be-governed. Whilst *ecological fit* – which refers to a technical approach to ecological problems – is well established, *social fit* is less so (Epstein et al., 2015; Turner et al., 2018). *Social fit* is essentially concerned with understanding local resource user's perceptions of, and thus the social acceptability of governance arrangements (DeCaro & Stokes, 2013; Epstein et al., 2015; Turner et al., 2018). Consequently, increasing the alignment of formal and informal governance networks and decision-making processes with local social behavioural patterns and rules can improve social fit (DeCaro & Stokes, 2013; Pittman et al., 2015). Though more concerned with social fit, this book acknowledges both ecological and social fit are equally important for 'good' governance (Epstein et al., 2015).

Consequently, building upon discussions thus far, and in accordance with Armitage et al. (2019: p523), environmental governance can be considered, "how institutions and social norms shape culture and societal behaviour and decisions; inform who is authorized to make decisions about and take action on natural resources; and influence what will be conceived as politically, economically, and environmentally acceptable."

3.2.4. Protected Area Governance

Whilst this book incorporates broadly defined CBC initiatives, which are not all 'area-based', since the three cases chosen all represent CCAs a brief introduction to PA types and governance is now provided.

3.2.4.1. Defining and Categorising Protected Areas

PA governance is about *who* defines the overall objectives of conserved areas (Borrini-Feyerabend & Hill, 2015). Nevertheless, PA governance has been shaped by numerous international conservation agencies, who have sought to develop common PA categories. Therefore, before discussing the various types of PA governance, a brief introduction to the history and recognized categories of PAs follows.

The International Union for Conservation of Nature (IUCN) defines a PA as, "[a] clearly defined geographical space, recognized, dedicated and managed, through legal or other effective means, to achieve the long-term conservation of nature with associated ecosystem services and cultural values" (Dudley, 2008: p9). However, this definition was accompanied by the statement that, "For IUCN, only those areas where the main objective is conserving nature can be considered protected areas; this can include many areas with other goals as well, at the same level, but in the case of conflict, nature conservation will be the priority" (Dudley, 2008: p10).

Early efforts to establish common terminology on PA categories led to the *World List of National Parks and Equivalent Reserves* at the first World Conference on National Parks in 1962, by the now *World Commission on Protected Areas* (WCPA - Brockman, 1962). Furthermore, the 1972 World Parks Conference urged the IUCN to define the various purposes of PAs, and develop suitable standards and nomenclature (Dudley, 2008; Brockington et al., 2010). Since 1994, the IUCN has recognized six PA categories with associated management objectives, which range from strictly controlled access to human inhabited cultural landscapes or seascapes encouraging sustainable natural resource use (*Table 3.1.*). Whilst, these management categories originally strived for common understanding of PAs (Dudley, 2008), notable limitations include differentiating between, and applying distinctive categories within different contexts (Brockington et al., 2010). Furthermore, 'critics' (e.g. Terborgh, 1999; Locke & Deardon, 2005; Terborgh & Peres, 2017), and 'promoters' (e.g. Mallarach et al., 2008; Ferraro et al., 2013) of less restrictive categories prevail (i.e. *Categories V* and *VI*). Nonetheless, greater acceptance of the contribution of all categories to conservation now exists, and all increasingly inform global PA systems (Bertzky et al., 2012).

Table 3.1.: IUCN categorisation of protected areas and associated management objectives. **Source:** Borrini-Feyerabend et al. (2013: p9)

Protected Area Category	Management Objectives
Ia. Strict nature reserve	Strictly protected for biodiversity and also possibly geological/ geomorphological features, where human visitation, use and impacts are controlled and limited to ensure protection of the conservation values
Ib. Wilderness area	Usually large unmodified or slightly modified areas, retaining their natural character and influence, without permanent or significant human habitation, protected and managed to preserve their natural condition
II. National park	Large natural or near-natural areas protecting large-scale ecological processes with characteristic species and ecosystems, which also have environmentally and culturally compatible spiritual, scientific, educational, recreational and visitor opportunities
III. Natural monument or feature	Areas set aside to protect a specific natural monument, which can be a landform, sea mount, marine cavern, geological feature such as a cave, or a living feature such as an ancient grove
IV. Habitat/species management area	Areas to protect particular species or habitats, where management reflects this priority. Many will need regular, active interventions to meet the needs of particular species or habitats, but this is not a requirement of the category
V. Protected landscape or seascape	Where the interaction of people and nature over time has produced a distinct character with significant ecological, biological, cultural and scenic value: and where safeguarding the integrity of this interaction is vital to protecting and sustaining the area and its associated nature conservation and other values
VI. Protected areas with sustainable use of natural resources	Areas which conserve ecosystems, together with associated cultural values and traditional natural resource management systems. Generally large, mainly in a natural condition, with a proportion under sustainable natural resource management and where low-level non-industrial natural resource use compatible with nature conservation is seen as one of the main aims

3.2.4.2. *The History of Protected Areas: from an African perspective*

People-free PAs were the favoured conservation approach of the nineteenth and twentieth century (Hutton et al., 2005). This approach was rationalized by a longstanding argument of the need to protect natural resources from local and indigenous communities who were heavily reliant on resources for their livelihoods, and therefore concerns existed that they would exploit them without restraint (Neumann, 1998; Agrawal & Gibson, 1999).

The first people-free PAs were established in Yosemite and Yellowstone in the United States of America in the late nineteenth century (Runte, 1987; 1990), and became the *de facto* model for many subsequent PAs (Nash, 1967; Igoe, 2005). This included the establishment of several South African forest reserves and colonial nature reserves in the late nineteenth century (Grove, 1987; Masuku Van Damme & Meskell, 2009). Rapid PA expansion followed World War II, with Africa in particular experiencing a "conservation boom" at the time (Neumann, 2002). This period also saw the founding of the IUCN, which subsequently established the present-day WCPA in 1958, the legacy of which is the influential World Parks Congress (Holdgate, 1999; WCPA, 2010).

People-free PAs were and still are conventionally controlled by governments and rationalized through criteria based upon wilderness, 'charismatic megafauna', endangered species, high species richness and uniqueness, and eliminating human impact (Adams, 2004b; Kalamandeen & Gillson, 2007; Dudley et al., 2014). Accordingly, local community evictions were rationalized based on a need to create "a new spatial order of nature and human occupation" (Neumann, 2001: p662), and became the mainstay of African colonial conservation (Neumann, 1996; 2001; Wolmer, 2005). This displacement of people for conservation is often referred to as *Fortress Conservation*, which prioritizes preservation of nature, through forceful exclusion, over the needs of those excluded (Rolston, 1996; Brockington & Igoe, 2006). More specifically, Brockington and Igoe (2006: p442) describe PA displacement in South Africa and Namibia at the time as "particularly thorough" in removing local communities for the cause of nature and the *Apartheid* regime (i.e. a complex set of laws enforcing racial segregation). However, whilst people-free PAs represent well-established and important tools to address biodiversity loss (see Coad et al., 2019; Visconti et al., 2019), the inseparable and inextricable connection between *nature* and *humans* suggest that categorizing landscapes as 'natural' or 'human-influenced' is both a false dichotomy and a historic anomaly as humans have greatly modified ecosystems for millennia (Agrawal & Gibson, 1999; Roe et al., 2000).

3.2.4.3. People-free PAs: shortcomings and the emergence of CBC discourse

People-free PAs are increasingly considered contrary to environmental and social justice imperatives, and endeavours to mainstream conservation into the wider landscape (Brechin et al., 2003; Brockington et al., 2010; Vucetich et al., 2018). Furthermore, these PAs have negatively altered local community perceptions of, and engagement with 'nature' (see Goldman, 2003), and resulted in poverty and resentment, due to denial of rights to historic lands and waters, notably in Africa (Brockington & Igoe, 2006). However, people-free PA assumptions began to be challenged in the 1960s in southern Africa with the rise of the sustainable use approach (Suich & Child, 2009; Child & Barnes, 2010). Furthermore, increasing evidence of the negative social impacts associated with PAs (e.g. restricted/ no access to traditional livelihood sources or areas of socio-cultural significance), was documented and condemned by numerous scholars both globally and in Africa (e.g. Brechin et al., 2003; Brockington & Igoe, 2006; West et al., 2006; Brockington et al., 2010; Kepe, 2018).

Not surprisingly, the past three decades witnessed widespread and increasing demands for rights-based approaches and socio-economic inclusiveness in conservation thinking, and conservation policy formulation and management (e.g. Brechin et al., 2003; Campese et al., 2009; Lele et al., 2010; Charles, 2013; Kashwan, 2013; Krause & Nielsen, 2014; Charles et al., 2016; Bennett et al., 2017; Singleton et al., 2017; Westlund et al., 2017). Recognition of the rights and needs of local communities began to emerge as a key consideration at the Third and Fourth World Parks Congresses (McNeely & Miller, 1984; McNeely, 1993). Furthermore, the 'Durban Accord' – agreed to at the Fifth World Parks Congress in 2003 – defined a new paradigm for PAs and the integration of interests of "all affected people" (IUCN, 2005: p220). Moreover, greater social consideration in PAs also led to the *Convention on Biological Diversity's* (CBD) establishment of the *Programme of Work on Protected Areas* in 2004, which describes the need for PAs to "... be integrated into the wider landscape and seascape, and into the concerns of the wider society, if they are to be successful in the long term" (Borrini-Feyerabend et al., 2013: p5). More recently, the focus continues to be placed upon acknowledging social inequity and injustice in conservation by the recent 2014 World Parks Congress in Sydney (IUCN, 2014a), and the proposed *Post-2020 Global Biodiversity Framework* (CBD, 2020). Nevertheless, the legitimacy of people-free conservation has been (and in many cases, continues to be) forcibly reproduced through modern PAs (Brockington et al, 2010), perhaps best illustrated by continued 'military-style' PA control, notably within the African context (Duffy et al., 2019). More specifically, this continues to be prevalent in South African PAs (Büscher & Ramutsindela, 2015; Annecke & Masubele, 2016).

Therefore, notwithstanding the above progress, the implementation of policy regarding greater social considerations in PAs remains problematic (Knight et al., 2008; Chandra & Idrisova, 2011; Calfucura, 2018; Stone et al., 2020). Furthermore, concerns exist that centralized PA managers solely work with local communities whose objectives align with *their* views on biodiversity conservation (Holt, 2005). Accordingly, ongoing concern exists for socially unjust conservation efforts resulting from the establishment of PAs (Brechin et al., 2003; Brockington et al., 2010; Vucetich et al., 2018). Nevertheless, recent decades have witnessed people-free PAs evolving to encompass a greater awareness of the importance of local conservation initiatives and interests in PA management, and the need to address the opportunity costs of conservation among the rural poor (Rands et al., 2010; Charles et al., 2016; Westlund et al., 2017). Additionally, growing recognition of the connection and interdependence of biological and cultural diversity for the future resilience of both ecosystems and local communities has increased (e.g. Pretty et al., 2010; Maffi & Woodley, 2012; Sterling et al., 2017).

3.2.4.4. Types of PA Governance

Globally PA governance regimes differ, but both the IUCN and the CBD recognize four broad 'modes' or 'types' of governance (*Table 3.2.*). Nevertheless, these governance types should be considered 'ideal types' in that they are "not a description of reality but … [aim] to give unambiguous means of expression to such a description" (Weber, 2017: p90). These governance types are based upon who possesses the authority and responsibility for key PA management decisions (Borrini-Feyerabend et al., 2013). However, three key issues include the difficulty of assigning governance types to PAs (i.e. which may be 'nested' and combine features of several governance types); that governance types may be perceived differently by different actors; and that governance arrangements may change over time (Borrini-Feyerabend et al., 2013). Therefore, in accordance with Borrini-Feyerabend et al. (2013), these PA governance types may be better considered on a continuum (*Figure 2.1.*), where the position on the continuum could vary for different kinds of PA governance approaches and decisions, and once again be perceived differently by different actors. A very brief introduction to each PA governance type follows.

Table 3.2.: The four broad protected area governance types. **Source:** Borrini-Feyerabend et al. (2013: p29).

Governance Type	Description
Type A. Governance by government (at various levels)	• Federal or national ministry or agency in charge • Sub-national ministry or agency in charge (e.g., at regional, provincial, municipal level) • Government-delegated management (e.g., to an NGO)
Type B. Shared Governance (governance by various rights-holders and stakeholders together)	• Transboundary governance (formal arrangements between one or more sovereign States or Territories) • Collaborative governance (through various ways in which diverse actors and institutions work together)· • Joint governance (pluralist board or other multi-party governing body)
Type C. Governance by private individuals and organizations	• Conserved areas established and run by: ▪ individual landowners ▪ non-profit organizations (e.g., NGOs, universities) ▪ for-profit organizations (e.g., corporate landowners)
Type D. Governance by indigenous peoples and/or local communities.	• Indigenous peoples' conserved territories and areas – established and run by indigenous peoples • Community conserved areas and territories – established and run by local communities

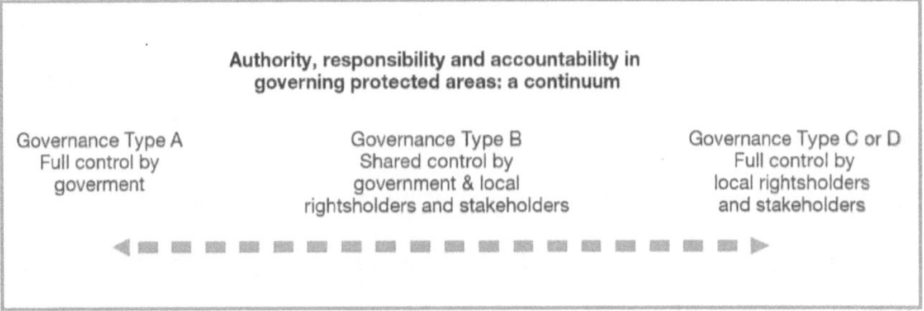

Figure 3.1.: The protected area governance continuum. **Source:** Borrini-Feyerabend et al. (2013: p102).

Type A governance refers to one or more state organizations holding authority, responsibility and accountability for PA management, and determining conservation objectives, and is usually associated with PA categories I and II (refer to *section 3.2.4.1.*). Furthermore, Type C, i.e. privately governed PAs, encompass PAs where individual(s), non-governmental organizations (NGOs) or corporations possess control and/or ownership (Borrini-Feyerabend et al., 2013). Whilst the importance of these PA governance types is acknowledged, specifically in South Africa (Langholz & Lassoie, 2001; Langholz & Krug, 2004), like all other types of PA governance, they have their strengths (e.g. the ability to conserve biodiversity and economic profitability), and weaknesses (e.g. forming "islands of elites" by concentrating land ownership with the wealthy).

In accordance with the introduction to CBC in *Chapter 1*, CBC governance is considered within this book to incorporate both Type B and Type D governance. Type B governance refers to *shared governance*, and is also known as, *collaborative governance*, *co-management* and *joint management*. Furthermore, Type D governance can be defined as, "protected areas where the management authority and responsibility rest with indigenous peoples and/or local communities through various forms of customary or legal, formal or informal, institutions and rules" (Dudley, 2008: p26). The term *co-management* became popular in the 1990's and can be describe as "a situation in which two or more social actors negotiate, define and guarantee amongst themselves a fair sharing of the management functions, entitlements and responsibilities for a given territory, area or set of natural resources" (Borrini-Feyerabend et al., 2000: p1). Therefore, co-management ideally synergistically combines the strengths and overcomes the weaknesses of each actor within the partnership, although achieving this balance and evaluating progress is often problematic (e.g. Carlsson & Berkes, 2005; Kothari et al., 2013; Cheng & Randall-Parker, 2017). This sharing of governance responsibility is the common *de facto* governance type represented in CBC initiatives. Consequently, Type B governance is a well-established strategy in modern conservation efforts globally (e.g. Borrini-Feyerabend, 2000; Pomeroy et al., 2001), in other African nations (e.g. Matose, 2006), and more specifically in South Africa (e.g. Hauck & Sowman, 2001; Koning, 2009).

Notwithstanding the above notion of shared governance in CBC initiatives, these governing systems are the expression of the relevant accumulated experience and/ or their direct interest in the protection, restoration and/or sustainable natural resource use in specific natural sites by local peoples (Borrini-Feyerabend et al., 2013). Therefore, Borrini-Feyerabend et al. (2013: p40) suggest that "[d]espite the real or perceived complexity" customary community institutions have been shown to function effectively, although this is not always the case due to erosion of customary governance systems (e.g. Cinner & Aswani, 2007). Nevertheless, Type D-specific governance contributions to conservation are increasingly recognized (see Levine & Richmond, 2014). Consequently, as both Type B and D PA governance are established approaches, it should be acknowledged that this book does not view CBC governance as new to communities or a new conservation concept.

3.3. Commons Theory

3.3.1. A brief introduction to Commons Theory

Since the 1990's empirical evidence of successful self-organising community's managing their *Common Property Resources* (CPRs), has generated interest in the applicability of *commons theory* to the challenges facing contemporary conservation governance (Dietz et al., 2003; van Laerhoven & Ostrom, 2007; Lejano et al., 2014; DeCaro, 2019). CPRs, in accordance with McKean (2000), are considered resources accessed, used and managed by a group of resource users who share rights and duties toward the resource. Accordingly, these challenges predominantly stem from the use of and management by multiple-users, and their rights to access, use and manage the CPRs, especially within cases of CPRs already being in a degraded state (Ostrom et al., 1999; Herzog & Ingold, 2019). Commons theory examines the management of both 'natural' (i.e. wild harvested resources), and 'human-constructed' (i.e. cultivated resources) CPRs (Ostrom et al., 1999), the focus of this book placed on the former.

Two key *Common Property Resource Management* (CPRM) problems are *exclusion* (i.e. controlling access) and *subtractability* (i.e. users subtracting from the welfare of others) (Ostrom et al., 1999). In reference to the latter Hardin's commonly cited "Tragedy of the Commons" predicts self-interested individual resource use at the expense of the common good (Hardin, 1968). However, Hardin has been criticized for two reasons; firstly, for assuming only two state-established institutional arrangements (i.e. state-property and private-property) can sustain CPRs in the long-term; and secondly, that resource-users are trapped in a 'commons dilemma' and unable to create solutions (Ostrom et al., 2002; Dietz et al., 2003; Acheson, 2011). Yet, Hardin's two aforementioned assumptions are refuted by extensive empirical research highlighting that communities, under certain conditions, can organise and manage CPRs successfully (e.g. Bromley et al., 1992; McKean, 2000, Ostrom, 1990, 1999; Agrawal, 2002; Berge & Van Laerhoven, 2011). Consequently, many scholars suggest Hardin remains widely cited to rationalise ongoing centralized-control of CPRs, and to perpetuate perceptions of local resource user's inability to regulate *their* natural resources (Berkes et al., 1989; Ludwig et al., 1993).

Commons theory considers four property regimes, common-property, open-access, private-property, and state-property, within which different forms of governance play out. The former two are of greatest relevance to CBC, where a common-property regime refers to resource-users sharing the rights and duties of a resource, whilst open-access regimes possess no rules regulating individual rights of use, though these may be implicit (Bromley et al., 1992; McKean, 2000). Furthermore, of relevance to the present context, state-property regimes are often considered to 'disincentivize' local control of resource use patterns due to the removal of local resource-user rights, and often result in *de facto* open-access regimes characterized by self-interested resource use, and thus the very 'tragedy' they attempt to avoid (Nagendra & Gokhale, 2008; Acheson, 2011). Lastly, while exclusion from resource harvesting is perhaps strongest in private-property regimes (Bromley et al., 1992), they too are not always successful (Acheson & McCloskey, 2008). Nevertheless, it should also be acknowledged that while Hardin's argument is not justified in relation to many 'small-scale' commons, this may not always be true for 'larger-scale' commons (Stern, 2011; Araral, 2014). Consequently, it is important to acknowledge no property regime works "efficiently, fairly, and sustainably in relation to all CPRs" (Ostrom et al., 1999: p279).

3.3.2. Lessons for CBC

3.3.2.1. Enabling CBC: lessons from commons research

CPRM is commonly viewed as a collective action problem (e.g. Ostrom et al., 1999; Ostrom, 1990, 2010b; Bodin, 2017; Herzog & Ingold, 2019). Accordingly, many have sought to frame the collective action of communities, and their partners, as *environmental stewards* (e.g. Bennett, N. et al., 2018a; Cockburn et al., 2018). The term *stewardship* is commonly associated with the collective action of a social group acting, "to protect, care for or responsibly use the environment in pursuit of environmental and/or social outcomes in diverse social–ecological contexts" (Bennett, N. et al., 2018a: p1).

Numerous scholars have examined the enabling factors, conditions and processes (i.e. "enablers') promoting 'successful' collective action in CPRM, most notably Ostrom (1990), Wade (1987), Baland and Platteau (1996), Pomeroy et al. (2001), and Agrawal (2002). I define an *enabler* here in accordance with Ostrom's (1990: p90) definition of a design principle, as "an essential element or condition that helps to account for the success of [CBC] institutions in sustaining the common property resources." Despite commons theory enhancing understanding of the factors supporting CBC, there remain numerous complex and enduring challenges associated with enabling collective action in highly diverse social, ecological and institutional contexts (Ostrom et al., 1999; Saunders, 2014; DeCaro, 2019; Herzog & Ingold, 2019). Nevertheless, valuable insights have emerged from commons scholars related to communities managing their natural resources, which are briefly discussed below.

Elinor Ostrom is arguably the most notable commons scholar, and her eight design principles associated with 'robust' CPRM institutions have been widely cited (Ostrom, 1990) *(Table 3.3.)*. Her design principles seek to explain conditions required "to sustain collective action in the face of social dilemmas posed by CPRs" (Cox et al., 2010: p2). These design principles have been extensively analysed for their usefulness and validity within diverse natural resource management contexts (e.g. Cox et al., 2010; Mills et al., 2013; Levine & Richmond, 2015; Baggio et al., 2016; Collen et al., 2016). More specific to the present context, they have been employed to investigate numerous African CBC-related initiatives (e.g. Crook & Mann, 2002; Quinn et al., 2007; Cinner, et al., 2009a; Biggs et al., 2019; Child, 2019).

Table 3.3.: Ostrom's Eight Design Principles for Sustainable Commons. **Source:** Ostrom (1990).

Ostrom's Eight Design Principles:

1. Define clear group boundaries.
2. Match rules governing use of common goods to local needs and conditions.
3. Ensure that those affected by the rules can participate in modifying the rules.
4. Make sure the rule-making rights of community members are respected by outside authorities.
5. Develop a system, carried out by community members, for monitoring members' behaviour.
6. Use graduated sanctions for rule violators.
7. Provide accessible, low-cost means for dispute resolution.
8. Build responsibility for governing the common resource in nested tiers from the lowest level up to the entire interconnected system.

Notwithstanding the contribution of Ostrom's design principles, they have also been subjected to extensive and often contentious discussion (e.g. Araral, 2014, 2016; Cox et al., 2016). More specifically, some have criticized Ostrom's overgeneralization of cases and contexts (e.g. Stein et al., 2000; Young, 2002; Blaikie, 2006). Nevertheless, extant literature emphasizes how her design principles remain relevant for sustaining CPRs by local actors, and thus enabling change towards a CBC mode of governance. For example, Cox et al. (2010) analysed 91 global case studies using Ostrom's design principles. Their findings reinforced Ostrom's, but did lead to an amended list of ten, deemed to better capture important social variables, including more clearly defining social boundaries within groups, the degree of alignment of rules with local conditions, and the accountability of monitors to resource users.

Of specific relevance to the present coastal and marine context, Pomeroy et al. (2001), in reference to their research in Asia, identify 18 conditions at three levels that they consider to be of importance to the success of community-based fisheries co-management initiatives (*Table 3.4.*). They specifically highlight the context-specific, and interactional nature of these conditions stating how "each supports and links to another to make the complex process and arrangements for co-management work" (Pomeroy et al., 2011: p2015). Furthermore, they strongly emphasize the need for planning and implementation to be conducted at multiple levels, and specifically, the importance of multi-actor partnerships.

Table 3.4.: Pomeroy et al.'s conditions for successful fisheries co-management. **Source:** Pomeroy et al. (2001).

Category	Attributes
Supra-community level	1. Enabling policies and legislation 2. External agents
Community level	1. Appropriate scale and defined boundaries 2. Membership is clearly defined 3. Group homogeneity 4. Participation by those affected 5. Leadership 6. Empowerment, capacity building, and social preparation 7. Community organizations 8. Long-term support of the local government unit 9. Property rights over the resource 10. Adequate financial resources/budget 11. Partnerships and partner sense of ownership of the co-management process 12. Accountability 13. Conflict management mechanisms 14. Clear objectives from a well-defined set of issues 15. Management rules enforced
Individual & household level	1. Individual incentive structure

In addition to Pomeroy et al.'s valuable insights into community-based fisheries management institutions, Agrawal (2002) – specifically building on the work of Ostrom (1990), as well as Wade (1987) and Baland and Platteau (1996) – offers an expanded set of 33 critical enabling factors for successful CPRM. He groups these into four categories: resource system, users, institutional framework, and externalities (*Table 3.5.*). However, in accordance with Pomeroy et al. (2001), Agrawal (2002) explicitly emphasizes the highly interdependent nature of these enablers, and the specific need to consider their presence, or ability to be present, based on this interdependence (Agrawal, 2002). Agrawal's factors have been used to evaluate diverse community-based governance systems, and explain possible reasons for governance problems and proposed solutions (e.g. Moore & Rodger, 2010; Hearn, 2008; Mackelworth et al., 2008; Gollwitzer, 2014).

Table 3.5.: Agrawal's critical enabling factors for Sustainable Commons. **Source:** Agrawal (2002).

Category		Attributes
Resource	System	1. Small size 2. Well-defined boundaries 3. Low levels of mobility 4. Possibility of storage of benefits from the resources 5. Predictability
	Users	6. Small groups 7. Clearly defined boundaries 8. Shared norms 9. Good leadership 10. Past successful experiences 11. Interdependence between group members 12. Similarities in identities and interests 13. Low levels of poverty 14. Overlap between user group residential location and resource location 15. High levels of dependence on the resource system 16. Fairness in allocation of benefits from common resources 17. Low levels of user demand 18. Gradual changes in level of demand
Institutional	Framework	19. Rules are simple and easy to understand 20. Locally devised access and management rules 21. Ease in enforcement of rules 22. Graduated sanctions 23. Availability of low-cost adjudication 24. Accountability of low-cost adjudication 25. Match restrictions on harvests to regeneration of resources
	Externalities	26. Low-cost exclusion technology 27. Time for adaptation of new technologies related to the commons 28. Low levels of articulation with external markets 29. Gradual changes in articulation with external markets 30. No undermining of local authorities by central government 31. Supporting external sanctioning institutions 32. Appropriate levels of external aid to compensate local users for conservation actions 33. Nested levels of appropriation, provision, enforcement, governance

Several African-specific CBC studies, three examples of which I discuss below, have provided valuable new insights into CBC enablers, and reinforce the findings of Ostrom (1990), Pomeroy et al. (2001), and Agrawal (2002 (see *Table 3.6.*). For example, Cinner et al. (2009a) underline how applying terrestrial conservation frameworks into marine realms, using examples from the Western Indian Ocean region, will likely result in inflexibility and institutional mismatches that will constrain effective management of these marine resource systems (*Table 3.6.*). Furthermore, they reiterate the negative effect when certain design principles are not fully present, notably: monitoring and the monitoring of monitors (i.e. accountability), clearly defined boundaries, and collective choice arrangements (i.e. locally devised rules). In another African example, Galvin et al. (2018) recently conducted a systematic review of African CBC initiatives, from which they propose ten institutional processes key to the success of these initiatives (*Table 3.6.*). They emphasize, among other conditions, the need for the presence of both key players/ leaders, the support of bridging organizations (e.g. government and NGO partners), and partnerships characterized by collaboration amongst all CBC actors (Galvin et al., 2018). More recently, Biggs et al. (2019), in reviewing Zimbabwe's CAMPFIRE program (i.e. a decentralized community wildlife management program), develop a list of ten conditions of emergence for robust conservation initiatives. Their key findings include the need to know who the beneficiary group is, and the ability of this group to participate and change rules, which should be perceived to be the result of legitimate decision-making processes, the need for adequate conflict resolution, and the need for multiple levels of external support (*Table 3.6.*). *Appendix 8* depicts further examples of African-specific studies that have investigated enablers for CBC in diverse contexts, and which informed the 14 CBC enablers proposed in the subsequent section.

Table 3.6.: African-specific CBC enablers identified by Cinner et al. (2009a), Galvin et al. (2018) and Biggs et al. (2019).

Scholars	List of enablers		
Cinner et al.'s (2009a) institutional design principles based upon community-based management of inshore marine resources in the Western Indian Ocean	1) Clearly defined membership rights 2) Congruence between the scale and scope of rules and local conditions 3) Resource users have rights to make, enforce and change rules 4) Conflict resolution mechanisms 5) Nested Enterprises 6) Monitoring of monitors 7) Clearly defined geographic boundaries 8) Collective choice arrangements with affected individuals able to participate in changing rules 9) Graduated sanctions 10) Monitoring of resources 11) Monitoring of resource users		
Galvin et al.'s (2018) ten institutional processes for African CBC initiatives	1) Length of establishment 2) Presence of key players/ leaders 3) Presence of supporting bridging organizations 4) Presence of diverse and multiple partnerships 5) Presence of collaboration in decision-making 6) Presence of social learning including monitoring and assessment 7) Devolution or rights to local community 8) Presence of monetary incentives for participating members 9) Presence of non-monetary incentives for participating members 10) Conservation model is in-line with cultural worldviews and practices		
Biggs et al.'s (2019) ten conditions of emergence for robust conservation based on Zimbabwe's CBC CAMPFIRE program	*Theme A: Recognising the need for change*	1) Collective recognition of the problem 2) Shared understanding of the problem
	Theme B: Expectations of positive outcomes	3) Collective interest in adopting new rules 4) High expectation and value of future benefits
	Theme C: Context that facilitates experimentation and collective learning among actors	5) Presence of policy entrepreneurs to champion the rule and advocate for its adoption 6) Context allows for collective learning within and outside the member group 7) Social norms that favour collaboration: Reciprocity, trust and cooperation should be valued by the actors in the system 8) Expectations that the group appropriating and benefiting from the new actions and rules will be stable.
	Theme D: Legitimate local scale governance	9) Perceived legitimate decision-making structure 10) Opportunity to generate new norms internally

Lastly, several systematic reviews have proposed complimentary and supplementary factors or conditions leading to CBC success or failure (and which informed the enablers identified in the subsequent section), and thus make valuable contributions to the CBC knowledge base (e.g. Pagdee et al., 2006; Waylen et al., 2010; Brooks et al., 2010, 2013; Brooks, 2016). Perhaps most notable in the present context the aforementioned recent African-specific CBC systematic review by Galvin et al. (2018). The next section proposes a list of 14 CBC enablers based upon the above scholars, and an extensive review of further relevant literature.

3.3.2.2. Identifying key CBC 'enablers': with an African emphasis

The presence of the list of 14 CBC enablers proposed below is assumed to positively influence the 'success' of a shift to a CBC mode of governance (*Cf.* Ostrom, 1990; Pomeroy et al., 2001; Agrawal, 2002).

Whilst acknowledging the importance, and interactional nature of ecological and social factors, in accordance with other 'design principle' studies (e.g. Cinner et al., 2009a; Levine & Richmond, 2015; Collen et al., 2016), the enablers proposed below focus mainly on 'potentially' enabling social-institutional factors, conditions and processes for CBC initiatives. More specifically, the focus is placed upon socio-economic, socio-cultural, and institutional factors and conditions within the CBC governing system. This does not attempt to undermine the importance of ecological factors, but is merely beyond the scope of the present research objectives. Accordingly, I explicitly acknowledge that ecological conditions of the resource system affect CBC governance regimes. Consequently, I briefly introduce some key ecological factors and conditions below which are assumed influential within this book.

Ostrom et al. (1999) effectively summarize key ecological attributes affecting a resource system, which they suggest include: clearly defined resource boundaries; the size and carrying capacity of the resource system; the measurability of the resource; the temporal and spatial availability of resource flows; the amount of storage in the system; resource mobility (e.g. water, wildlife, and fish) or stationarity (e.g. trees and medicinal plants); and how fast resources can regenerate. High resource mobility is especially important (and taken as a given) within the present context of coastal and marine conservation (e.g. wide-ranging fish stocks harvested along regional coastlines), and is also emphasized by Agrawal (2002). Consequently, while I acknowledge the importance of the aforementioned ecological conditions to governing any CBC initiative, within the scope of this research I only consider *clearly defined resource system boundaries* in the case study analyses.

Therefore, based upon an extensive review of literature, as discussed subsequently, and in accordance with the current research focus, I propose a consolidated list of 14 socio-institutional enablers, the presence of which is hypothesized to be key for CBC initiation, implementation and governance (*Table 3.7.*). Nevertheless, it should be acknowledged from the outset that compiling a list of enablers for CBC initiatives is extremely challenging, since factors and conditions affecting CBC initiatives are highly complex, and enablers are very context-specific, as has been acknowledged by the prominent commons scholars discussed above. For example, high levels of dependence upon a natural resource may result in greater resource harvesting activity and environmental degradation, which is exacerbated by contexts in which few alternative livelihood sources exist for the resource user (Barrett et al., 2011; Roe et al., 2015). Yet high levels of resource dependence may also promote greater sustainable harvesting practices in an effort to conserve resources of importance to the resource user and their future (Stern et al., 2002; Wilson et al., 2016). Moreover, Wilson et al. (2016: p227) suggest one may assume a potential "trend in which conservation behaviors peak at a moderate level of resource dependence." Consequently, this example serves to emphasize not only the complexity, and ultimately context-specific nature, but also the directionality and nonlinearity of CBC enablers.

Table 3.7.: An adapted and consolidated set of socio-institutional enablers for successful CBC implementation and governance based upon the literature reviewed.

	Enablers	Key Supporting Literature
Resource System & Users	1. Clearly defined resource system and user boundaries	Ostrom (1990); Pomeroy et al. (2001); Agrawal (2002); Cox et al. (2010); Cinner et al. (2009a); Child (2019)
	2. Shared norms, values, interests & identities	Pomeroy et al. (2001); Agrawal (2002); Biggs et al. (2019); Child (2019b); Herzog & Ingold (2019)
	3. Strong community ties	Agrawal (2002); Alexander et al. (2016); Infield et al. (2017); Crona et al. (2017); Biggs et al. (2019)
	4. Strong local leadership	Pomeroy et al. (2001); Agrawal (2002); Ostrom (2010b); Galvin et al. (2018); Biggs et al. (2019)
	5. Low levels of poverty	Ostrom (1990); Agrawal (2002); Barrett et al., (2011); Brooks et al. (2012); Raycraft (2019)
	6. High levels of dependence on resource	Agrawal (2002); Robbins et al. (2009); Wilson et al. (2016); Gilliam & Charles (2018)
	7. Equitable distribution of benefits from common property resources	Agrawal (2002); Pomeroy et al. (2001); Cox et al. (2010); Galvin et al. (2018); Biggs et al. (2019); Child (2019b)
Institutional Structures & Externalities	8. The presence of a community institutions	Pomeroy et al. (2001); Cinner et al. (2009a); Western et al. (2015); Kawaka et al. (2017); Bluwstein & Lund (2018)
	9. Locally devised natural resource access and management rules	Ostrom (1990); Pomeroy et al. (2001); Agrawal (2002); Cinner et al. (2009a); Cox et al. (2010); Galvin et al. (2018); Biggs et al. (2019)
	10. Rules strongly align with local priorities/ needs	Ostrom (1990); Pomeroy et al. (2001); Agrawal (2002); Cox et al. (2010); Cinner et al. (2009a); Galvin et al. (2018); Biggs et al. (2019); Child (2019)
	11. Ease in enforcement of rules, and conflict resolution	Ostrom (1990); Pomeroy et al. (2001); Agrawal (2002); Cinner et al. (2009a); Cox et al. (2010); Armitage et al. (2017); Biggs et al. (2019); Charles et al. (2020)
	12. High levels of accountability	Pomeroy et al. (2001); Agrawal (2002); Cinner et al. (2009a); Cox et al. (2010); Lockwood et al. (2010); Bluwstein et al. (2016); Galvin et al. (2018); Biggs et al. (2019)
	13. Low levels of articulation with external markets	Agrawal (2002); Stern (2002); Brewer et al. (2012); Bersaglio & Cleaver (2018)
	14. The presence of 'nested' governance with high levels of initial external support, including, enabling legislation and political will, and external financial and technical support	Ostrom (1990); Pomeroy et al. (2001); Agrawal (2002); Cox et al. (2010); Cinner et al. (2009); Galvin et al. (2018); Biggs et al. (2019); Child (2019); Armitage et al. (2020)

Like Ostrom (1990, 2002), Pomeroy et al. (2001), and Agrawal (2002), I acknowledge that these enablers are often highly interconnected and interdependent, and therefore, recognize that the presence of one enabler may promote the presence of another (see also Ostrom, 2009). *Figure 3.2.* depicts, using a social network analysis approach, the potentially highly interconnected nature of this dissertation's proposed 14 CBC enablers, as I see it. This social network map represents the number of proposed ties or connections between enablers (i.e. an enablers proposed *degree of centrality*). More specifically, it is bidirectional as it depicts how the presence of an enabler affects, or is affected by, the presence of other enablers. *Enabler 11* (i.e. the ease of rule enforcement & conflict resolution) provides a notable example of an enabler's potentially highly interconnected nature (*Figure 3.2.*). It is plausible to expect that the ease with which enforcement takes place within a CBC initiative should improve when: the resource system and user boundaries are clearly defined (i.e. *enabler 1*); and a community possesses shared norms, values, interests and identities (i.e. *enabler 2*), strong community ties (i.e. *enabler 4*), and community institutions (i.e. *enabler 8*), characterized by strong local leadership (i.e. *enabler 3*). Furthermore, when a community possesses low levels of poverty (i.e. *enabler 5*), and benefits that emerge from the initiative are equitably distributed (i.e. *enabler 7*), it is likely community members would have less need to overharvest the specific natural resource of concern, and hence not break the resource access and use rules. Moreover, the ease with which rule enforcement is carried out would plausibly be positively affected when high levels of community support for a CBC initiative exist as a result of rules having been locally devised (i.e. *enabler 9*), and when these rules strongly align with a community's priorities or needs (i.e. *enabler 10* – which also links back to *enabler*

3, as this commonly includes striving for lower local poverty levels). Additionally,
enforcement should be positively influenced by high levels of accountability
amongst actors (i.e. *enabler 12*), and when the initiative is supported by external
partners (i.e. *enabler 14*). These two enablers perhaps most notably associated
with generating greater required legitimacy for enforcement of a CBC initiative,
which is frequently as a consequence of legal recognition and effective monitoring
by a government department (*Figure 3.2.*).

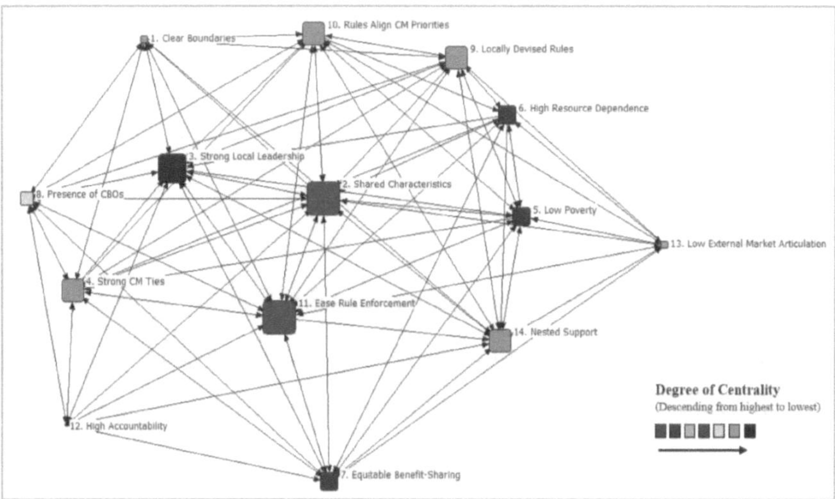

***Figure 3.2.*:** A social network map depicting the interactional nature of the 14 proposed CBC enablers. **Note:** Enablers
have been 'mapped' by their *degree of centrality* (i.e. the number of proposed ties of each enabler to other enablers),
which is indicated by size (the bigger the icon the higher the centrality – i.e. more ties to other enablers) and colour (see
legend). **Source:** Designed by author.

Consequently, many recognized commons scholars have acknowledged the
difficulties experienced by attempts to contend with the aforementioned
complexity of enablers (e.g. Ostrom et al., 1999, 2002; Pomeroy et al., 2001;
Agrawal, 2002, 2014; Araral, 2014, 2016; Cox et al., 2016). In addition to the above

issues, challenges associated with wording of the proposed enablers is acknowledged. This notably concerns the ability of the choice of wording to influence the analysis of these enablers, which is an acknowledged limitation here. While every effort has be made to make the clarity of wording of these enablers specific enough to facilitate their analysis within the three case studies, it should also be acknowledged – in accordance with numerous notable commons scholars who have proposed lists of enablers (e.g. Ostrom, 1990; Pomeroy et al., 2001; Agrawal, 2002; Cox et al., 2010) – that wording should also be as simple, and as broadly encompassing of relevant factors and conditions, as possible (see *Tables 3.3., 3.4.,* and *3.5.* above).

The 14 proposed CBC enablers should not be viewed as 'set-in-stone' but as a 'null hypothesis' for what factors and conditions may enable successful CBC implementation and governance. Accordingly, the subsequent analysis within the contexts of the three selected cases (i.e. in *Chapters 6, 7* and *8)* seeks to investigate these 'hypotheses', and ultimately suggest modifications and/ or additional enablers where necessary within the South African CBC context, at the culmination of the book in *Chapter 9.* This case study enabler analysis makes use of a *four-point rating scale of presence*, namely: *absent; partially present; partially absent;* and *present.* Accordingly, this 'rating scale' can be considered as a continuum ranging from the complete *absence,* to the complete *presence* of an enabler (*Figure 3.3.*). Furthermore, this *rating scale* was specifically chosen to provide greater insight into the degree of 'partial presence' of an *enabler.* While a five-point Likert scale was initially proposed, it was determined that the data collected lack sufficient detail to confidently assess the enabler at that level of differentiation. Nevertheless, the *four-point scale* is considered to sufficiently address the present scope of research, and thus achieve the aim and objectives. Consequently, based upon the empirical findings that emerge from this case study analysis, and the South African specific CBC enablers identified in *Chapter 4*, this list of 14 enablers is amended in *Chapter 9*. The subsequent section endeavours to explain the importance of, and the rationale for the inclusion of these 14 enablers within the proposed list. Inspired by the work of Agarwal (2002), the 14 proposed CBC enablers are discussed here under two categories: *resource systems and users;* and *institutional structures and externalities.*

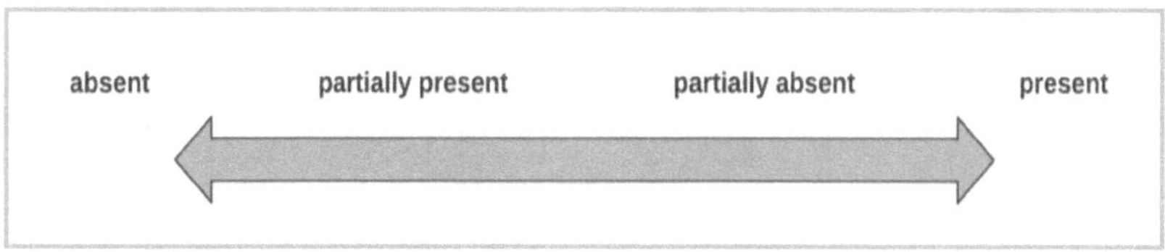

Figure 3.3.: A four-point rating scale for the case study analysis of the presence of the 14 proposed CBC enablers.

The Resource System & Users
Enabler 1: Clearly defined resource system and user boundaries
In accordance with numerous past CBC studies (e.g. Pomeroy et al., 2001; Cinner et al., 2009a; Child, 2019), I consider *clearly defined resource system boundaries* useful to assisting CBC initiatives to avoid the two aforementioned major CPRM 'dilemmas', i.e. *exclusion* and *subtractability* (refer to *section 3.3.1.*). Similarly, like Cox et al. (2010) I propose *clearly defined resource-user boundaries* as important for the creation of exclusion rules ideally considered to reward the 'conservation investment' made by 'insiders', by preventing 'free-riding' by 'outsiders'. This is considered especially relevant to natural resources with high mobility such as fish

(Cinner et al., 2009a; Almeida et al., 2009; Baggio et al., 2016). Defined resource-user boundaries is considered here to be two-fold concerning both socio-cultural and geographical boundaries (*Cf.* Baggio et al., 2016).

Many scholars caution against identifying communities as 'clear entities' (e.g. Agrawal & Gibson, 1999; Titz et al., 2018). Extant CBC literature emphasizes the problematic nature of defining clear socio-cultural community boundaries, as community composition and perceptions are often wrongly assumed homogeneous (Agrawal & Gibson, 1999; Bray et al., 2012). Accordingly, Titz et al. (2018: p2) suggest the use of the term 'community' to refer to "'local' or 'place-based' communities displays a rather one-dimensional and static understanding of community, ignoring social dynamics and the multiple, sometimes conflicting, layers of meaning that are embedded in the term." Not surprisingly challenges in defining 'community' are widely acknowledged, notably in the present context within both African (e.g. Mandondo, 2001; Blaikie, 2006; Klein et al., 2008), and South African studies (e.g. King & Peralvo, 2010). The topic of community homogeneity is discussed further in the subsequent enabler relating to *shared norms, values, interests and identities*.

<u>Enabler 2: Shared norms, values, interests & identities</u>
CBC success is commonly considered heavily dependent upon a community with shared cultural values, as this has extensively been shown to positively influence governance of CBC initiatives both globally (e.g. Levine & Richmond, 2014; Alcántara-Salinas et al., 2015; Delgado-Serrano, 2017; Infield et al., 2017), and in Africa specifically (Harris, 2007; Scanlon & Kull, 2009; Brown & Lassoie, 2010; Holmes et al., 2018; Biggs et al., 2019; Child, 2019). In this book, I refer to

culture as "the system of shared beliefs, values, customs, behaviors…members of society use to cope with their world and with one another" (Bates & Plog, 1990: p7). Furthermore, Biggs et al. (2019) note the need for collective recognition and shared understanding of the 'problem', and a collective interest in adopting new rules. Therefore, whilst some scholars have shown homogeneity increases community engagement in conservation governance without the need for economic incentives (e.g. Berkes 2009b; Delgado-Serrano, 2017), it should be acknowledged that cultural erosion – due to colonial influences or present day political and market forces – has potential negative implications for CBC (Cinner & Aswani, 2007; Ibarra et al., 2011). More specifically, the negative effects of cultural 'erosion' have been shown within African CBC contexts (e.g. Cinner, 2007; Walters et al., 2015; Fernández-Llamazares et al., 2018).

However, scholars have also emphasized that the notion of homogeneity ignores differences in power, interests, norms, and values found within communities (Agrawal & Gibson, 1999; Leach et al., 2011; Ojha et al., 2016). Furthermore, community heterogeneity has specifically been shown in certain contexts to possesses no consistent relationship with levels of collective action and is subject to the type of heterogeneity (i.e. economic; political, and cultural) (Adhikari & Lovett, 2006; Andersson & Agrawal, 2011; Brooks et al., 2013). Consequently, notwithstanding the contentious nature of the perceived influence of socio-cultural homogeneity/ heterogeneity (see also Mudliar & Koontz, 2018), I propose here that the presence of *shared norms, values, interests and identities* will positively influence the (potential) success of CBC initiatives.

<u>Enabler 3: Strong local leadership</u>

Leadership has been described as "the most important attribute in the toolbox of conservation science" (Dietz et al., 2004: p274). Accordingly, numerous scholars have emphasized the importance of the presence of key players or leaders for positive social and ecological CBC outcomes (Pomeroy et al., 2001; Galvin et al., 2018; Biggs et al., 2109). More specifically, the presence of strong *local* leadership plays particularly central and influential role in social and institutional processes considered crucial for CBC, as has been shown extensively within both diverse global (e.g. Brooks et al., 2012; Cinner et al., 2012a; Warren & Visser, 2016; Lyon & Cavaye, 2016; Crona et al., 2017), and African contexts (e.g. Zulu, 2012; Nelson, 2012; Hauzer et al., 2013; Mbaru & Barnes, 2017; Barnes et al., 2019; Hebinck et al., 2020). Nevertheless, strong local leadership may both enable and constrain CBC governance, since the power of local leaders may advance CBC initiatives or fulfil vested interests, i.e. result in *elite-capture* (Warren & Visser, 2016; García-López, 2019). *Elite-capture* of CBC initiatives is especially prevalent within community's unequal to begin with (Agrawal, 2002; Platteau, 2004;), and consistently shown to constrain successful CBC outcomes (e.g. Saito-Jensen et al., 2010; Uberhuaga, et al., 2011; Hebinck et al., 2020). Nevertheless, many CBC studies indicate the risk of elite-capture is not an inevitable and long-term outcome as it can be minimized (Campbell et al., 2007; Saito-Jensen et al., 2010; Lund & Saito-Jensen, 2013; Warren & Visser, 2016). Consequently, the need exists to understand and acknowledge the complexity of leadership and the specific motivations of leaders (Barnes-Mauthe et al., 2015; Steenbergen, 2016; Musgrave & Wong, 2016; Steenbergen & Warren, 2018).

In accordance with *enabler 2*, representation of diverse community interests and concerns is central to local leadership and its ability to improve CBC governance (e.g. Ribot, 2007; Ratner et al., 2013; Steenbergen, 2016; Crona et al., 2017). Furthermore, Crona et al. (2017) emphasize both "leadership engagement" and "leadership legitimacy" as key elements of social capital likely to increase positive social and ecological CBC outcomes. Moreover, Englefield et al. (2019: p20-21) identify various "leadership competencies" which may enable improved conservation governance, including: building trust amongst followers; the ability to create a vision that is inspirational; the ability to demonstrate the behaviours they expect to see in others; and displaying experience with technical skills. Lastly, many note strong local leadership is more about stimulating communication and 'bridging' within communities, as well as between communities and external partners (Berdej & Armitage, 2016; Crona et al., 2017; Mbaru & Barnes, 2017; Steenbergen & Warren, 2018). Therefore, I propose the presence of *strong local leadership* will positively influence the (potential) success of CBC initiatives.

<u>Enabler 4: Strong community ties</u>

This enabler overlaps substantially with *enabler 2* above since strong community ties based on shared socio-cultural values can result in increased levels of intra-community trust, reciprocity and respect for both formal and informal (notably customary) CBC-related institutions and practices (Biggs et al., 2019). This in turn can prevent conflicts related to opportunistic 'free-riding' natural resource harvesting behaviour, and thus promote sustainable use of natural resources (Berkes 2009a; Ruiz-Mallén et al., 2015; Ulambayar et al., 2017). Furthermore, Crona et al. (2017) emphasize the importance of communication and bridging among community members for successful local CBC leadership, a topic discussed under *enabler 3* above.

However, whilst strong community ties can motivate collective action, they may be used to either exploit or conserve a natural resource (Pretty & Smith, 2004; Ostrom & Ahn, 2009). Accordingly, a lack of coordination among community actors, especially exclusion of marginalized community sub-groups – such as women and youth – can lead to conflicting objectives that constrain CBC governance (see Alexander et al., 2016). In contrast, communities which maintain frequent face-to-face communication, and which possess high levels of social capital and a desire to maintain one's reputation within small groups, increase the potential for trust and thereby can induce compliance and decrease the cost of monitoring (Pretty, 2003; Young et al., 2016; Dressel et al., 2020). However, strong family ties in particular can still make reporting and enforcing CBC compliance problematic (Delgado-Serrano, 2017). Nevertheless, strong community and family ties motivated by a desire to maintain one's cultural identity and resist outside threats (e.g. State and private-sector resource capture), can also promote positive CBC outcomes, although this is increasingly constrained by increased modernization (Ruiz-Mallén et al., 2015; Infield et al., 2017). Consequently, I assume the presence of *strong community ties* will positively influence the (potential) success of CBC initiatives.

Enabler 5: Low levels of poverty

Poverty is commonly linked to environmental degradation (due to high resource-dependence – discussed subsequently in *enabler 5*), and emphasized by many within CBC literature as a primary motivating factor for non-compliance (e.g.

Cinner, 2009; Barrett et al., 2011; Brooks et al., 2012). Accordingly, poverty may lead resource users to overexploit resources as it may 'discredit' future incomes (Ostrom, 1990), and can stimulate elite-capture (García-López, 2019). Furthermore, poor households – especially those headed by females – have been shown to benefit less from commons resources, and therefore, levels of poverty may further exacerbate local (gender-based) inequalities (Adhikari, 2005; Robbins et al., 2009; Barbier, 2010; Angelsen et al., 2014; Thondhlana & Muchapondwa, 2014). That's said high levels of poverty may also promote higher levels of 'care' amongst resource users for resources they are highly dependent upon, and therefore, the effect of poverty levels on CBC outcomes is debatable (Raycraft, 2019) (this topic is discussed further in *enabler 6* below).

Conservation initiatives often attempt to link poverty alleviation with conservation (e.g. Roe et al., 2013, 2015; Gurney et al., 2014). However, some scholars have emphasized the threat of this strategy to biodiversity conservation (e.g. Sanderson & Redford, 2003). Nonetheless, linking poverty alleviation objectives and biodiversity conservation efforts is widely considered central to the success of CBC initiatives, and specifically relates to *enabler 10*, i.e. the alignment of CBC governance to local priorities. That said it should be acknowledged that low poverty levels are very problematic to obtain since CBC initiatives commonly occur in developing world. Therefore, although the effect of poverty levels is context-specific, I operate here on the premise that the presence of *low levels of poverty* will enable the success of CBC initiatives.

<u>Enabler 6: High levels of dependence on natural resources</u>
Communities in developing countries often possess high levels of dependence on natural resources, and are therefore often considered more vulnerable to change. This has specifically been shown within both African (Cinner et al., 2012b; McClanahan et al., 2015), and South African CBC contexts (Shackleton & Shackleton, 2004; Thondhlana & Muchapondwa, 2014). As introduced above in *enabler 5*, a higher reliance on natural resources among poor communities is often associated with environmental degradation. However, as discussed previously, this 'poverty-environment trap' is more complex, and subject to social and ecological context (Agrawal & Angelsen, 2009; Barbier, 2010; Thondhlana & Muchapondwa, 2014; Raycraft, 2019). Some scholars have shown how a relation of well-being linked to a resource can promote recovery from an environmental disaster (Gilliam & Charles, 2018).

Resource-dependency influences interactions between different conservation actors (Agrawal & Gibson, 1999; Robbins et al., 2009). Robbins et al. (2009) specifically note how community members will generally approve rules allowing *their* continued resource harvesting practices, and restricting others. Nonetheless, there are "paradoxical implications" for natural resource-dependence, since while high resource-dependency may ostensibly increase the value of the resource to the user, and thus motivate conservation behaviours, it may also inhibit a community's ability to conserve a resource due to a lack of alternative livelihoods. Nevertheless, broader socio-economically inclusive planning of conservation areas that address factors affecting resource-dependency are considered more effective than attempting to manage resource-harvesting activity itself (Andrew et al., 2007; Peterson & Stead, 2011). As introduced previously, Wilson et al. (2016:

p227) suggest one may assume a potential "trend in which conservation behaviors peak at a moderate level of resource dependence." Therefore, based on the aforementioned issues this enabler is very context-specific, and may not be enabling in certain contexts. Consequently, while this enable may require modification or omission, subject to contextual findings in the three cases, in the final list of enablers proposed at the culmination of this book, I propose the presence of *high levels of resource-dependence* to enable the success of CBC initiatives by motivating resource-users to conserve *their* resources.

Enabler 7: Equitable distribution of benefits from common property resources

A key concern for CBC initiatives is the equitable distribution of benefits amongst community members, and thus as discussed previously, the occurrence of elite-capture. Therefore, I refer here to equitable distribution as the ability of all community members to access and use natural resources to derive both tangible and intangible benefits. Much research has shown that providing economic incentives, which outweigh the consequences of changed conservation behaviours, may motivate conservation action (Nilsson et al., 2016; Fletcher et al., 2016; Biggs et al., 2019; Stone et al., 2020). However, scholars have specifically shown that southern African CBC initiatives have often resulted in local-level elite-capture (e.g. Balint & Mashinya, 2006; Blaikie, 2006; Nelson, 2012; Zulu, 2012; Matenga, 2015; Child, 2019; Hebinck et al., 2020). However, as introduced in *enabler 3*, many CBC studies indicate the risk of elite-capture is not inevitable, nor a long-term outcome of CBC initiatives. As with *enabler 6*, I propose that the presence of *equitably distributed benefits* will positively influence the (potential) success of CBC initiatives by motivating resource-users to conserve their resources.

<u>Enabler 8: The presence of a community institutions</u>

The presence of a community-based management organization (CBO) is recognized as crucial to a community-based mode of governance (Pomeroy et al., 2001). This has especially been shown in both African (e.g. Cinner et al., 2009a; Aheto et al., 2016; Kawaka et al., 2017; Bluwstein & Lund, 2018), and more specifically South African CBC research (e.g. Fabricius & Collins, 2007). CBOs are required to perform crucial functions such as knowledge dissemination, community deliberation, project implementation, monitoring and enforcement, benefit-sharing and conflict resolution (Olsson et al., 2004; Western et al., 2015; Aheto et al., 2016; Kawaka et al., 2017). In accordance with *enabler 3*, the legitimacy of local CBOs often depends on the leadership structures, as well as their relationship to other local/ customary management organisations, and their legal recognition by the State.

Whilst a community may already possess customary leaders, these leaders may lack the necessary legitimacy, capacity or time to manage local natural resources, often necessitating a more bottom-up process facilitated by other leaders deemed more acceptable to local resource users (Pomeroy et al., 2001; Fabricius & Collins, 2007; Cinner et al., 2009a; Crona et al., 2017). However, it has been shown that a strong relationship and mutual understanding needs to be established (preferably within the planning phase) between CBOs and customary leadership for CBC initiatives to be 'successful' (e.g. Aheto et al., 2016; Bluwstein & Lund, 2018). Lastly, strong local-level partnerships between conservation-specific CBOs and other local CBOs, for example women's organizations, can positively influence conservation outcomes (Radel, 2012). Consequently, I propose here that the mere *presence of a CBO* in the community, especially when a CBC-dedicated CBO is present, will positively influence the success of CBC initiatives.

<u>Enabler 9: Locally devised natural resource access and management rules</u>
Local participation in rule making is widely considered crucial to CBC initiatives (e.g. Campbell et al., 2010; Cox et al., 2010; Brooks et al., 2012; Brooks, 2016; Biggs et al., 2019; Armitage et al., 2020). Accordingly, the recently proposed *Post-2020 Global Biodiversity Framework* specifically calls for the recognition and participation of local and indigenous communities (CBD, 2020). Nevertheless, many conservation initiatives have retained a state-centric approach to rule design (as discussed previously in *section 3.2.4.3.* above).

Agrawal (2005: p22) specifically states that local participation is necessary to "generate the concern for conservation that renders environmental protection a moral act." Furthermore, local participation in conservation management can increase legitimacy, and social cohesion and acceptance for CBC initiatives, i.e. the *social institutional fit* (Jentoft, 2000; Pretty, 2003; Kawaka et al., 2017; Alexander et al., 2018a&b). However, local participation in rule-making has been shown to both positively and negatively affect CBC outcomes, depending on who participates and their position in the local social network (Agrawal, 2009a; Campbell et al., 2010; Miller et al., 2012; Alexander et al., 2016). Therefore, encouraging community participation without regard for community inequalities and power relations defining *who* participates, the *type* of participation, and its outcomes, increases the risk of elite-capture (Cooke & Kothari, 2001; Calfucura, 2018). Lastly, the inclusion of women in rule-making is considered particularly key to ensuring access and management rules represent all resource users (e.g.

Agarwal, 2009a; Horwich et al., 2011; Zanotti, 2013; Mermet, 2018). Therefore, notwithstanding the aforementioned challenges associated with increased community participation in rule design, I propose that when *natural resource access and management rules* are *devised locally* this will enable success of CBC initiatives.

Enabler 10: Rules strongly align with local priorities/ needs

This *enabler* supports efforts to achieve greater *social institutional fit* within CBC governance. Aligning natural resources access and use rules with local priorities is commonly shown to positively affect CBC governance (Granek & Brown, 2005; Cox et al., 2010; Waylen et al., 2010; Brooks, 2016; Charles et al., 2016), as has specifically been shown in African CBC initiatives (Harris, 2007, 2011; Cinner et al., 2009a; Galvin et al., 2018; Biggs et al., 2019; Child, 2019). Levels of rule-alignment should consider both cultural and socio-economic alignment. For example, Galvin et al. (2018) suggest, based upon their detailed analysis of 65 African CBC initiatives, that the alignment of the community's cultural worldviews and practices to these initiatives is key. Furthermore, rule-alignment with local priorities is considered especially relevant when 'Western' scientific conservation discourse is of limited relevance to a community's lived realities (Blaikie, 2006; West et al., 2006; Golden et al., 2014a; Alcántara-Salinas et al., 2015). Accordingly, extensive research shows CBC is more likely to garner community support, if local knowledge, customs and priorities are taken into consideration (Infield et al., 2017; Aswani et al., 2018). A commonly cited example is the presence of a sacred natural site, which represents a form of local motivation for conservation within many communities (Dudley et al., 2009), and has been shown specifically influential within diverse African CBC contexts (e.g. Metcalfe et al., 2010). However, it should be acknowledged that efforts to align conservation with customary institutions and practices, may be affected by 'erosion' of customary institutions (as discussed in *enabler 2*).

In addition to cultural alignment, the ability to CBC initiatives to provide tangible benefits and ultimately contribute toward poverty alleviation, and thus align with local socio-economic priorities, is considered especially key (as discussed in relation to *enablers 5* and *6*). Therefore, aligning rules with both local socio-economic and cultural priorities (in accordance with *enabler 9*) remains an important and highly influential strategy for CBC governance. Consequently, I propose that when natural resource access and use rules are *strongly aligned with local socio-economic and cultural priorities* this will positively influence the (potential) 'success' of CBC initiatives.

<u>Enabler 11: Ease of enforcing rules & resolving conflict</u>
As introduced above this enabler is highly interdependent on and positively influenced by the presence of numerous other enablers. Accordingly, enforcement is influenced by numerous factors within CBC initiatives, including perceived legitimacy of regulations, social taboos and exclusion, and moral obligations (Pomeroy et al., 2001; Cinner et al., 2009a; Hayes & Persha, 2010). Furthermore, as introduced above ecological factors and conditions affect the ability to exclude 'outsiders', which can increase resource-user support for effective monitoring and rule enforcement and thus effective CBC governance (Burger et al., 2001; Gibson et al., 2005; Almeida et al., 2009; Nilsson et al., 2016; Biggs et al., 2017).

Some scholars emphasize that local users should enforce natural resource access and use rules (e.g. Padgee et al., 2006; Chhatre & Agrawal, 2008; Cinner et al., 2009a). Furthermore, levels of community involvement in monitoring have been shown to influence the potential for positive CBC outcomes (Campbell et al., 2007; Hauzer et al., 2013; Andersson et al., 2014; Armitage et al., 2017; Turreira-García et al. 2018; Charles et al., 2020), and is considered particularly effective for enforcement when local peer-pressure and monitoring are coupled (e.g. Harris et al., 2003; Child & Child, 2015).

A key reason for conflict within CBC initiatives is the ability of individuals to source livelihoods, which is often linked to externalities such as the presence of markets and levels of external support (see *enablers 13 & 14 below*). Consequently, conflict over natural resource access and use is often linked to the 'commodification' of natural resources (see Aswani et al., 2007; Büscher & Dressler, 2012). However, clearly defined resource-use rights, the use of and the ability to enforce graduated sanctions, and recognition of diverse actor interests, values and contexts, can promote greater transparency, build trust, and subsequently reduce conflicts and facilitate effective CBC governance (Pomeroy et al., 2001; Cox et al., 2010; Redpath et al., 2013; Young et al., 2016; Baynham-Herd et al., 2018; Dressel et al., 2020). However, enforcement and conflict-resolution also rely on the *accountability* of the 'enforcer', discussed next. Consequently, I propose that the *ease with which rules can be enforced and conflict resolved* positively influences the (potential) 'success' of CBC initiatives.

<u>Enabler 12: High levels of accountability</u>

As introduced in *section 3.2.3.*, accountability has long been emphasized as crucial to 'good' governance. According to Lockwood et al. (2010: p993), accountability can be defined as "(a) the allocation and acceptance of responsibility for decisions and actions and (b) the demonstration of whether and how these responsibilities have been met." Furthermore, Bierman & Gupta (2011: p1857) emphasize the importance of clearly defining expected behaviour, linking those accountable and those with the right to hold account, and the judgement and ability of the governing actor to sanction deviant behaviour of those held accountable. Therefore, accountability is strongly linked to the concepts of transparency and legitimacy, high levels of which promote improved environmental governance (Lockwood et al. 2010; Biermann & Gupta, 2011). This links back strongly to *enablers 3* (i.e. strong local leadership) and *enabler 11* (i.e. ease of enforcement).

It is important to acknowledge that accountability can refer to both horizontal (i.e. amongst actors at the same governance level – e.g. within a local community) and vertical accountability (i.e. upward/ downward governance accountability across different levels – e.g. between the State and a local community). A lack of either constrains effective community-based governance (e.g. Ribot et al., 2010; Bluwstein et al., 2016; Wright, 2017). Furthermore, Larson (2003: p221) specifically notes how local 'interactive capacities', notably the ability to build support and navigate power relations and networks across scales among diverse actors, is key to keeping the State downwardly accountable (i.e. to communities) within decentralized conservation governance. Accordingly, local community representation (i.e. enabler 3 - strong local leadership) forms a key component of both horizontal and upward CBC accountability, though this may be problematic because of elite-capture (Mufune, 2015; Ece et al., 2017). A further key horizontal accountability consideration is the participation of women, since their exclusion from CBC management activities has been linked to poor outcomes (e.g. Leisher et al., 2018; Mermet, 2018). Yet, perhaps most notable to CBC is how local perceptions of accountability correlate with low levels of support for conservation (Bennett et al., 2019). Consequently, I propose that *high levels of accountability*, both upward and downward accountability, will positively influence the (potential) 'success' of CBC initiatives.

<u>Enabler 13: Low levels of articulation with external markets</u>
Agrawal (2002) suggests that low-levels and gradual changes in articulation with external markets are key to the success of community-based institutions. In the current globalized world, external market forces are increasingly playing a greater role in influencing local-level natural resource use practices, and the livelihoods of impoverished and resource-dependent communities (Brewer et al., 2012; Wilson et al., 2016; Ojha et al., 2016; Bersaglio & Cleaver, 2018; Koot, 2019). However, as is the case with resource-dependence (i.e. *enabler 6*), the influence of market forces is complex and may either increase or restrain natural resource use, and therefore CBC success (e.g. Ojha et al., 2016; Brewer et al., 2012; Bersaglio & Cleaver, 2018), the latter potentially out of concern for sustainable future livelihoods (Stern et al., 2002). For example, while Ojha et al. (2016) and Bersaglio & Cleaver (2018) note examples of increased local benefits emerging from increased market access, they also note how market-based approaches have altered the structures of CBOs. Other scholars like Brewer et al. (2012) emphasize how such market-based approaches have had negative effects on natural resources. Notwithstanding the lack of consensus and context-specific nature of

market articulation, I propose that *low levels of articulation with external markets will* positively influence the (potential) 'success' of CBC initiatives.

<u>Enabler 14: The presence of 'nested' governance institutions with high levels of initial external support</u>
It is widely acknowledged that CBC initiatives continue to require partnerships between CBOs and external partners (i.e. actors outside of the community) as 'nested' tiers of governance responsibilities and support increase the legitimacy of CBC initiatives (e.g. Ribot et al., 2006; Biggs et al., 2019; Armitage et al., 2020), and specifically in coastal CBC initiatives (e.g. Pollnac et al., 2001; Pomeroy et al., 2001; Cinner et al., 2009a; Mayol et al., 2013; Levine & Richmond, 2015; Gurney et al., 2016). In particular, the need for external partner support is often due to the presence of low levels of community governance capacity, and specifically the inability of communities to independently control 'outsider' natural resource harvesting (Nagendra et al., 2005; Hayes & Persha, 2010; Kothari et al., 2013).

From the perspective of state support, a lack of state support – notably a lack of enabling legislation providing secure rights to community managers to make decisions relevant to their unique contexts – can undermine the legitimacy of CBC initiatives (Pomeroy et al., 2001; Cinner et al., 2009a Govan et al., 2009; Uberhuaga et al., 2011). Furthermore, many note CBC needs to overcome a lack of political will from state leaders to be more effective (see Carbonetti et al., 2014; Stone et al., 2020).

However, external institutional support goes beyond the State, as partners such as NGOs are often the primary 'catalysts' of change in bottom up development (Kamat, 2004; Hearn, 2007). However, the notion that NGOs are inherently superior to other types of organizations when it comes to supporting community-based governance is often challenged (e.g. Tole, 2010; Anderson, 2013). Furthermore, NGOs may also be powerless to hold state actors accountable (Brass, 2012; Nuesiri, 2018). Therefore, robust and productive relationships between States and NGOs, in addition to their support of communities, may better serve the interests of society (Barr et al., 2005; Singleton et al., 2017).

Lastly, a key factor associated with external support of CBC initiatives is financial support. Between 1980 and 2008, nearly 75% of the $18 billion in international biodiversity aid was devoted to projects falling under the CBC umbrella (Miller, 2014). However, a lack of long-term funding is considered especially constraining to developing robust community-based governance institutions (e.g. Ostrom, 2000; Wells et al., 2010). Yet at the same time some suggest continued dependence of developing nations on aid is counter-productive, and can undermine 'good' governance (e.g. Bare et al., 2015). Nevertheless, external funding remains a key component of required external support for CBC initiatives. Consequently, I propose the *presence of 'nested' governance institutions with high levels of initial external support, including, enabling legislation and political will, and external financial and technical support*, will enable the (potential) 'success' of CBC initiatives.

3.4. Closing Remarks

No 'ideal' governance arrangement exists for all conservation contexts. Therefore, contextually appropriate governance is critical for effective, equitable and socially just conservation, which may require greater flexibility and even the combining of different governance approaches (see Jones et al., 2019a). Changes to governance structures resulting from changes to "the role of actors, instruments and powers at their disposal, and the decision-making levels at which they engage," dictate that governance might more accurately be considered "a process than ... a fixed state of affairs" (Borrini-Feyerabend et al. (2013: p10). Consequently, a key objective is to devise institutional arrangements that assist in iteratively establishing enabling conditions, or meeting key obstacles to governance in the absence of ideal conditions (Ostrom et al., 2002; Meinzen-Dick et al., 2002). Accordingly, one of the main focuses of this book is to better understand enabling and constraining factors, conditions and processes affecting the ability of actors to initiate, implement and sustain CBC governance in the selected African cases. The first step toward exploring this has been taken with the identification of the 14 proposed CBC enablers above. Accordingly, the next step requires exploring the presence (or absence) of these 14 enablers within the selected case studies in *Chapters 6-8*, where a shift to CBC governance occurred. However, *Chapter 4* firstly explores CBC progress, as well as enabling and constraining factors, conditions and processes, within the South African context.

Review of South African Community-Based Conservation Progress

4.1. Introduction and context

This chapter examines progress with CBC in South Africa, and in doing so identifies common 'constrainers' and 'enablers' for its implementation and governance. The term *constrainers* refers to factors, conditions and processes related to the resource or resource-user that constrain decisions and actions regarding managing that resource (*Cf.* Ostrom, 2010a). Consequently, this chapter serves to address **objective 2** (*Box 4.1.*).

Box 4.1.:

> **Objective 2:** To review progress with CBC in South Africa, within the context of national conservation legislation, and identify the current enabling and constraining factors, conditions and processes to its implementation

4.1.1. National Context: apartheid & South African conservation's socio-political past

South Africa possesses high levels of poverty, income inequality and child malnutrition, and low levels of education (SSA, 2019). Furthermore, the country continues to perform poorly on global State corruption and capacity indices (e.g. Transparency International, 2019; World Justice Project, 2019). Accordingly, some suggest 'post-apartheid' has only further accumulated wealth within 'upper societal-echelons' (e.g. Adato et al., 2006; Southall, 2016). *Apartheid* existed between the years of 1948 and 1994, and involved the implementation of a complex set of laws and regulations aimed at the spatial and ethnic separation of races into a power hierarchy, with all races subservient to 'white' rule (Clark & Worger,

2016). This system of racial separation and oppression has resulted in immense social, political, economic and ecological impacts (Cock & Fig, 2000; Dubow, 2014; Butler, 2017). *Figure 4.1* depicts some of the country's key national statistics in order to better contextualize the findings of this chapter.

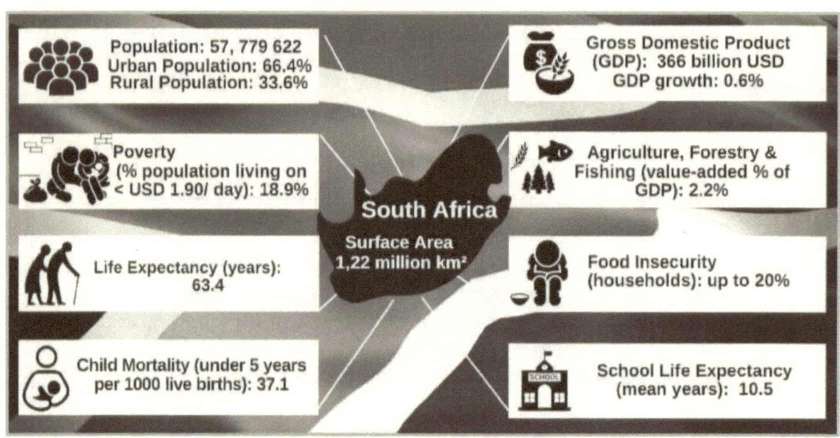

Figure 4.1.: A summary of key South African socio-economic statistics. **Source:** World Bank (2019); SSA (2019).

South Africa is a biologically diverse country, possessing three of 34 global conservation hotspots (Mittermeier et al., 2004); however, it has a 'storied' conservation past. Whilst, historical conservation records depict the 'remarkable' contribution of colonial environmentalism (Grove, 1987), the period is associated with colonial protected areas (PAs) perpetuating erstwhile political ideologies, values and identities to the detriment of local communities and their customary and local conservation practices (van Sittert, 1998; Cock & Fig, 2000; DeGeorges & Reilly, 2009; Sunde, 2014). The twentieth century witnessed South Africa adopting a fortress conservation approach, in accordance with many other African nations, which initially focused on establishing game reserves (for political reasons) to protect 'prized' colonial hunting species, and later transforming these into national parks to attract tourists at the start of the Apartheid era (Adams, 2003; Child, 2004; Carruthers, J., 2008). This resulted in widespread extreme 'environmental racism' and displacement of people (Cock & Fig, 2000; Ruiters, 2001; Fabricius & De Wet, 2002). South Africa's most famous conservation evictions include local communities in and adjacent to the *Kruger* and *Kalahari Gemsbok National Parks* (discussed further in *section 4.2.1.*). Furthermore, across the country local community ownership, access to, and use of land and resources for traditional livelihoods was prevented or restricted, resulting in food- and livelihood-insecurity and conflicts with conservation agencies, as many local communities perceived these restrictions to place nature ahead of their needs (Cock & Fig, 2000; Kepe et al., 2005). Consequently, South African conservation necessitated urgent redress within the 'post-apartheid' State (Wynberg, 2002).

4.1.2. The Emergence of a South African CBC discourse

Amongst criticisms of the past exclusionary conservation regime, many called for a more people-centred, community-based conservation approach (e.g. Cock & Fig, 2002; Shackleton & Shackleton, 2004). This was envisaged through land restitution processes and co-management agreements, and increased funding for conservation and development programmes (Cock & Fig, 2000; Kepe, 2018). As mentioned in *Chapter 3,* co-management refers to the sharing of management and decision-making power amongst governmental and or non-governmental organizations, and local communities (Cundill et al., 2013). The newly elected democratic State enacted policies and legislation to enable redress by including provisions for restoring land and resource-use rights to communities living in and adjacent to PAs, and promoting increased participation in PA decision-making and management activities (Kepe et al., 2005; Cundill et al., 2013).

Therefore, it can be argued that in terms of discourse and policy, 'post-apartheid' South Africa increasingly exhibited a shift away from its colonial natural protectionist model, to a more people-centred approach to conservation, and greater recognition of socio-political and economic implications of conservation actions (Cock & Fig, 2000; Fabricius et al., 2001; Kepe, 2008). This shift in discourse and policy has manifested in several economic development initiatives targeting previously marginalised communities, and largely focused on international ecotourism which has increasingly been endorsed by conservation agencies (Davies, 2000; Magome & Murombedzi, 2003). Two State programmes implemented in 1996 to this end were firstly, the TRANSFORM programme (Training and Support for Resource Management) and GEAR strategy (Growth, Employment and Redistribution). The former initiated several initiatives guided by principles of 'good practice' to strengthen Community-Based Natural Resource Management (CBNRM - Fabricius & Collins, 2007; Collins & Snel, 2008). The latter focused on tourism associated poverty alleviation programmes to create jobs and generate local economic growth (Streak, 2004). This supported the country's neo-liberal approach to economic development, which sought to encourage foreign investment and 'private-sector-driven' development (see Büscher et al., 2012). CBNRM is another term commonly used in southern Africa, to describe devolved management authority over natural resources to the local community level in an attempt to promote human well-being and their ability to derive lasting benefits from natural resource use (Fabricius & Collins, 2007; Matose & Watts, 2010; De

Beer, 2012). De Beer (2012) provides a valuable introduction to CBNRM within a South African CBNRM context, specifically emphasizing the problem of defining 'community'.

Therefore, some scholars have described CBC in the country as "neoliberalization from above" (Büscher & Dressler, 2012: p369). Furthermore, Kepe (2009) notes that despite significant shifts in conservation policies, conservation practices remain unchanged. Accordingly, some describe the role of community in CBC in the country as "little more than a token role" (De Beer, 2012: p555). Consequently, many suggest conservation agencies pursue 'so-called' CBC initiatives in the country solely to generate local support for conservation, but in practice restrict local access to natural resources, and consider community concerns merely an 'add-on' and 'exogenous' to top-down conservation agendas, rather than a paradigmatic shift (Els & Bothma, 2000; King, 2007; Dressler et al., 2010). Not surprisingly, substantial literature describes the challenges encountered, and specifically emphasizes the problematic roles of, and the relations between State and communities, and their coordination in conservation co-management in the country (Fabricius et al., 2007; Kepe, 2012, 2018; Cundill et al., 2013).

Notwithstanding the above challenges, there has been some progress with CBC initiatives in South Africa which should be acknowledged, including advances in enabling legislation advocating for CBC initiatives (see *section 4.1.3.*), and conservation planning in general (RSA, 2010). However, poor implementation and governance of the aforementioned change in conservation agenda is well-documented (e.g. Algotsson, 2006; Knight et al., 2008; Cundill et al., 2013). Therefore, some argue that successful implementation of CBC in South Africa, both within and outside PAs, requires securing local community rights to make decisions on rules of access, modes of usage, and the manner of benefit-distribution for resources they have traditionally harvested (Paterson, 2011, 2015). Accordingly, I now discuss the current legislation enabling a change toward a CBC mode of governance in South Africa.

4.1.3. South African Environmental Law

4.1.3.1. South Africa's International and Regional Conservation Commitments

Numerous international and regional conservation commitments shape South Africa's biodiversity conservation legislation, notably the country's 1995 ratification of *The Convention on Biological Diversity of 1992* (hereafter CBD – CBD, 2011), which represented a fundamental shift toward a CBC approach (Paterson, 2011). In particular, the *CBD Programme of Work* (CBD, 1996) calls for "the full and effective participation of indigenous and local communities" in conservation management and promotes legal recognition of "indigenous and local community conserved areas" (*Supra* Elements 1 and 2). CBD commitments promoted the promulgation of various national policy documents, most notably the *1997 White Paper on the Conservation and Sustainable Use of South Africa's Biological Diversity*. Further national legislation emerging from CBD commitments include the *National Environmental Management* Act of 1998 (hereafter NEMA – RSA, 1998a), the *National Environmental Management: Biodiversity Act 10 of 2004* (hereafter NEMBA - RSA, 2004a), and the *National Environmental Management: Protected Areas Act 57 of 2003* (hereafter NEMPAA - RSA, 2004b). Moreover, the *National Biodiversity Strategy and Action Plan* (most recently the 2015-2025 version) promotes community involvement and empowerment in conservation, and is in direct response to the country's CBD commitments (RSA, 2016).

Regionally, South Africa is also party to numerous African Union (AU) model laws established to improve and standardize the regulation of various biodiversity issues, including the rights of local communities (Strydom, 2015). For example, *Article XVII* of *The Revised (1966 Algiers) Convention on the Conservation of Nature and Natural Resources (2003)* includes provisions for recognising traditional rights of local communities and indigenous knowledge (Strydom, 2015). Furthermore, the country is party to several *Southern African Development Community* (SADC) *Protocols*, for example, the *SADC Protocol on Forestry (2002)*, which specifically calls for the adoption of "national policies and mechanisms to enable local people and communities to benefit collectively from the use of forest resources and to ensure their participation in forest management activities" (*Supra* Article 12(a)). This is particularly relevant to South Africa's Eastern Seaboard where several communities rely on forest products for various purposes (e.g. *Dukuduku Forest* in KwaZulu-Natal province - Sundnes, 2013).

<u>4.1.3.2. South Africa's Environmental Legal Framework</u>
South Africa's contemporary environmental legal framework is essentially three-pronged. It firstly concerns allocation of legislative and executive competence to the three spheres of government (i.e. local, provincial and national) to produce and administer laws, as required by *The Constitution of the Republic of South Africa of 1996* (hereafter *Constitution*). The *Constitution* emphasizes everyone's right to have their environment protected "through measures that secure sustainable development and the use of natural resources" whilst "promoting social development" (*Section 24(b) (iii)*). Secondly, it concerns the applicable legal and policy framework, which emerges through exercising the aforementioned

authority; and thirdly, the institutions tasked with its implementation (see Van der Linde & Feris, 2010).

Since democracy in 1994, a broad range of policies and laws has been introduced relevant to biodiversity conservation of both terrestrial and marine environments, and enforceable by national and provincial authorities. For example, the aforementioned *White Paper on the Conservation and Use of South Africa's Biodiversity (1997)*, which subsequently informed various national laws (introduced above). This *white paper* also led to the publication of *The Guidelines for the Implementation of Community-Based Natural Resource Management (CBNRM) in South Africa* in 2003 (Fabricius et al., 2003); and the *National Co-Management Framework* in 2010 (DEA, 2010). The former sought to promote a shared understanding and purpose of CBNRM, emphasize its main challenges, and promote principles and processes related to its implementation, as well as clarity on the roles and responsibilities of actors to improve co-operation, and ultimately increase the chances of success (Fabricius et al., 2003: p5 & 6). The latter prescribes how co-management should be implemented through the development of co-management agreements, specifically allowing "communities to play a critical role in the management of protected areas" (DEA, 2010: p9). However, various overlapping laws regulating other sectors that directly or indirectly affect biodiversity conservation complicate this legal framework. This most notably includes firstly, the land reform agenda, which focuses on restoring rights to land and resources to communities that were dispossessed of these rights during Apartheid, which influences conservation through for example the *Restitution of Land Rights Act 22 of 1994* and its various amendments leading up to its *Amendment*

Act 15 of 2014 (hereafter RLRA – RSA, 1994, 2014a). Secondly, this concerns legislation pertaining to land-use planning, notably the *Spatial Planning and Land Use Management Act (SPLUMA) 16 of 2013,* which attempts to promote greater consistency and uniformity in the application procedures and decision-making by authorities responsible for land use decisions and development applications.

The institutions tasked with implementing the aforementioned legal requirements are considered as convoluted as the legal landscape, and include an array of State departments and statutory authorities spanning all three spheres of government (Paterson, 2015). At the national sphere the *Department of Environmental Affairs* (DEA) oversees the implementation of NEMA, NEMBA, NEMPAA, as well as *the National Environmental Management: Integrated Coastal Management Act 24 of 2008 and its Amendment Act 36 of 2014* (hereafter NEM: ICMA – RSA, 2009; 2014b). Furthermore, the *South African National Biodiversity Institute* (SANBI), established in terms of NEMBA, and the *South African National Parks* (SANParks), governed in terms of NEMPAA, are two key statutory authorities aiding implementation of these laws. Moreover, in addition to DEA, the implementation of coastal specific legislation (notably NEM:ICMA), involves the *Department of Agriculture, Forestry and Fisheries* (DAFF)[1]. DAFF also administers fisheries-related legislation such as the *Marine Living Resources Act 18 of 1998* and its *Amendment Act 5 of 2014* (hereafter MLRA – RSA, 1998b & 2014c), and forestry-related legislation such as the *National Forestry Act 84 of 1998* (hereafter NFA - RSA, 1998c). Consequently, the PA

---

[1] It should be noted that certain functions of DEA and DAFF recently became amalgamated. This occurred after the completion of national interviews.

implementation involves a complex legal and institutional landscape (see also Goosen & Blackmore, 2019).

In addition to national legislation administered by the aforementioned State departments, the enactment of provincial biodiversity conservation legislation often involves provincial parastatal conservation agencies, specifically responsible for PA management. *Table 4.1.* provides a summary of the provincial conservation institutions and the relevant legislation they are required to administer. Accordingly, provincial legislation and management increases the complexity of conservation governance in some provinces, since in many cases this legislation overlaps or contradicts national legislation. For example, this has been noted regarding coastal management, specifically in the Eastern Cape after the inclusion of former 'homelands'[2] and their legislation, such as Ciskei and Transkei, into the country (Goble et al., 2014).

---

[2] 'Homelands' refer to areas reserved for residence of 'black' African persons during the Apartheid era.

Table 4.1.: A summary of the provincial conservation authorities and legislation. **Source:** Paterson (2015)

Province	Institutional Authority	Provincial legislation
Eastern Cape	• Department of Economic Development, Environmental Affairs & Tourism • Eastern Cape Parks and Tourism Agency *(management of protected areas)*	• Eastern Cape Parks and Tourism Act 2 of 2010 • Provincial Parks Board Act (Eastern Cape) 12 of 2003 • Transkei Environmental Conservation Decree 9 of 1992 • Nature Conservation Act (Ciskei) 10 of 1987 • Nature Conservation Ordinance (Cape) 19 of 1974
Free State	• Department of Economic Development, Tourism, Environmental Affairs & Small Business	• Nature Conservation Ordinance (OFS) 8 of 1969
Gauteng	• Department of Agriculture and Rural Development	• Nature Conservation Ordinance (Transvaal) 12 of 1983
Kwazulu-Natal	• Department of Economic Development, Tourism & Environmental Affairs • Ezemvelo KZN Wildlife *(management of protected areas)*	• KwaZulu-Natal Nature Conservation Management Act 9 of 1997 • KwaZulu-Natal Nature Conservation Act 29 of 1992 • Transkei Environmental Conservation Decree 9 of 1992
Limpopo	• Department of Economic Development, Environment and Tourism • Limpopo Tourism Agency *(management of tourism activities within protected areas)*	• Limpopo Environmental Management Act 7 of 2003 • Limpopo Tourism and Parks Board Act 8 of 2001
Mpumalanga	• Department of Economic Development, Environment & Tourism • Mpumalanga Tourism and Parks Agency *(management of protected areas)*	• Mpumalanga Tourism and Parks Agency Act 5 of 2005 • Mpumalanga Nature Conservation Act 10 of 1998
North West	• Department of Rural, Environment & Agricultural Development • North West Parks and Tourism Board *(management of protected areas)*	• Protected Areas Act (Bophuthatswana) 24 of 1987 • Nature Conservation Ordinance (Transvaal) 12 of 1983 • Nature Conservation Ordinance (Cape) 19 of 1974 • Bophuthatswana Nature Conservation Act 3 of 1973
Northern Cape	• Department of Environment & Nature Conservation	• Northern Cape Nature Conservation Act 9 of 2009
Western Cape	• Department of Environmental Affairs & Development Planning • Cape Nature *(management of protected areas)*	• Western Cape Biosphere Reserve Act 6 of 2011 • Nature Conservation Ordinance (Cape) 19 of 1974

<u>4.1.3.3. Enabling Legislation for CBC in South Africa</u>

National legislation enables the shift toward a community-based mode of governance, including the implementation of community-conserved areas (CCAs). NEMA (RSA, 1998a) possesses several provisions for CBC including enhanced community access to environmental benefits and resources (*Supra* Section 2(4) (d)), recognition of traditional ecological knowledge (*Supra* Section 2(4) (g)), and facilitates necessary capacity-building for equitable and effective community participation (*Supra* Section 2(4) (f)), and empowerment in governance (*Supra* Section 2(4) (h)).

NEMBA (RSA, 2004a) provides an integrated, co-ordinated and uniform approach to biodiversity management by all spheres of government, communities, the private sector and the public. Of particular relevance to CBC are *Biodiversity Management Plans* (BMP) – which are informed by the DEA's *Norms and Standards for Biodiversity Management Plans* (RSA, 2014d) – which must identify a suitable person, organization or organ of state responsible for implementing the plan (i.e. inclusive of local community members or organizations (NEMBA Section 43(2)). Furthermore, BMPs must assign responsibility to this person or entity for doing so (*Supra* Section 43(2) (c)), and in the case of an ecosystem BMP, may enter into a *Biodiversity Management Agreement* (BMA) with this person or entity to regulate its practical implementation (*Supra* Section 44). To date several BMPs of this nature have been introduced (DEA, 2019a), however, since these are voluntary mechanisms and subject to high costs and subject to developing a BMA (with none developed to date), the practical implementation of BMPs has been constrained (Paterson, 2015; CER, 2019). An addition concern is the lack of recognition of BMPs, most notably for example by the *Department of Mineral Resources* (DMR) when granting mining rights on land included in a BMP (CER, 2019).

BMAs provide a potentially useful legal mechanism to promote CCAs as they allow for various tenure relationships with vast implications for the nature, number and geographical extent of contracting parties, and are therefore flexible to both conservation authorities, and local community needs (Paterson, 2015). Furthermore, provincial level mechanisms complement NEMBA's national CBC enabling regime (refer to *section 4.1.3.2.*). Moreover, the establishment of provincial biodiversity stewardship programmes aim to create alternative conservation mechanisms for securing, and encouraging commitment to, and the implementation of good biodiversity management practices on private and communal land for conservation; and provide landowners tangible rewards for doing so (see Barendse et al., 2016; Wright et al., 2018). Moreover, these biodiversity stewardship agreements afford the same level of protection as other formally declared PAs (CER, 2019). An example of a biodiversity stewardship agreement on communal land can be found in the Mgundeni community of north-west KwaZulu-Natal province, where the *Mabaso Community Biodiversity Agreement* was signed between the community, WWF, and Ezemvelo KZNWildlife (i.e. the provincial parastatal conservation agency) to protect 455 hectares of grasslands (Jonas & MacKinnon, 2016; WWF, 2019). Lastly, like BMAs, the NFA explicitly expresses the intention to promote the role of communities in both the use and management of South Africa's natural forests through *Community Forest Agreements* (CFAs – NFA Section 29(1)). However, despite their potential, like

BMAs no CFAs have been concluded to date, with many institutional challenges prevailing since their inception (Paterson, 2018a).

NEMPAA (RSA, 2004b) introduces a fundamental shift in how PAs are established and managed in the country, and specifically calls for its implementation to take place "in partnership with the people" (*Supra* Section 3(b)), with provisions for the incorporation of communal land within PAs; and the participation of local communities in PA management (*Supra* Section 2). Due to its obligations to reach international CBD targets, and a lack of suitable State-owned land, the State has sought to expand the national PA network on private and communal land (Paterson, 2015). Therefore, like NEMBA, NEMPAA enables devolution of PA management by the State to suitable persons, organizations and organs of state (NEMPAA Section 39(1)), and therefore potentially local communities. When land is owned by a community, they would usually be governed by a *Communal Property Association* (CPA), however the devolution of management authority to other forms of *Community-Based Organizations* (CBOs) with no tenurial relationship to the land situated in PAs is not precluded (Paterson, 2015). A CPA represents an association of juristic persons formed in order to acquire, hold, and manage property on a basis agreed to by members of that community in terms of a written constitution (RSA, 1996). Of direct relevance to CBC, NEMPAA prescribes mandatory content of management plans including the implementation of CBNRM "where appropriate" (*Supra* Section 41(2) [f]), and financial and technical support in implementation of a co-management agreement (*Supra* Section 41(3) [c]). Consequently, whilst NEMPAA provides legal mechanisms enabling collaborative governance and therefore the implementation of CCAs, its use to establish CBC initiatives is limited, with the exception of some wildlife CCAs (Paterson, 2015; Novellie et al., 2016– discussed further in *section 4.2.1.*).

Lastly, NEM: ICMA (RSA, 2009a&b) facilitates community management of coastal resources through the establishment of a *Special Management Area* (*Supra* Sections 23 & 24). An appointed manager may be among others a juristic person constituted for that purpose or a traditional council (*Supra* Section 24(2) (a) & (c)). Therefore, like NEMBA, NEMPAA and NFA, NEM: ICMA enables devolution of management authority and responsibilities to local communities or a local CBO. *Table 4.2.* provides a summary of the aforementioned key CBC enabling legislation.

Table 4.2.: A summary of key CBC enabling national legislation. **Sources:** RSA (1998a,b,c; 2004a&b; 2009; 2014b; 2016).

Legislation	CBC enabling stipulation/ provisions
National Environmental Management Act 107 of 1998 **(NEMA)**	**Key provisions:** • Promotes enhanced *community access* to environmental benefits and resources • Promotes recognition of *traditional ecological knowledge* • Facilitates community *empowerment in conservation management*
National Environmental Management: Biodiversity Act 10 of 2004 (NEMBA)	**Key provisions:** • *Biodiversity Management Plans* (BMPs) enable devolution to suitable persons or organizations inclusive local community member or community-based organization (CBO) • *Biodiversity Management Agreements* (BMAs) allow for various tenure relationships flexible to natural resource and diverse actor objectives
National Environmental Management: Protected Areas Act 57 of 2003 **(NEMPAA)**	**Key provisions:** • Promotes *incorporation of communal land* within PAs • Promotes *participation of local communities* in PA management • Enables *devolution* to suitable persons or organizations inclusive local community member or community-based organization (CBO) • Enables implementation of *Community-Based Natural Resource Management* (CBNRM)
National Environmental Management: Integrated Coastal Management Act 24 of 2008 **and** *Amendment Act No. 36 of 2014* **(NEMICMA)**	**Key provisions:** • *Special Management Areas* enable devolution to suitable persons or organizations inclusive local community member or community-based organization (CBO)
National Forests Act 84 of 1998 **(NFA)**	**Key provisions:** • Promotes *Participatory Forestry Management* (PFM) • *Community Forest Agreements* (CFAs) promote role of communities in both the use and management of natural forests
National Biodiversity Strategy and Action Plan **(NBSAP)**	**Key stipulations:** • Promotes mainstreaming of biodiversity considerations into development planning, *capacity building* and *community empowerment*

Notwithstanding, the country's aforementioned CBC enabling legislation, implementation of this legislation takes place within a complex, and recently reformed legal and political system. The land reform agenda is especially relevant to biodiversity conservation since many PAs fall on land claimed or is under claim by communities that were historically displaced (Cundill et al., 2013, 2017). Apartheid was characterised by racially based land dispossession grounded in several diverse laws beginning with the *Native Land Act 27 of 1913,* which resulted in the relocation of approximately 3.5 million 'black' South Africans (and their descendants) by precluding their ability to purchase, hire or otherwise acquire land outside of 'homelands' (Wissink, 2019). This displacement also led to disorderly institutional land administration, and the decay of traditional communal land-tenure regimes (McCusker et al., 2016). Moreover, these areas remain zones of contestation, vulnerability and poverty (Noble & Wright, 2013; Zenker & Jensen, 2018). To reverse this historic dispossession, the State initiated a comprehensive land reform programme under the aforementioned RLRA (RSA, 1994; 2014a), which consisted of *land redistribution, tenure security*, and *land restitution (Supra* Section 25(5-7)). The RLRA established the legislative framework necessary for a person, community or direct descendants of a person dispossessed of land after 19 June 1913 to be entitled to restoration of land rights representing either the entire piece of land, or any other right in land; or the payment of compensation to the claimant; or both (*Supra* Section 42D(1)(a)-(f)).

When a "community" is the land claimant, the RLRA provides for all community members to have access to the land and compensation in question, on a fair and equitable basis (*Supra* Section 42D (2)). This process usually involves the formation of land trusts or CPAs. Accordingly, the *Communal Property Associations Act 28 of 1996* (hereafter CPAA – RSA, 1996) prescribes procedures for establishing CPAs to hold land rights, and is thus key to CCA implementation (Paterson, 2011). The CPAA's main objective is to ensure CPAs are established and managed in a participatory and nondiscriminatory manner which is accountable to their members (*Supra* Sections 9 (d) (i) & (e) (i)). However, several challenges related to establishing CPAs include: cumbersome procedures governing the establishment and operation of CPAs; the fact that customary practices are frequently not taken into account by those tasked with assisting communities in drafting their CPA constitutions; requiring fragmented and multi-layered community structures to form a single CPA; and the failure of the State to provide adequate support during the pre- and post CPA establishment phase (see Lahiff, 2008). Furthermore, many CPAs are characterised by "self-help, elite capture, uncontrolled use of the resources and internal conflict" (DLA, 2007: p160). These challenges are acute within the context of conservation initiatives in the country (e.g. Beyers & Fay, 2015; Koot & Büscher, 2019). Consequently, land reform has not been without serious challenges, with some labelling the process 'recolonisation', and describing it as a failure for social justice (see Cousins, 2016; Hall & Kepe, 2017; Kepe & Hall, 2018), particularly in conservation contexts (Cundill et al., 2013, 2017; Clements et al., 2021).

4.2. Findings

This section discusses the progress with CBC in South Africa since 1994, and identifies the constraining and enabling factors for CBC to be initiated, implemented and governed. These findings are based on firstly, an extensive review of relevant literature (*section 4.2.1.*), and secondly, the perceptions of various South African conservation governance actors gathered from semi-structured interviews (*section 4.2.2.*).

4.2.1. Review of South African CBC literature

An extensive review of South African literature spanning wildlife, forestry, and coastal CBC initiatives was undertaken to appraise national progress with CBC, and in doing so, identify common 'constrainers' and 'enablers' to its implementation and governance. Whilst the wildlife sector possesses examples of CCAs, no forestry, or coastal CCAs exist to date. Nevertheless, all national conservation initiatives falling under the umbrella term of *CBC* (i.e. as broadly defined in *Chapter 1*), are included here. Fabricius and Collins (2007) provide a useful starting point for the identification of 'constrainers' and 'enablers' for CBC in the country. I seek here to build on their initial work by incorporating studies that are more recent, and apply an expanded focus that includes other sectors, notably forestry, and coastal conservation. In doing so I propose an updated list at the culmination of this section, and compare these with those that emerge from interviews with national CBC actors in *section 4.2.2.*, to propose a revised consolidated list in *section 4.2.3. Figure 4.2.* provides a map depicting the locations of all CBC examples discussed throughout this chapter.

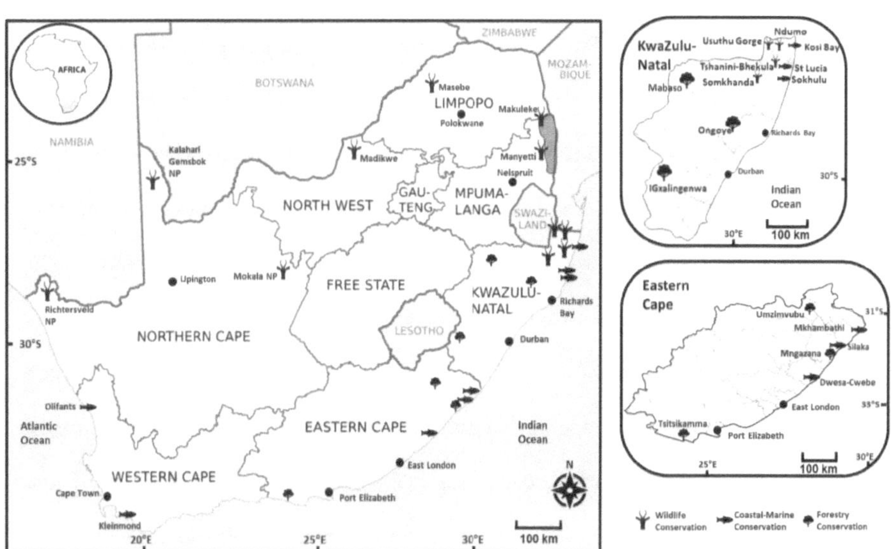

Figure 4.2.: Map depicting the localities of South African CBC examples discussed. Inset indicates the location of South Africa on the southern tip of Africa. Note Kruger National Park, located in the northeast outlined in green. Due to numerous examples from the KwaZulu-Natal and Eastern Cape provinces, these areas expanded.

Whilst much of the discussion below focuses on past and present socio-institutional challenges to achieving CBC initiation, implementation and governance in South Africa, in many cases both positive social and ecological outcomes have been identified. For example, the CCAs of *Somkhanda* and *Tshanini-Bhekula Game Reserves* have made positive contributions toward threatened biodiversity, the former to African Wild Dog and Black Rhino populations (McCann et al., 2015), and the latter to bird assemblages and rare endemic eastern sand forest vegetation (van Eerden et al., 2006). Furthermore, some consider the *Sokhulu Mussel Project* to have reduced the negative ecological effects of mussel harvesting (Harris et al., 2003). Moreover, certain CBC initiatives are considered to have produced tangible benefits in the form of increased employment, access to wild game (as a source of protein), and monetary benefits from ecotourism initiatives (e.g. Relly, 2008; McCann et al., 2015). In addition, some CBC initiatives are thought to have strengthened intra-community and community-partner relations through improved community representation and increased multi-actor collaboration, for example at the *Mngazana Mangrove Forests* (Traynor & Hill, 2008) and the *Olifants estuary* (Sowman, 2009). Nonetheless, whilst much work remains to improve South African CBC implementation and governance, progress and in particular 'failures' thus far provide valuable lessons for future efforts (*Cf.* Catalano et al., 2019). Accordingly, what follows is a discussion of the commonly identified 'enablers' and 'constrainers' that emerged from South African CBC literature.

<u>4.2.1.1. The Land Reform Process</u>

A central emergent topic regarding CBC implementation (especially regarding CCAs) is the land reform process (Kepe, 2018; Ramutsindela & Shabangu, 2018; Clements et al., 2021). Land reform is a major driver for enabling a people-centred and 'socially just' conservation approach in South Africa (Paterson, 2011; Clements et al., 2021). However, some describe this process as "in flux – and, arguably, in crisis" (Hall & Kepe, 2017: p1), with a poor track record of land settlement processes in conservation areas in particular (Paterson & Mkhulisi, 2014). Nevertheless, socio-economic transformation of conservation initiatives in South Africa – targeting previously marginalized communities – via land reform and co-management agreements has resulted in two (potential) forms of community-owned and or -managed PAs namely: *contractual parks* (i.e. incorporated into established State PAs); and more recently *CCAs. Contractual parks* involve the State (commonly through a parastatal conservation agency) and the community (as the new landowner subject to a successful land claim) entering into a co-management agreement to share power and 'jointly' establish natural resource access, use and benefit-sharing arrangements (Cundill et al., 2013, 2017). Arguably the country's most famous example – and the first settled land claim of its type – is associated with the Makuleke community and SANParks (i.e. the national parastatal conservation agency) in the *Kruger National Park* (Ramutsindela & Shabangu, 2013, 2018). However, some describe how the community was "coerced" into pursuing eco-tourism upon successfully obtaining land title (Ramutsindela & Shabangu, 2018: p77). Furthermore, whilst claimants believed agreements included the ability to hunt, SANParks ceased hunting activities and instead encouraged non-consumptive eco-tourism initiatives through community-private sector partnerships (Ramutsindela & Shabangu, 2013, 2018). However, the success of these partnerships is heavily contested (Ramutsindela & Shabangu, 2013, 2018). Moreover, some have noted how SANParks has since voiced disapproval of these partnerships as it restricts its own control over decision-making regarding natural resource use (Ramutsindela & Shabangu, 2013, 2018).

Similar co-management arrangements were reached between SANParks and the Khomani San and Meir communities (i.e. displaced by the *Kalahari Gemsbok National Park*), and the Nama community (i.e. displaced by the *Richtersveld National Park*). Like the Makuleke agreement, both these initiatives aim to provide community benefits through eco-tourism. The former agreement allows for cultural, historical and ceremonial use of natural resources (Thondhlana et al., 2011), whilst the latter includes provisions for sustainable livestock grazing (Magome & Fabricius, 2004; Michler et al., 2019). However, continued top-down and science-based approaches in both have resulted in conflict over community participation and representation, and equitable benefit-distribution (Magome & Fabricius, 2004; Thondhlana et al., 2011; Michler et al., 2019). Furthermore, these two contractual parks are now part of *Trans-Frontier Conservation Areas* (TFCAs) (Fabricius & Collins, 2007; Thondhlana et al., 2011). Whilst TFCAs represent an increasingly popular approach to 'bridging' conservation and development in southern Africa, they are commonly criticized for their complex and often inequitable institutional arrangements (Büscher, 2013; Hanks & Myburgh, 2015; Thondhlana et al., 2015).

Several wildlife CCAs have now legally declared, predominantly through NEMPAA legislation (refer to *section 4.1.3.3.*). Examples include *Somkhanda, Usuthu Gorge, Ndumo, Tshanini-Bhekula, Masebe* and *Manyeleti Game/Nature Reserves* (refer to *Figure 4.2.* for locations), which all originate from the conclusion of a community land claim. Like the above contractual parks, many of these CCAs have since been incorporated into TFCAs, e.g. *Usuthu Gorge, Ndumo, Tshanini-Bhekula*.

Therefore, whilst land claims can result in communities (as the new landowners) agreeing to maintain their land under conservation in perpetuity, such agreements are commonly considered 'socially unjust' (Thondhlana et al., 2011; Kepe, 2012, 2018). Furthermore, since CCAs predominantly originate with a land claim, this often delays their implementation (Obiri & Lawes, 2002; Cock, 2007). Consequently, I identify the slow and complex land claims, and subsequently CCA declaration processes as a 'constrainer'.

4.2.1.2. Capacity, Coordination and Collaboration

The aforementioned slow and complex land claims processes, and subsequent declaration of CCAs, is commonly depicted as the result of corrupt, under-resourced and incapacitated, and often unwilling, local, national and provincial state officials, amongst high staff turnover (Fabricius & Collins, 2007; Paterson & Mkhulisi, 2014; Coetzee & Nell, 2019). Consequently, this represents a further 'constrainer'. Consequently, national CBC studies emphasize the continued need for long-term external technical, as well as financial, institutional support and flexibility, inclusive of improved support from the State and other partners (Hauck & Sowman, 2001; Fabricius & Collins, 2007; Cundill & Fabricius, 2010; Krüger et al., 2016; Sanders et al., 2019). This represents an important 'enabler'.

The above institutional processes are often further constrained by unproductive multi-actor relations. South African CBC is often characterized by limited direct actor interaction (notably between State and other actors), and thus communication, resulting in a lack of trust (Hauck & Sowman, 2003; Thondhlana et al., 2011, 2015). These 'strained' relations cause frustration and a loss of local motivation and support for CBC, as observed for example in *Ndumo Game Reserve* and *Masebe Nature Reserve* (Boonzaaier, 2012; Meer & Schnurr, 2013), as well as some southern and Eastern Cape *Participatory Forest Management* (PFM) initiatives (e.g. Holmes-Watts & Watts, 2008; Brown, 2009). Consequently, a current lack of coordination and collaboration amongst diverse CBC actors (and thus a lack of consideration for diverse interests), represents an additional 'constrainer'. Accordingly, Boonzaaier (2010, 2012) notes how limited communication between State, community leaders, and community members, essentially resulted in a *de facto* top-down conservation approach in *Masebe Nature Reserve*. Furthermore, the rapid implementation process at *Madikwe Game Reserve* is considered to have prohibited true community participation and communication in decision-making (Davies, 2000; Relly, 2008). This 'constrainer' has also emerged in other cases such as PFM attempts in *Tsitsikamma Forest* (Holmes-Watts & Watts, 2008), and three co-managed coastal initiatives in *Sokhulu*, *Kosi Bay* and *Lake St Lucia* (Hauck & Sowman, 2001, 2003; Harris et al., 2003; Mann, 2003). A lack of community involvement in decision-making has at times even led to violence, as witnessed in *Ndumo Game Reserve* and *Silaka Nature Reserve* (Meer & Schnurr, 2013; Thondhlana et al., 2016). Therefore, a failure to devolve 'true' decision-making power to community members within CBC governance institutions (see Boonzaaier, 2012; Cundill et al., 2013), represents an additional 'constrainer'.

The above constraints are often explicitly the result of 'community-partner' relations characterized by mistrust (Kepe, 2012; Sunde, 2014; Thondhlana et al., 2016). Consequently, fostering relations of trust between actors for improved communication and coordination to promote a clear, shared and formalised vision for CBC, represents a key 'enabler' (Hauck & Sowman, 2001; Napier et al., 2005; Matose & Watts, 2010). The presence of a local 'champion' (i.e. who possesses qualities making them influential) can improve communication and coordination, and subsequently improve levels of trust. These 'champions' may include local community leaders, as well as members of State or parastatal conservation agencies and/ or other external partners such as academics or NGOs. Furthermore, their presence is often especially key to keeping local communities informed, motivating community participation, and providing necessary support through the aforementioned complex and onerous institutional processes (Hauck & Sowman, 2001; Fabricius & Cundill, 2010; Pool-Stanvliet et al., 2018). Nevertheless, these 'champions' may also cause community conflict if perceived to be capturing benefits (Fabricius & Collins, 2007; Sutton & Rudd, 2014). Moreover, Shackleton (2009) suggests a lack of State 'champions' focused on community engagement, plagues sustainable natural resource use in the country. Consequently, the presence of 'champions' to motivate actors and drive CBC implementation and governance processes, is another key 'enabler'. This finding in accordance with global CBC literature reviewed previously in *Chapter 3: section 3.3.2.1.* (e.g. Pomeroy et al., 2001; Agrawal, 2002; Galvin et al. 2018; Biggs et al., 2019)

4.2.1.3. Socio-political past

Negative feelings towards past colonial and apartheid conservation agencies and practices often leads to community resistance toward conservation initiatives, and strains collaborative conservation governance between community members and partners (Dressler et al., 2010; Thondhlana et al., 2016; Kepe, 2018). Accordingly, a lack of consideration for both historical and current social, as well as ecological, contexts when initiating CBC represents a further 'constrainer'. This notably includes cultural recognition (Cocks et al., 2012; Thondhlana & Shackleton, 2015; Boonzaaier & Wels, 2016). Failure to recognize local cultural contexts may even lead to conflict, as widely observed in several CBC initiatives across all three conservation sectors (e.g. Holmes-Watts & Watts, 2008; Meer & Schnurr, 2013; Boonzaaier & Wels, 2016; Thondhlana et al., 2016; Sowman, 2017). Thus, the need for increased understanding and recognition of the socio-ecological (notably cultural) context represents another key 'enabler' for CBC in the country.

4.2.1.4. Community Rights and Participation

A fundamental 'constrainer' central to other 'constrainers' highlighted thus far, is a lack of community rights to access, use and manage natural resources, and thus the ability to derive both cultural and monetary benefits. Of particular relevance to this constraint is the major challenge of resolving the highly contentious, complex and political issue of land rights and biodiversity conservation (Paterson & Mkhulisi, 2014; McCusker et al., 2016; Kepe, 2018). Numerous national studies show community motivation for CBC is often predicated on a desire to consolidate land and natural resource access and use rights, which continues to be constrained by a lack of recognition and internal conflicts over rightful beneficiaries (Robertson & Lawes, 2005; Cundill et al., 2013, 2017; Sunde, 2014; Thondhlana et al., 2016).

A common cited CBC 'enabler' linked to community rights to benefits, is the need to establish sustainable and tangible incentives for continued community participation and commitment, including the provision of alternative or supplementary livelihoods. Accordingly, a lack of, conflict over, and slow realization of tangible benefits has been widely observed to result in frustration in diverse CBC initiatives across all three sectors within the country (e.g. Hauck & Sowman, 2001; Matose & Watts, 2010; Thondhlana et al., 2015, 2016). This necessitates clearly defining and legitimising conflict resolution strategies in conjunction with communities, since reducing conflict and promoting 'pro-conservation' behaviours forms the basis for good governance (as introduced in *Chapter 3*), as shown necessary for example in *Mokala National Park*, and *Dwesa-Cwebe* and *Mkhambathi Nature Reserves* (Kepe, 2008; Ntshona et al., 2010; Sunde, 2014; Krüger et al., 2016). Nonetheless, local motivation for CBC is not solely concerned with deriving monetary benefits (Brown, 2009; Meer & Schnurr, 2013). Consequently, lack of effective participation and recognition of cultural values, have been shown to constrain South African CBC initiatives (Brown, 2009; Sunde, 2014; Boonzaaier & Wels, 2016; Thondhlana et al., 2016).

<u>4.2.1.5. Local Governance</u>
In addition to State incapacity, a key constraint commonly cited by CBC initiatives in the country is weak, and in some cases incapacitated, local governance institutions, inclusive of customary authorities and other CBOs (e.g. Sanders et al., 2019). This notably concerns inequitable management systems, leading to elite-capture (Fabricius & Collins, 2007; Thondhlana et al., 2015; Coetzee & Nell, 2019). Ineffective local governance institutions are commonly characterised by poor community representation due to poor relations between customary or local authorities and their constituencies. This often stems from assuming communities are homogenous (i.e. possess shared values, interests and identities), as well as a lack of consideration of social inequalities and diverse intra-community interests and objectives (Koning, 2010; Boonzaaier, 2012; McCann et al., 2015; Thondhlana

et al., 2015; Coetzee & Nell, 2019). One factor constraining effective local institutions is limited knowledge and capacity, which in some cases has even led to a community perceived dependency on State institutional structures, as witnessed in the two PFM cases of *Mngazana Mangrove* and *Tsitsikamma Forests* (Traynor & Hill, 2008; Matose & Watts, 2010). However, local community skills development has occurred in certain cases, for example at *Madikwe* and *Somkhanda Game Reserves* (Relly, 2008; McCann et al., 2015). Furthermore, the *Sokhulu Mussel Project* showed improved community knowledge and conservation attitudes toward their natural resources (which incorporated LEK) increased compliance and respect for community monitors and the provincial parastatal conservation agency (Harris et al., 2003; Napier et al., 2005). Consequently, the need to identify institutional strengths and weaknesses and collaboratively develop not only community but also State knowledge and management capacity is a key 'enabler' for South African CBC. Moreover, an additional 'enabler' is the need to continuously monitor, learn and adapt through an iterative and community inclusive process, so as to develop the required knowledge and capacity at all levels for improved CBC governance (Grundy & Michell, 2004; Holmes-Watts & Watts, 2008; Cundill & Fabricius, 2010).

4.2.1.6. State and Local Institutional Alignment

A lack of alignment of State, and local and customary institutions emerged as another key 'constrainer' to CBC in the country (Fabricius & Collins, 2007). This in accordance the aforementioned need to recognize local socio-cultural and ecological context. For example, in *Masebe Nature Reserve* Boonzaaier and Wels (2016) describe the "juxtaposition" of established institutional boundaries (i.e. fencing of the reserve) and the cultural landscape, since the former did not account for the latter, and they suggest that this type of social injustice promotes opposition to conservation. Similarly, others found the inflexibility of reserve officials concerning restricted access to natural resources and cultural sites negatively influences local perceptions of conservation institutions (e.g. Thondhlana et al., 2015, 2016; Thondhlana & Cundill, 2017). However, customary institutions have been at least partially eroded over time due to colonial and apartheid era discriminatory systems of law. For example, Obiri and Lawes (2002) showed a decrease in traditional authority recognition has negatively affected several community forestry initiatives. Similarly, Meer and Schnuur (2013) explain how issues with local traditional authorities affected management of *Ndumo Game Reserve*. Furthermore, they describe the problematic nature of relations between the two local communities and Ezemvelo KZNWildlife (i.e. a provincial parastatal conservation agency - Meer & Schnuur, 2013). Consequently, greater alignment of customary and State institutions represents a central 'enabler' for CBC in the country. However, whilst many customary practices and systems of governance are still functioning and compatible with CBC objectives (e.g. Sunde et al., 2013; Sunde, 2014; Thondhlana & Shackleton, 2015), CBC initiatives continue to require support from external institutions.

Consequently, a consolidated list of 'constrainers' and 'enablers' for CBC implementation and governance in South Africa, as they emerge from the literature, can be depicted by *Tables 4.3.* and *4.4.*, respectively. I separate these two tables since whilst at times 'enablers' are the inverse of 'constrainers', this is not always the case. See *Appendix 9* for a comprehensive table depicting South

African CBC case-studies and references that have informed the selection of these 'constrainers' and 'enablers'.

Table 4.3.: A summary of the key literature-based 'constrainers' for South African CBC implementation and governance.

Constrainers
1. Slow land claims and CBC proclamation, planning & implementation processes
2. High turnover & weak participation by under-capacitated local, national & provincial government
3. Poor coordination and collaboration amongst diverse actors
4. Lack of devolution of decision-making power and weak community participation
5. Lack of consideration for historical and current socio-political, cultural and ecological contexts
6. Lack of community rights to access, use and manage natural resources to derive both cultural and monetary benefits
7. Capture of and conflict over benefits
8. Weak & incapacitated local governance institutions
9. Lack of alignment of state and customary institution

Table 4.4.: A summary of the key literature-based 'enablers' for South African CBC implementation and governance.

Enablers
1. External financial and technical support for the CBC initiative
2. Relations of trust between actors for improved communication and coordination toward a clear, shared and formalised vision
3. Formal decision-making structures jointly creating and enforcing rules for natural resource access and use in collaboration with the community
4. Presence of 'champions' to motivate actors and drive CBC implementation and governance processes
5. Understanding of the social-ecological context (including recognition of livelihood and culturally significant practices)
6. Sustainable and tangible incentives for continued community participation and commitment
7. Clearly identified and legitimised conflict resolution strategies
8. Collaboratively developed knowledge & management capacity of community and partners
9. Continuous monitoring, learning and adapting of CBC initiatives through an iterative and community inclusive process

4.2.2. Actor Perceptions of South African CBC
<u>4.2.2.1. CBC & the State:</u>

Respondents described the current South African CBC implementation 'landscape' as "vexed, uncertain, stuck and not moving", and "riddled with politics" (SA15), and in particular emphasized how "political instability plagues South African CBC" (SA13). Accordingly, two major 'State-centric' constraints to CBC implementation and governance identified by respondents include CBC legislative concerns, and a lack of political will and capacity. Many considered South African CBC-related legislation enabling, but 75% of respondents emphasized it is not always used accurately, with most citing its complexity as the reason (reaffirming *constrainer one* above – *Table 4.3.*). Accordingly, approximately 82% of respondents noted the complexity of CBC-related legislation as a constraint to implementation, and emphasized how the resultant 'drawn-out' process commonly results in continued disillusionment amongst local communities regarding CBC initiatives. Responses on legislative complexity encompassed, firstly, the unclear articulation of legislation, and secondly, the overlapping nature of legislation and the responsible State authorities. Most respondents noted how diverse institutional mandates by different State departments constrains CBC implementation progress, especially in the context of land restitution in conservation areas (e.g. land reform by DRDLR and conservation management by DEA and parastatal agencies). Additionally, one respondent highlighted how "[CBC] policy and legislation interpretation is different for different actors" (SA11), and noted this further exacerbates institutional complexity. Consequently, most respondents strongly emphasized that streamlined legislative processes would *enable* more effective CBC implementation. This in accordance with previous discussion under the proposed *enabler 14* in *Chapter 3: section 3.3.2.2.*

An additional State-centric 'constrainer' highlighted by 71% of respondents is a continued lack of political will for CBC initiatives. Respondents noted in particular that this manifests in a lack of attendance by state representatives. Accordingly, two respondents specifically described State inaction regarding CBC initiatives as "doing window-dressing" (SA15), "feet-dragging" (SA20), and "it's a talk show" (SA26). One respondents went as far as to state that while, "NGOs and researchers are open [to CBC], but the State is not" (SA10), and this requires urgent attention. More specifically, 60% of all and 75% of non-state respondents perceived a reluctance by the State to devolve secure rights and powers, and a failure to recognize local communities as the management authority, as especially constraining to progress with CBC in the country. In particular, one respondent emphasized that, "the biggest challenge is a lack of initial support when communities get land handed-over or land rights" (SA25). Consequently, many respondents acknowledged increased devolution of authority and decision-making power to local communities. That said approximately 86% of respondents noted that external support, especially in the initial stages, was still necessary and a key 'enabler' for CBC implementation to succeed (reaffirming the *enablers* above - *Table 4.4.*). Furthermore, most respondents acknowledged community motivation for conservation often stems from a desire to consolidate their rights and meet their livelihood and cultural needs, which resonates with the South African literature reviewed above, and the global literature reviewed in *Chapter 3: section 3.3.2.2.* Accordingly, approximately 92% of respondents specifically emphasized how slow progress, and notably the delivery of benefits, and the difficulty in navigating these onerous institutional processes leads to community frustrations and constrains CBC implementation. Therefore, many respondents specifically emphasized the need for government support, for example as two respondents stated there is a need for "strong leadership from high politics" (SA15), and, "[CBC] is only going to work if you get government 'buy-in'" (SA9). However, certain respondents did acknowledge that a lack of political will is perhaps more an issue of capacity, emphasizing there is a "total lack of [government] understanding of how CBC works" and therefore, an "absolute lack of ability of government to move forward [with CBC]" (SA25). Consequently, both greater clarity and streamlining of legislation, and State political will and capacity emerged as key 'enablers' to facilitate CBC initiation, implementation and governance in the country.

In accordance with the literature reviewed above, approximately 86% of respondents noted a lack of State funding, which they described as difficult to source, or when available is short-term, and does not generate long-term sustainable opportunities for CBC initiatives. However, others suggested, "there is enough money available but [it is] not been used wisely" (SA9). Two major funding-related challenges that emerged from responses include, firstly, maintaining community interest and support whilst waiting for State funding, and secondly, the difficulty managing community expectations of potential benefits once funding is received, since many CBC initiatives do not live up to high community expectations. Lastly, some respondents acknowledged a specific lack of funding from the private sector, and stated that this sector commonly perceives CBC initiatives as high-risk investments.

<u>4.2.2.2. CBC & the Local Context</u>

A key 'constrainer' identified in the literature is the lack of alignment of the interests of the State and communities. This was confirmed by approximately 89% of respondents. In particular, many respondents emphasized the lack of legislative alignment with the local context. For example, respondents stressed that, "policy-makers don't understand the context of the people they are working with" (SA10), and that, "policy is drawn up for the people not by the people" (SA16). Four aspects highlighted by respondents regarding this alignment included a lack of consideration for the country's socio-political past (i.e. specifically mentioned by approximately 57% of respondents); the need to align CBC initiatives with poverty and livelihood needs (i.e. specifically mentioned by approximately 86% of respondents); a lack of State and customary institutional alignment (i.e. specifically mentioned by approximately 42% of respondents); and the need for targeted local capacity building when required (i.e. specifically mentioned by approximately 82% of respondents). Therefore, as one respondent stated, "if the institution fits with local objectives then implementation is good" (SA6). Accordingly, the importance of CBC alignment with the local context emerged as an important 'enabler' among respondents (reaffirming *enabler five* above – *Table 4.4.*). This strongly aligns with the global literature, and specifically *enabler 10* proposed in *Chapter 3, section 3.3.2.2.*

<u>*Consideration for the Socio-Political Past*</u>

Linked to the above 'enabler', many respondents suggested that whilst CBC is attractive to both the State and local communities, they emphasized these interventions take place within a socio-political context linked to South Africa's colonial-apartheid conservation legacy. Accordingly, as one respondent emphasized, it can be "difficult to sell the conservation agenda due to past experiences" (SA16). Furthermore, another respondent specifically noted, conservation partners "haven't been able to see the extent of apartheid undermining epistemological approaches [to conservation]" (SA11). Therefore, in accordance with discussions throughout, all respondents specifically acknowledged greater recognition of the socio-political context is required for CBC governance moving forward. Once again the importance of consideration for past experiences strongly resonates with Agrawal's (2002) who described "past successful experiences" as a critical factor for success in CPRM.

Alleviating Poverty

All respondents acknowledged poor socio-economic circumstances of many communities as a major 'constrainer' for CBC in the country. More specifically, responses noted poverty forces, "communities [to] think of today not tomorrow" (SA22), and changes local perceptions for nature. For example, respondent SA11 emphasized, in reference to the latter in *Dwesa-Cwebe Nature Reserve*, that while, "people believe in customary rules of 'care for nature' [they] are forced to break them as they can't afford to survive". Nevertheless, many other respondents noted that communities are often in favour of CBC initiatives, since as some stated, "areas with little economic opportunities and good biodiversity are a driver for CBC" (SA7), and specifically noted how, "rock bottom communities are open to improvements and therefore willing to engage in CBC" (SA13). Nonetheless, all respondents acknowledged persistent widespread poverty continues to constrain 'pro-conservation' attitudes and behaviours, since community priorities are focused on survival. Moreover, approximately 64% of respondents acknowledged that CBC initiatives are seldom able to live up to the high economic expectations, resulting in community disillusionment. This emerged particularly strongly from established wildlife CCAs, and therefore, questions whether the goals of such initiatives are realistic. However, Biggs et al. (2019: p3) recently identified based upon their review of Zimbabwe's CBC program CAMPFIRE, that, "High expectation and value of future benefits" is a key condition required to enable the adoption of new rules by a group. Therefore, while high community economic expectations may be difficult to manage, they may provide the impetus needed, but still need to be realistic. Additionally, this calls for improved communication between actors regarding realistic expectations of a CBC initiative, perhaps most notably local leaders and their constituents as this is commonly the only source of information for community members (both notions are discussed further below under *section 4.2.2.3.*).

Therefore, the ability to alleviate poverty emerged as a major 'enabler' for promoting CBC, since as respondent SA4 stated "socio-economic issues drive natural resource use." This echoes the findings related to *enabler 5* in *Chapter 3: section 3.3.2.2*. Accordingly, responses noted the need exists to figure out "how we can unlock socio-economic opportunities [in pursuing CBC]" (SA27). Accordingly, ecotourism was a common proposition amongst respondents as a 'solution' to socio-economic concerns of previously marginalized and impoverished communities, however, some also noted implementation and management of such initiatives is problematic. Therefore, the need to incentivise CBC through the delivery of tangible benefits was specifically emphasized by approximately 85%

of respondents, stating that if communities see value in a CBC initiative it increases their support and participation. Nevertheless, many respondents also acknowledged that communities value CBC initiatives for both monetary and socio-cultural benefits, which must be considered. Consequently, the need to poverty alleviation, and especially incentives, strongly correlate with the findings in *Chapter 3: section 3.3.2.1.* (e.g. Pomeroy et al., 2001; Galvin et al., 2018; Biggs et al., 2019).

<u>*Cultural Alignment*</u>

As stated throughout cultural alignment consistently emerged key to enabling CBC in the country, which also strongly correlates with the findings related to *enabler 10* in *Chapter 3: section 3.3.2.2.* For example, as one respondents specifically noted there is a need "to emphasize the cultural history of conservation" and "create linkages to living landscapes and cultural heritage" (SA9). Accordingly, many respondents specifically acknowledged the importance of recognition for cultural practices, and noted that cultural value is key to the sense of identity and belonging local communities attach to their environment and natural resources, which are turnkey to promoting positive conservation outcomes. However, all respondents acknowledged these have been at least partially eroded. For example, respondent SA11 notes – in reference to cultural values of the isiXhosa people (in particular in the Eastern Cape former 'homelands' as described in *section 4.1.3.3.*) – that "isiXhosa involves knowing to care for nature", however, they noted that in many who have grown up outside reserves customary practices have now been eroded. Furthermore, another respondent noted the erosion of customary practices in KwaZulu-Natal province, stating that, "there is now pressure on young people to develop and be educated," and, "the youth are no longer interested in learning from their elders" (SA4).

Most respondents (i.e. approximately 86%) also acknowledged the complexity of community heterogeneity for cultural practices, and noted the negative impact of establishing CBC institutions and practices in communities wrongly assumed to be homogenous. This also echoes findings related to *enabler 3* from *Chapter 3: section 3.3.2.2*. Respondent SA3 provided the example of the Meir and San people of the Kalahari Gemsbok National Park stating, "[they] are portrayed as a homogeneous community but they are not", and further noted that they are not only divided between the *Meir* and the *San* people but the 'traditionalists' and the 'modernists', emphasizing that, "since their displacement they don't want to use customary practices, the community has evolved", and they have "lost connection with the land" (SA3). Likewise, respondent SA25 noted 'modern' communities possess different value systems, and even LEK is not always endorsed by the whole community. Therefore, some suggested there is perhaps a growing "need to nurture community pride in the environment" (SA10), including LEK, and "reinforce relations with the natural resource" (SA16). Notwithstanding the above concerns, as respondent SA14 stated, "culture is important [for CBC] to succeed!" Nevertheless, all respondents acknowledged customary erosion often plagues local institutions and inhibits their effectiveness especially where there is no external support. Consequently, these findings resonate strongly with Galvin et al., (2018) who emphasize the need for African CBC initiatives to be better aligned with the cultural worldviews and practices of communities (in *Chapter 3: section 3.3.2.1.*).

Local Capacity

In connection with cultural alignment, numerous respondents specifically noted, "[CBC] enabling legislation is not building on the cultural and customary foundation [found in communities]" (SA11), including tenure and resource governance systems. Accordingly, approximately 91% of respondents suggested, in accordance with above discussions of the literature, that this necessitates greater alignment and integration of State and local conservation practices and institutions. However, all respondents acknowledged a lack of local governance capacity for CBC – including limited capacity in customary institutions and other CBOs – and approximately 82% of respondents specifically emphasized the need to increase awareness. Moreover, approximately 29% of respondents specifically mentioned the need for effective use of LEK systems in CBC institutions, and to empower its dissemination. Lastly, as established throughout, all respondents emphasized the need for external institutional support, particularly in the initial stages of developing a CBC initiative, and thereafter build local capacity to improve CBC implementation and governance. This aligns with discussion related to *enabler 14* in *Chapter 3: section 3.3.2.2.* As respondent SA22 noted, "the community needs to get to the point of managing the initiative themselves before the end of a project cycle, otherwise it collapses." Consequently, as introduced above approximately 82% of respondents emphasized the above issues necessitate targeted local capacity building.

4.2.2.3. Social Relations in CBC: Diverse Objectives, Power Dynamics & Collaboration

All respondents emphasized the complexity of CBC implementation and governance in the country, and most notably approximately 86% of respondents described 'complexities of interest' (i.e. diverse intra-community and external partner objectives). Consequently, the difficulty of addressing diverse CBC actor objectives emerged as a major challenge for progressing CBC initiatives in South Africa. Accordingly, the importance of social relations within CBC institutions was an overarching theme among many responses, which was captured by respondent SA7 who highlighted that "CBC success hinges on relations."

Power Dynamics in CBC

All respondents identified *power*, and specifically the importance of 'who' has *de facto* decision-making authority, as a key factor, in addition to the effect of the country's socio-political past, affecting social relations in South African CBC institutions. All respondents expressed strong concerns about power relations at multiple levels ranging from the upper-echelons of conservation management (largely concerning State and parastatal conservation agencies) to local-level concerns regarding community representation by CBOs and traditional authorities. Additional some respondents identified the constraints of the 'power of science', for example two respondents expressed specific concern that, "science has never been about communities!" (SA4), and emphasized that conservation in South Africa, "was treated originally as a science question, but it is a societal question" (SA10).

All respondents also raised concerns about intra-community power relations, and specifically local elite-capture, as introduced in *Chapter 3* and specifically emphasized by *enabler 3* (i.e. strong local leadership) in *section 3.3.2.2*. Responses relate to two common assumptions. Firstly, many conservation actors assume communities are homogeneous, cohesive, and benefits are equitably shared (the latter notion captured *by enabler 7* in *Chapter 3: section 3.3.2.2.*). Secondly, that community members talk to each other. As respondent SA25 stated, "a community is not just this big unicellular organism, you can't assume all know or agree with what's going on." Accordingly, as discussed above approximately 86% of respondents referred to the complexities of communities, and respondents emphasized that a common and persistent lack of understanding of local social dynamics within CBC initiatives requires attention, as this will influence the entry point for CBC initiatives into each community, since all are different. Consequently, respondents emphasized the need "to be conscious of power relations" (SA12), and, "to understand power! [and] who controls and based on what relations" (SA14).

Traditional Authorities

Though some respondents suggested, "communities are motivated, and CBC can work" (SA22), they strongly emphasized the necessity to involve the right people in local CBC institutions. This notably regarding the need for *strong local leaders* (in accordance with *enabler 3* from *Chapter 3: section 3.3.2.2.*). Therefore, as one respondent stated, local leaders "can be seen as gate-keepers" (SA4), as they have the potential to either enable or constrain CBC initiatives. Accordingly, respondent SA16 provided the example of how a traditional authority not consulted in the inception of an MPA in *Mkhambathi*, promoted fishing in the MPA in defiance of the State-imposed laws. Consequently, some respondents noted that, "the community may be excited by [the CBC initiative], but traditional authorities may be closed to it" (SA15), and whilst traditional authorities may be, "open to communication [they] have the potential to 'mutiny'" (SA7).

Relations between traditional authorities and other community leaders emerged as especially key to enabling CBC. Accordingly, with specific reference to the land claim process, respondent SA25 emphasized the potential for conflict between newly established CPAs and well-established customary institutions – which aligns with discussions related to *enabler 8* in *Chapter 3: section 3.3.2.2* (i.e. the presence of community institutions). Accordingly, respondents noted that a misalignment between CBOs within communities can potentially take value away from the initiative. Therefore, numerous respondents noted the importance of CBC initiatives accommodating and working better with customary institutions, and highlighted the need for the presence of committed and strong local leaders or champions in communities. Consequently, these two conditions emerged as key 'enablers' for facilitating CBC initiation, implementation and governance.

Collaboration and Communication

A key 'constrainer' identified by numerous respondents concerning social relations in CBC initiatives is a continued lack of communication and collaboration between communities and their local representatives, and amongst various actors including State departments, local communities and other non-state partners. In accordance with discussions throughout, most respondents specifically emphasized a lack of intergovernmental coordination by overlapping responsible State authorities. This finding strongly correlates with *enabler 14* (i.e. the presence of nested governance) as identified in *Chapter 3: section 3.3.2.2*. One State respondent characterized intergovernmental coordination as, "not much talk, and even less doing, [and this] needs to be the other way around" (SA20). Furthermore, in specific reference to concerns within coastal conservation, respondent SA16 emphasized, "DAFF and DEA need to co-operate." Accordingly, respondent SA27 emphasized the "need to unlock the opportunities different institutions bring to the table." Moreover, many respondents, including those from the State, also expressed concerns regarding 'corrupt' relations between the State and private sector, specifically noting the conflicting interests of conservation and other industries, most notably commercial fishing and mining. In reference to commercial fishing, respondent SA11 voiced concerns that, "power relations between DAFF and industry mean there are no fish for the small-scale fisher's basket." In addition, numerous responses expressed concerns about the impacts of mining, stating there are "issues with mining and getting the Department of Mineral Resources to hold companies accountable" (SA20). This statement was in reference to environmental management mandates such as impact assessments and post-project restoration. Additionally, it was emphasized by some that, "mining is a priority of the State, not the people!" (SA28). Consequently, as respondent SA16 stated there is a specific "need to avoid private capture in [South African] conservation."

All respondents emphasized the specific importance of community-partner relations, especially those with the State, for successful facilitation of CBC initiation, implementation and governance in the country. However, the approximately 63% of responses characterized these relations as *average* to *poor*. Furthermore, it could be inferred from many responses that the aforementioned slow institutional processes associated with land claims and CCA declaration, and a lack of tangible benefits, contribute equally to deterioration of these relations. Moreover, a few respondents noted a risk of CBC 'fatigue' in communities due to previous perceived 'failed' experiences. This once again in accordance with findings in

Chapter 3. For example, in reference to the *Usuthu Gorge CCA*, respondent SA25 noted how numerous previous projects have failed to produce outcomes for the community, which has generated a sense of disillusionment, and a lack of trust within the community for all partners. Consequently, a major consensus among respondents was the need for greater collaboration between all CBC actors. As respondent SA27 stated, "we need to crack the nut together."

<u>*Respect & Trust*</u>

Therefore, based upon the above concerns a central 'constrainer' to emerge from respondents was a lack of respect and trust between partners notably between the State and communities. Not surprisingly, approximately 64% of respondents specifically emphasized the need to re-build these relations, since "Trust is key!" (SA27). Furthermore, respondent SA12 suggested, "conservation can't win battles if [the State] goes to war with the people, [the State] needs to reach compromises." Respondent SA11 specifically noted the presence of "adversary relationships in the marine environment", which are constraining community participation in coastal CBC initiatives. This was reiterated by respondent SA16 who noted, "Government, researchers, and scientists don't respect the ability of communities, we need to change [their] perceptions of communities." In relation to building necessary relationships of trust, some respondents highlighted the necessity "for champions to believe in the community" (SA11), especially emphasizing the need for State champions (reaffirming *enabler four* in Table 4.4 above). This again strongly correlates with findings from the global literature in *Chapter 3*.

Respondents emphasized that to build relations of trust and respect in CBC institutions requires a long-term presence of external governance actors within communities. Consequently, as one NGO respondent noted:

> "If you start working with a community don't expect quick results ... to work with communities is not an easy task ... [partners] need to commit time and effort to the community ... keep working at it" (SA26).

Numerous respondents specifically described the need for partners to take community concerns and suggestions seriously, and follow through on their promises. Furthermore, some respondents specifically noted "a desperate need for greater support from parastatal conservation agencies" (SA17). Accordingly, as a SANParks respondent noted, "SANParks is not in touch with the needs of local communities, but (we) want to now change perceptions of fortress conservation," and "have an objective of reaching out to neighbours to establish their needs" (SA2). Accordingly, respondent SA27 suggested that whilst SANParks and other partners have their mandates, "the challenge is to explore the opportunities these mandates bring to the table for the beneficiaries", noting local communities as the beneficiaries (SA27). Nevertheless, some respondents also noted that whilst "community value having partners coming in to support them" (SA25), at times they have "unrealistic expectations of partners" (SA16). These issues once again mirror those emerging from South African literature (*section 4.2.1*), and findings in *Chapter 3*: section 3.3.2.1., most notably those emerging from African CBC studies (e.g. Galvin et al., 2018; Biggs et al., 2019).

4.2.3. A consolidated list of South African CBC 'constrainers' and 'enablers'

Consequently, based upon the above discussions in *sections 4.2.1.* and *4.2.2.*, I propose a consolidated list of 'constrainers' (*Table 4.5.*) and 'enablers' (*Table 4.6.*) for CBC implementation and governance in the country. These are revisited and consolidated with the case study findings later in *Chapter 9*.

Table 4.5.: A summary of the key emergent 'constrainers' for South African CBC implementation and governance, based on both a review of relevant literature and national conservation actor interviews.

Constrainers
1. Slow land claims and CBC proclamation, planning & implementation processes
2. Lack of State capacity & political will for CBC
3. Lack of devolution of decision-making power and weak community participation
4. Lack of community rights to access, use and manage natural resources to derive both cultural and monetary benefits
5. Lack of consideration for historical and current socio-political, cultural and ecological contexts
6. Capture of and conflict over benefits
7. Poor coordination and collaboration amongst diverse actors
8. Weak & incapacitated local governance institutions
9. Lack of alignment of state and customary institutions
10. Failed past CBC experiences

Table 4.6.: A summary of the key emergent 'enablers' for South African CBC implementation and governance, based on both a review of relevant literature and national conservation actor interviews.

Enablers
1. Streamlined CBC-related legislative processes
2. State capacity and political will for CBC initiatives
3. Devolution of authority and decision-making power to local communities
4. External financial and technical support to the CBC initiative
5. Understanding and alignment of CBC initiatives for social-ecological context to address local priorities
6. Presence of sustainable and tangible incentives to alleviate poverty and encourage community participation and commitment to CBC initiatives
7. Awareness of power dynamics and presence of strategies to legitimise conflict resolution
8. Strong local leadership characterized by supportive traditional authorities
9. High levels of recognition of customary institutions, knowledge and practices
10. Strong relations of trust between actors for improved communication and coordination toward a clear, shared and formalised vision
11. Presence of local 'champions' to motivate actors and drive CBC implementation and governance processes
12. Ability to continuously monitor, learn and adapt through an iterative and community inclusive process

4.3. Conclusion

This chapter reviewed CBC progress in South Africa. Despite a plethora of legislation enabling CBC in the country, CCA implementation in particular remains limited terrestrially to a few wildlife examples, with no forestry or coastal equivalents. The first step to better addressing the present CBC policy-praxis disjuncture is the need to learn lessons from experiences. Accordingly, numerous insights emerged here, and led to a proposed list of common 'constrainers' and 'enablers' for CBC implementation and governance in the country.

Notwithstanding a change toward a pro-CBC rhetoric, a lack of State support persists, as highlighted by other recent studies (e.g. Sanders et al., 2019). This perhaps best depicted by a continued lack of *de facto* devolution of decision-making power to local community-level institutions. Therefore, whilst national CBC legislation enables *de jure* devolution to the community level, complexity and a lack of political will, continue to constrain CBC implementation. Consequently, the streamlining of legislation and improved political will are key 'enablers'. Moreover, a lack of both State and local capacity, and multi-actor communication and collaboration are further key 'constrainers'. This requires targeted institutional capacity building, and the fostering of relations of trust amongst all actors. In addition, a lack of consideration and alignment of CBC initiatives for local ecological, and socio-cultural and -economic context requires attention. Further key 'enablers' emerging include the need for strong local leadership, consideration of power dynamics at all levels, and the presence of 'champions' to drive the CBC implementation process. Consequently, these South African specific CBC enablers mirror many of the 14 proposed CBC enablers from *Chapter 3: section 3.3.2.2.* Most notably these include the enabling presence of *strong local leadership*, *strong rule alignment with local priorities*, and *external support* from partners. These findings are consolidated together with the findings of subsequent empirical chapters in the final discussion (*Chapter 9*).

Chapter 5

Theoretical Foundations Part II:

Theory of Change and its application to CBC

5.1. Introduction

This chapter builds upon the theoretical foundations presented in *Chapter 3*. It explores the theoretical underpinnings and application of the *Theory of Change* (ToC) approach in the context of attempting to understand *how* to make a change towards a community-based mode of conservation governance. Firstly, I introduce the ToC approach, and secondly, describe a basic sequential *ToC pathway development process*, before proposing a *ToC Pathway Design Framework*. The chapter then culminates with a proposed *Generic CBC ToC Pathway*. This generic pathway represents the conceptual framework for the study as it informs data collection and interpretation in the case study sites (in *Chapters 6-8*). Consequently, in doing so this chapter addresses **objective 3** (*Box 5.1.*). This generic ToC pathway is revisited and amended, based on the findings of the empirical chapters (i.e. *Chapters 6-8*), to ultimately propose a *South African Empirical CBC ToC pathway* in *Chapter 9*.

Box 5.1.:

Objective 3: To draw on theoretical ideas from Governance Theory, Commons Theory, and Theory of Change to develop a Generic Theory of Change Pathway that offers a theoretical understanding of factors, conditions and processes that enable change towards a community-based mode of conservation governance in the developing world

5.2. Theory of Change: Theoretical Underpinnings

Many scholars suggest that present challenges within social-ecological systems (SESs) – including climate change, ecological degradation and enduring 'poverty traps' – necessitate social and institutional change (e.g. Biggs et al., 2010; Olsson & Galaz, 2012; Chaffin et al., 2016; Barnes et al., 2017; Blythe et al., 2017). More specifically, as mentioned previously, many scholars and practitioners have acknowledged the shortcomings of conventional conservation approaches and the need for alternative approaches that include local resource users in management and decision-making, *when* contextually-appropriate (discussed in *Chapter 3: section 3.2.4.3.*). Therefore, this study recognizes the need to identify pathways that facilitate contextually appropriate change towards a CBC mode of governance (*Cf.* RARE, n.d.; Biggs et al., 2017; Blythe et al., 2017). At this stage it should be acknowledged that while the primary focus of this book is on change emerging from within the governance system, the ability to facilitate change towards a CBC mode of governance is also affected by external factors and conditions. This is expanded upon in *section 5.3.2.6.* below.

In response to greater recognition of the complexity in SESs, the need for innovative approaches guiding and assessing change pathways in CBC has emerged. Accordingly, ToC, in its various guises, is one such approach that has recently and increasingly been shown useful in this endeavour. More specifically, it has been employed to improve monitoring, evaluation, and decision-making processes, and their outcomes, within diverse conservation governance contexts, including 'community-based' conservation initiatives (RARE, n.d.; Margoluis et al., 2013; Bottrill et al., 2014; Mascia et al., 2014; Morrison, 2015; Biggs et al., 2017; Romero & Putz, 2018; Blue Ventures, 2019a). In particular, the use of ToC in developing world conservation contexts is especially relevant to this book (e.g. RARE, n.d.; Biggs et al., 2017; Romero & Putz, 2018). These developing world cases strongly emphasize the ToC development process needs to provide opportunities for initial and ongoing involvement of local resource users, the development of local capacity, and the align with local socio-economic and cultural priorities. Nevertheless, some suggest, "well-informed ToC...remain uncommon in conservation" (Romero & Putz, 2018: p547). Consequently, whilst ToC provides a potentially useful and flexible approach to better understand *how* to facilitate initiation, implementation and governance of CBC, greater understanding of *how* to develop 'well-informed' and 'robust' conservation ToC pathways persists. Accordingly, this chapter explores this goal, and entails, firstly, improved understanding of *what* ToC is, and secondly, *how* a ToC pathway is developed.

ToC is a theory-driven approach that assists in developing, managing, and evaluating interventions (Rogers, 2007, 2014a; Mayne, 2015, 2017). However, providing a common perspective or definition of ToC, and a generic methodology, is widely considered to be problematic (Coryn et al., 2011; Stein & Valters, 2012; Vogel, 2012; Mayne & Johnson, 2015). This is because ToC can simultaneously be considered a 'way of thinking', 'a process' and/ or 'a product/ representation' (Rogers, 2014b; van Es et al., 2015). Furthermore, ToC pathways (i.e. a product/ representation) can be depicted in numerous ways (Funnell & Rogers, 2011; Mayne & Johnson, 2015). As a result, debate exists over its usefulness, and 'best' method of use (Coryn et al., 2011; Funnell & Rogers, 2011; Mayne & Johnson, 2015; Prinsen & Nijhof, 2015). Nevertheless, this book proposes that ToC provides a flexible, interdisciplinary approach that can promote increased understanding of *how* and *why* an intervention works, and the processes that bring about change.

A fundamental quality of ToC is that it enables practically and backwardly 'mapping' the logical pathways and sequences of actions toward desired or expected outcomes (Connell & Klem, 2000; Stein & Valters, 2012; Valters, 2015). Furthermore, ToC can account for both how change is *expected* to happen (i.e. the initiation and implementation phases) or how change *has* happened (i.e. post-implementation - the evaluation and adaptation phase) (Rogers, 2014b; Mayne, 2017). In doing so ToC can inform the actions required to bring about change by considering multiple levels of change and learning from the intervention as it evolves. Therefore, ToC facilitates project implementation cognisant of multiple levels of change, and learns from the change process as it evolves (Mayne, 2017). This way of thinking resonates with the well-established conservation management concepts of *adaptive management* (see Holling, 1978; Armitage et al., 2010; Schultz et al., 2015), and *social learning* (see Keen et al., 2005; Armitage et al., 2011; Ernst, 2019). *Social learning* is particularly relevant to the CBC change process, and the development of ToC pathways, since it is considered a process of transformative and iterative social change involving actors critically challenging existing norms, values, institutions, and interests through iterative practice, evaluation and action modification, so as to pursue desirable collective actions in the context of achieving positive social and ecological outcomes (Mezirow, 1993; Keen et al., 2005; Wals, 2007; Biggs et al., 2019). Consequently, a foundational aspect of a ToC pathway is that it queries, "*what* it is about an intervention that works for *whom*, in *what* circumstances, in *what* respects, over *which* duration" (Pawson, 2013: p167 - *emphasis* added). I now discuss the process of developing a ToC pathway.

5.3. Developing a Conceptual Framework for CBC ToC Pathway Development

5.3.1. A ToC pathway development process

As Weiss (1997: p524) notes, ToC's, "very ambitiousness seems to tempt the gods." Consequently, developing a ToC pathway is not simple. Nevertheless, several scholars have offered criteria for developing 'robust' ToC pathways. For example, Connell and Klem (2000: p94-95) offer four criteria based on how *plausible*, *doable*, *testable*, and *meaningful* a ToC pathway is. More recently, building upon this premise, and the work of Davies (2012), Mayne (2017: p159) suggests a 'robust' ToC can be considered "structurally sound and plausible", if it "supports a solid and plausible intervention design", following which, "it is reasonable to expect that the intervention, if implemented as designed, will be able to contribute to the intended results."

Given the diverse ToC terms in use, I now define some key 'change elements' described in this dissertation's CBC ToC pathway development process. Firstly, whilst it is acknowledged that CBC initiatives vary greatly, the term *intervention* is used throughout to describe any CBC initiative as broadly defined previously in *Chapter 1*. Furthermore, *intervention design* refers to the proposed *actions* and their associated *assumptions*, designed to produce an identified *desired result*. Here I consider an *action* to be an event, a project or programme, a policy or strategy, or even formation of an organization (Rogers, 2014b; Mayne, 2015). Moreover, *assumptions* are a central determinant as to whether *actions* implemented will (potentially) achieve the *desired result* or not. This includes *rationale* and *causal assumptions*. The former being the fundamental hypotheses or premise(s) on which the intervention is grounded, and the latter the prominent and likely necessary events, factors or conditions for realising a particular causal link in a ToC pathway (Mayne, 2017). Lastly, *desired result* encompasses both the intervention's intermediary *desired outcomes and outputs*, and subsequently the *desired impact*. Accordingly, the *desired impact* refers to the final measurable outcome(s), which is itself dependent on achieving the intermediary *desired outputs and outcomes* (Mayne, 2015, 2017). Therefore, the above change elements can be represented by the basic ToC Pathway design schematic in *Figure 5.1*.

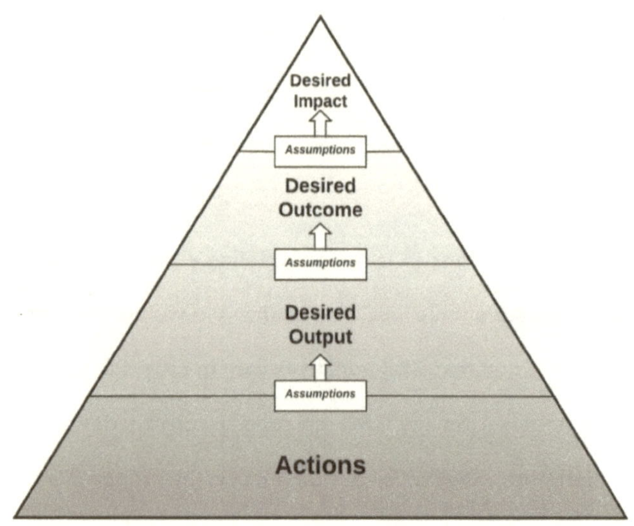

Figure 5.1.: A basic ToC pathway design schematic. Inspired by and adapted from Harries et al. (2014)

Based on accepted theoretical ideas underpinning ToC, as introduced in *section 5.2.*, and a review of CBC-related literature, I propose that the development of 'well-informed', 'robust' CBC ToC pathway comprises six core steps: 1) identify the intervention's main beneficiaries; 2) articulate the intervention's desired result; 3) define and analyse the contextual factors, conditions or events that may positively or negatively affect the intervention's implementation and management; 4) identify actions, and the associated assumptions that underpin these actions, to achieve the intervention's desired result; 5) implement and evaluate actions to identify persistent or newly arising issues; and 6) constantly adapt the intervention, most notably the actions, based upon evaluation to better achieve the intervention's identified desired result (*Figure 5.2.*). These steps are discussed further directly below, and subsequently, with specific reference to CBC literature in *section 5.3.2.*

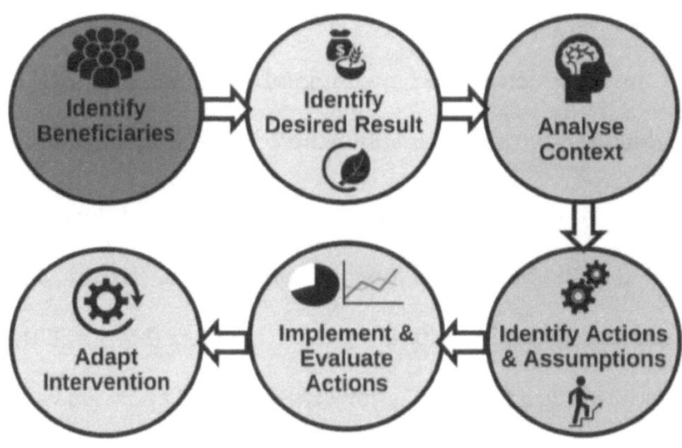

Figure 5.2.: A simple step-by-step ToC Pathway design sequence.

The first step involves identifying the intervention's *beneficiaries* (Harries et al., 2014). Accordingly, I consider *beneficiaries* to include all actors who may be affected by or have an interest in the interventions objectives, and in the case of

CBC initiatives notably includes local resource users. Therefore, ToC pathways should consider, "the needs, interests and behaviour of stakeholders and other key actors" (van Es et al., 2015: p16). As CBC initiatives involve multiple 'nested' actors (i.e. government and non-governmental partners and local communities) with multiple diverse interests and agendas (Baird et al., 2019a&b; Armitage et al., 2020); CBC ToC pathways should therefore be considered both "multitargeted" and "messy" (Mayne, 2015: p133). Mayne (2015: p122) suggest that a specific consideration is both the "reach" and "reaction" of these beneficiaries. In other words, *who* the intervention is reaching, and *how* they have or will react to such changes (Mayne, 2015). Consequently, developing a 'robust' CBC ToC pathway requires attempting to capture all interests and activities affecting various target groups. Lastly, identifying beneficiaries is crucial to all subsequent steps as these steps should involve active and meaningful participation and collaboration amongst the identified *beneficiaries* within the ToC pathway development process.

The second step requires identifying and articulating the intervention's desired result, which should be *clear*, *logical*, *measurable*, based upon previous *experience*, and be *achievable* with the planned actions (Mayne, 2017). However, it should be acknowledged that ToC pathways commonly possess various potential 'routes' to a desired result (e.g. RARE n.d.; Biggs et al., 2017; Balfour et al., 2019).

Once the intervention's desired result has been identified, step three involves considering the intervention's context to enable the formulation and implementation of appropriate actions to achieve the intervention's desired result. However, adequately accounting for context is one of the greatest challenges in developing a ToC pathway (Weiss, 1997; Blamey & McKenzie, 2007). This requires identifying what Blamey and McKenzie (2007: p446) refer to as the "causal and situational triggers for changes." Accordingly, both ecological and social factors, conditions and/ or events that (may potentially) stimulate initiation and/ or maintenance of a CBC initiative, should be considered. It should be acknowledged from the outset that the classification of 'change triggers' may represent *initial* contextual factors, conditions or events leading to both sudden change, as well as those considered 'motivators' for actions that may manifest over longer time periods, but nonetheless still stimulate the change process. Initial socio-institutional contextual issues (i.e. the primary focus of this book), which encompass a range of 'social dimensions' inclusive of issues related to, culture, gender, equity and institutional power relations, may stimulate the need for change, and thus the need for the intervention and its proposed actions (see van Es et al., 2015; Mayne, 2015, 2017). In addition to *initial contextual issues* that may stimulate the change process, the CBC literature specifically refers to "catalytic elements" and "trigger events" (Seixas & Davy, 2008). Seixas and Davy (2008: p103) define *catalytic elements* as factors contributing "to speeding up the process of organizing an initiative (initial catalytic elements) and those that maintain the initiative (continuing catalytic elements)." Furthermore, they define *trigger events* as "the motives or events, which led people to get mobilized around an initiative" (Seixas & Davy, 2008: p103). In additional to socio-institutional triggers, *ecological triggers* specifically include changes in ecological attributes, which serve as ecological system indicators (e.g. a decrease in species abundance), that subsequently trigger management decisions and actions in an intervention (Bie et al., 2018). Consequently, I use *contextual change triggers* as an overarching term within this book to incorporate the above concepts related to the context of intervention stimulating the change process.

Once a desired result has been articulated, and the intervention's context has been considered, this information can be used to formulate appropriate *actions*, i.e. step four. A key consideration in formulating appropriate actions is that they need to be broadly *acceptable*, *doable*, *measurable* and *sustainable* to bring about the desired result (Mayne, 2015, 2017). This has been especially noted within conservation interventions (e.g. Biggs et al., 2017; Romero & Putz, 2018; CBD, 2020). To be broadly acceptable ToC pathway development must consider the needs, interests, capacity, behaviour and visions of the identified beneficiaries for the conservation intervention, which will have begun in step one.

Formulating actions ultimately involves identifying underlying *assumptions*. These notably include *causal assumptions*, which are the likely necessary events, factors or conditions that determine whether a 'causal link' associated with a proposed action in the intervention's ToC pathway may be realized, and therefore, the intervention's ability to produce the desired result (Mayne, 2015, 2017). However, some assumptions are more or less likely to be realized. For example, if an assumption is related to a necessary action by a specific CBC actor (e.g. a conservation agency) not taken before, then such an assumption, and subsequent desired result, is "at-risk" or even implausible (Mayne, 2015, 2017). Furthermore, "counter pressures", for example elite-capture or a lack of political will in CBC interventions, may ensure the assumption remains unrealized or may even 'derail' a ToC pathway (Mayne, 2017). The concepts of *actions* and *assumptions* are discussed more comprehensively in *sections 5.3.2.3.* and *5.3.2.4.*

Step five involves the continuous monitoring and evaluation of the effectiveness of the implemented *actions to* achieve the *desired result.* This should be conducted through a set of social and ecological indicators, as emphasized by recent CBC studies that have employed a ToC approach (Romero & Putz, 2018; Béné, et al., 2020). Accordingly, systematic feedback is crucial to a ToC pathway, as it enables its adaptation so as to increase the chances of achieving its desired result. Larrosa et al. (2016: p318) describe systematic feedback within a conservation context as how, "results from some action travel through the system and eventually return in some form to the original action, potentially influencing future actions." This feedback process identifies 'issues arising' within the intervention. The term *issues arising* is used here in keeping with the definition of *initial contextual issues* described above. However, 'issues arising' refer specifically to those issues identified by after the CBC intervention's initial actions. Yet, the term 'issues arising' can include both newly arising issues and/ or persistent *initial contextual issues*. Therefore, there is much overlap between the 'issues' considered initially and those 'arising' post-implementation. For example, high levels of natural resource dependence may be identified as an *initial contextual issue*, and subsequently as a *persistent issue*. Furthermore, if it is only discovered to be constraining the *desired result* after implementing the intervention's initial actions, then it would be considered an '*issue arising*'.

Therefore, these 'issues arising' then feedback into the change process to allow actions to be reformulated and implemented to increase the chances of achieving the desired result in the theory of change pathway. Consequently, monitoring and evaluation are crucial to the development of a 'robust' ToC pathway. Hence, as

Pawson & Tilley (2004: p2) specifically note, a 'robust' ToC pathway should aim to identify a "perceived course whereby wrongs might be put to rights, deficiencies of behaviour corrected, [and] inequalities of condition alleviated".

Notwithstanding the need to account for the above *change elements*, one needs to acknowledge that *external influences* can also affect progression through a ToC pathway (Mayne, 2015). Accordingly, ToC pathways should always, and are considered here, merely as "a model of the *contribution to* and not cause per se of the intended result" (Mayne, 2015: p128 – *emphasis* in original). I consider *external influences* to refer to external factors, events and/ or conditions that either positively or negatively influence achievement of the desired result within a ToC pathway (Mayne, 2015). As Galvin et al. (2018: p41) emphasize, "There are myriad factors that determine the social and ecological outcomes of CBCs, which include exogenous socio-political, economic, historical, and biophysical factors." For example, a lack of political will, funding or supporting policy and diverse development/ commercial agendas of other sectors may constrain, while international conservation commitments (e.g. the *Aichi Targets* or the proposed *Post-2020 Global Biodiversity Framework*), and CBC compatible development in other sectors may stimulate and facilitate national and local action required to achieve the intervention's desired result (Sanders et al., 2019; CBD, 2020).

Therefore, based upon the six steps, and discussion of the associated *change elements* described above, I propose a *ToC Pathway Design Framework* (*Figure 5.3.*). This framework informs development of the *Generic CBC ToC Pathway* in the subsequent section.

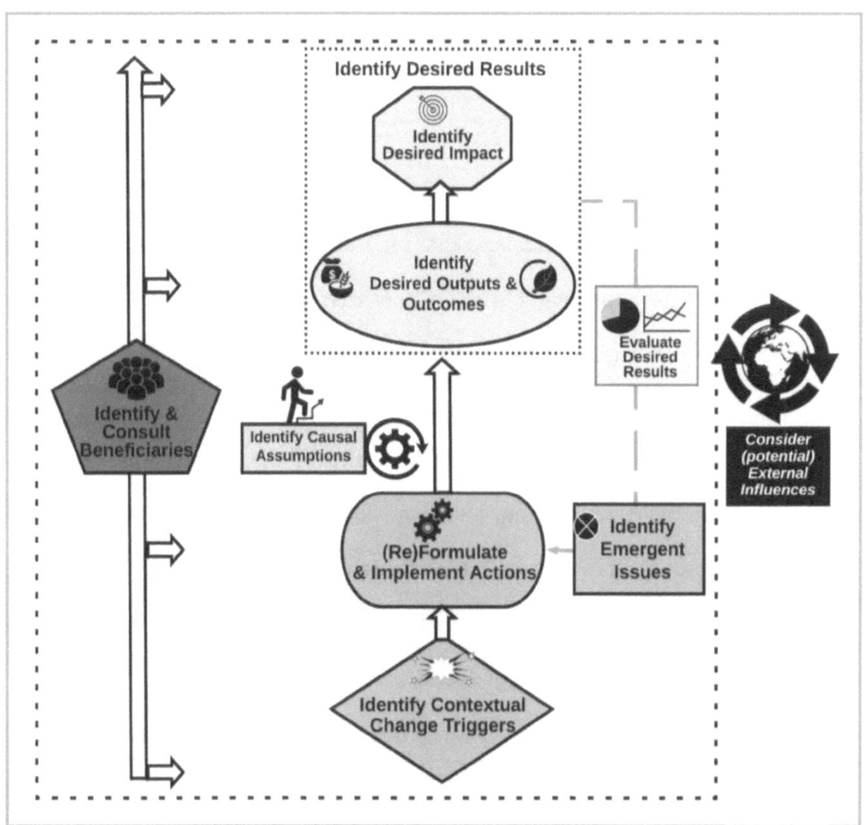

Figure 5.3.: A ToC Pathway Design Framework. Note: All change elements from 'contextual change triggers' through to 'desired results' require consultation with the intervention's beneficiaries. This is indicated with an 'arrow line'.

To sum up, multitargeted, multi-actor CBC interventions require ToC pathways that are able to assist in handling complexity effectively, whilst avoiding the 'pit-falls' of over-simplification, but should also avoid aiming for perfection, and instead be "good enough" and not overly complex (Mayne, 2015: p138). In doing so Mayne (2017: p160) suggests that a "reasonably robust" ToC pathway may improve: "(a) designing and planning an intervention, (b) managing an intervention, (c) assessing and evaluating an intervention, and (d) scaling up an intervention." This book focuses upon (a) and (b), as evaluation and scaling up are beyond the scope of this research. More specifically, I explore the CBC change process in each case study site to better understand *how* it unfolded. This entails identifying the *common change elements*, including the *contextual change triggers*, *actions* taken to facilitate change, the *assumptions* underpinning the change process, and the *issues arising* and *external influences* that may facilitate (or affect) achieving the *desired result*. The subsequent section now discusses these common change elements as they emerge from CBC-related literature.

5.3.2. *Developing a Generic CBC ToC Pathway*

In addition to the steps and considerations above, the 'robustness' of a ToC pathway is improved with testing an initial ToC pathway against the logic and assumptions proposed, and available evidence from previous research or evaluations, prior to its 'field-based' testing (Kautto & Silila, 2005; Brousselle & Champagne, 2011). Accordingly, this section seeks to propose a 'prior', 'literature-based' *Generic CBC ToC Pathway*, i.e. which is based upon an extensive review of the literature in the subsections below. Furthermore, this discussion compliments literature previously reviewed in *Chapter 3,* most notably the 14 proposed CBC enablers, the presence of which is hypothesised to increase the effectiveness of the generic pathway, and its actions, to achieve an intervention's desired result. Consequently, this generic pathway ultimately informs the development of a modified, 'field-based' *South African Empirical CBC ToC Pathway* – based upon the findings emerging in *Chapters 4, 6, 7* and *8* – at the culmination of this book in *Chapter 9.* Therefore, it should be acknowledged from the outset that the components for the generic pathway proposed in subsequent subsections – most notably the generic *action categories* and *actions* proposed – will require modification.

In developing the generic pathway the various aforementioned change elements are discussed in accordance with the proposed *ToC Pathway Design Framework* above (refer to *Figures 5.3.*). The discussion in the subsequent subsections provides common CBC-related examples of the change elements as they emerge from the literature. While I acknowledge that this is a non-exhaustive list of examples, it does serve to adequately propose a generic pathway, which can be considered to form the 'baseline' for further investigations into the three case studies. Furthermore, I will also refer in particular to examples of these change elements as they emerged from three CBC-related interventions where ToC has been directly used (*Table 5.1.*). In the first example Romero and Putz (2018) develop a country-specific ToC for evaluating sustained timber yields of natural forest management in Indonesia. Biggs et al. (2017) provide a second example where they develop a topic-specific ToC to guide actions of policy makers, practitioners, and donors tasked with decreasing illegal wildlife trade. Finally, the third example relates to the *Pride* campaigns of RARE (i.e. a non-profit organization), who provide a step-by-step guide to developing a ToC for local conservation interventions depicted using the case of Corazon Bay Marine Protected Area (MPA) in the Coral Triangle (RARE, n.d.). *Table 5.1.* provides a summary of key examples of the change elements in these three CBC-related interventions which will be referred back to throughout the following subsections. As established, step one of the CBC ToC pathway development process involves identifying the target beneficiaries of the desired result (i.e. notably the local community or resource-users – *Table 5.1.*). The next step involves identifying the desired result, which is now discussed

Table 5.1: A summary of key examples of the change elements that emerged from three CBC interventions making use of Theory of Change.

Intervention (Reference)	Main Beneficiaries	Desired Result(s)	Contextual Change Triggers	Actions	Assumptions	Issues Arising
Example 1: Evaluation of Forest Stewardship Council Certification in Indonesia (Romero & Putz, 2018)	Natural Forest Management Enterprises	• Sustained Timber Yield • Increased post-logging timber recovery rates	• Diminishing timber yields • Lack of markets • National policy approved increased harvesting intensity • Decline in certified natural forest management enterprises	• Train workers in reduced-impact logging techniques to avoid unnecessary damage • Increase understanding of the motivational context for worker performance for enabling environment for sustained timber yield decisions • Make decisions based on reliable harvesting data	• Workers will employ reduced-impact logging techniques • Presence of reliable data • Resources are available to adopt management practices • Presence of timber regulatory frameworks • Natural forest management enterprises will respect timber regulations and monitor annual tree growth	• Lack of political will and adequate policy and legislation • Concerns about worker and subcontractor training and supervision • Monitoring challenges • Lack of realized market benefits • Declining profits from subsequent harvests which produce small fraction of initial harvest • High opportunity costs of forest retention
Example 2: A community-based response to illegal wildlife trade (Biggs et al., 2017)	Local communities living with wildlife	• Decreased illegal wildlife trade	Escalating poaching and illegal wildlife trade High cost of living with wildlife for local community	• Strengthen disincentives for illegal behaviour • Increase incentives for wildlife stewardship • Decrease costs of living with wildlife • Support livelihoods that are not related to wildlife	• Community rangers use equipment and training to combat illegal wildlife trade and do not use them to poach themselves or for other purposes • Benefit sharing within communities is sufficiently equitable, and capture of benefits by elites does not undermine success. • Compensation does not lead to perverse behaviour (e.g., damage from wildlife is not actively induced to receive payments). • The value of wildlife products poached or traded in illegal markets is not so high that all other forms of income cannot come close to competing	• Local elite capture • Government resistance to decentralization of authority and community or individual ownership of wildlife • Government corruption leading to lack of trust in law enforcement authorities • Threats to community game guards enforcing laws • Risk of in-migration when benefits are perceived • If illegal wildlife trade decreases and wildlife populations increase, this may lead to increased human-wildlife conflict
Example 3: RARE's Pride conservation campaign. Case Study: Corazon Bay no-take fishing area (RARE, n.d.)	Local fishers and community members	• Increase white-spotted grouper population size • Reduce number of white-spotted grouper taken in no-take area	• Declining fish populations • Limited local income opportunities leading to overfishing, reef gleaning and dynamite fishing in the marine protected area and surrounds • Fish is important source of protein to local community	• Build a local management committee • Train local enforcement teams to increase regular enforcement • Increase awareness of the effects of negative fishing behaviour • Increase interpersonal communication among community members • Allocate exclusive fishing rights outside MPA	• Community will engage with the campaign • Fishers will respect local leaders and comply with no-take fishing regulations • Decreased fishing activity will increase white-spotted grouper population size • Provision of exclusive fishing rights will persuade fishers to fish outside the no-take area	• Family tradition of fishing in area • Local community don't see benefit of protecting no-take area to improve fishing • Enforcement teams are not respected and hesitant to arrest/ prosecute fellow community members • If some fishers (notably outsiders) fish in the no-take area, others feel like they should be able to too

<u>5.3.2.1. Identifying a Desired Result</u>

As mentioned above the *desired result* includes the *desired outputs and outcomes* and ultimately the final *desired impact*. Commonly identified conservation desired results include desired changes in attitudes and behaviour of policy makers, managers, partners, and resource users toward natural resource use and management. For example, the three interventions introduced above, as depicted in *Table 5.1.,* refer to desired results of: sustained timber yield and increased post-logging timber recovery rates (Romero & Putz, 2018), decreased illegal wildlife trade (Biggs et al., 2017), and increased population size of a local fish species and reduced catches an MPA (RARE, n.d.).

As established throughout evaluation of a ToC pathway is beyond the present scope, and therefore for the purposes of developing a *Generic CBC ToC Pathway* the *desired impact* is broadly stated as, to facilitate the successful initiation, implementation and governance of CBC interventions. Furthermore, the proposed intermediary desired outputs and outcomes, in accordance with emergent trends in the literature, are categorized based on three proposed 'desired output categories'. The core desired output category concerns *behaviour,* more specifically desired outputs in behaviour of CBC institutional actors that support conservation interventions and a willingness to work with other actors. The two further desired output categories relate to achieving desired *mindsets* (i.e. the influence of actor's the values, perceptions, attitudes and motivations towards the CBC initiative) and *institutions* (i.e. actor-relations, regarding decision-making processes, and actor's knowledge and capacity). Collectively these three categories, together with the desired impact, inform the generic desired results for the proposed *Generic CBC ToC pathway*. These three categories should be viewed as a guideline that would inform the identification of desired results within a specific CBC intervention. A brief discussion on the rationale for the selection of these categories and the generic desired results therein follows.

<u>*Behaviour*</u>

Behaviour has been described, in reference to evaluating conservation interventions, as "the only indicator that translates into real world impact" (Veríssimo, 2013: p29). Conservation is increasingly recognized as a behaviour change 'problem' by both conservation scholars (e.g. Schultz, 2011, 2014; Nilsson et al., 2016; Reddy et al., 2017; Olmedo et al., 2018; Cinner, 2018; Dobson et al., 2019; Schill et al., 2019), and practitioners (e.g. RARE, n.d.; WWF, 2017; RARE & BIT, 2019). More specifically, implementation of conservation interventions is widely considered a product of human decision-making processes, and therefore, changes in human behaviour are considered necessary for conservation 'success' (e.g. Mascia et al., 2003; Milner-Gulland, 2012; Veríssimo, 2013). Consequently, this has led some to suggest it requires "behaviorally-informed solutions" (RARE & BIT, 2019: p6). However, behaviour change involves a dynamic balancing of opposing forces which either 'drive' or 'restrain' change (Lewin, 1952). Furthermore, influencing 'pro-conservation' behaviour is problematic since a multitude of internal and external contextual factors affect behaviour change (Clayton et al., 2013; Gifford, 2014; Nilsson et al., 2016; Manfredo et al., 2016; Reddy et al., 2017; Schill et al., 2019). Not surprisingly influencing conservation behaviours is considered especially challenging for "ill-equipped" conservation professionals (Veríssimo, 2013: p29).

Behaviour change is largely a result of an actor's psychological and physical capacity to engage in actions, and their motivation and opportunity to do so (Michie et al., 2011: p4). Reddy et al. (2017) describe three primary approaches to encouraging 'pro-conservation' (or discouraging 'anti-conservation') behaviour, which include either *promoting* awareness and concern; *incentivizing* behaviour; and/ or *nudging* behaviour (i.e. making small changes to the decision-context targeting intuitive thinking). Essentially behaviour change requires understanding both individual and group conservation behaviours from the perspective of *what* is important to *who* and *why*, whilst also acknowledging an actor's context (Clayton & Brook, 2005; Clayton et al., 2013; Morrison, 2015; Reddy et al., 2017). Furthermore, behaviour change should be viewed as 'cyclical' and not linear, and thus allow for 'behavioural relapses', and the potential for learning, adaptation and action modification in the future (Prochaska & DiClemente, 1982, 1986). Consequently, based upon the above discussion and the present research focus, I propose a desired behaviour result simply as *increased pro-CBC behaviour*, though I acknowledge that this will be highly context-specific.

Mindsets

By identifying 'high-priority' behaviours in need of change, and aligning these with local priorities, conservation interventions may promote 'pro-conservation' *mindsets* (Ehrlich & Kennedy, 2005; Schultz, 2011; Waylen et al., 2013; Schultz, 2014). In turn, *conservation mindsets* will influence *conservation behaviours* (Dietz, 2015; Chan et al., 2016; Manfredo et al., 2016). These *mindsets* – including values, attitudes, perceptions and motivations – determine an individual's intention, openness and propensity and actual behaviour change (Petty, 1997[3]; Fishbein & Ajzen, 1975)[4]. Conservation-based values will influence an individual or organization's view on the purpose of conservation, which can include protecting nature for humans' (i.e. *instrumental values*) or for nature's sake (i.e. intrinsic values), and/ or a combination of the two (i.e. *relational values* – see Chan et al., 2016).

Key considerations for promoting 'pro-conservation' mindsets include encouraging a higher perceived-connectedness with nature (Pyle, 2003; Gosling & Williams, 2010); and improving local perceptions of (and thus collaboration with) external partners associated with the conservation intervention (Harris, 2007; Bennett & Dearden, 2014; Mahajan & Daw, 2016). Numerous scholars emphasize local conservation perceptions as particularly influential on CBC outcomes (Bennett & Dearden, 2014; Delgado et al., 2015; Mahajan & Daw, 2016; McClanahan & Abunge, 2016; Bennett et al, 2019). More specifically, local perceptions can strongly influence community participation in conservation governance, particularly amongst marginalized groups such as women (Nuggehali & Prokopy, 2009; Horwich et al., 2011; Zanotti, 2013). Therefore, as Morrison (2015: p960) states, "People are integral to any conservation outcome, and understanding motivations of potential supporters – and opponents – of conservation is essential in planning a theory of change." Consequently, I propose a desired mindset result simply and broadly as *increased pro-CBC mindsets*, and recognize that they are crucial to achieving *increased pro-CBC behaviours*.

---

[3] Although beyond the present scope, Petty (1997) provides a comprehensive discussion on 'attitude change theory', which is still highly relevant today and specifically to the topic of CBC.

[4] See also previous sections discussion related to "behaviour change"

Change in SESs requires 'systemic shifts' in institutional underpinnings such as 'mental models' (Biggs et al., 2011; van den Broek, 2018; Moon et al., 2019). Institutions are sets of formal and informal rules and norms that shape interactions and thoughts of organizations and individuals, including consideration of the possibility of change (North, 1990; Agrawal & Gibson, 1999; Redmond, 2005; Scott, 2014). Institutions are based on human behaviours and the differences between rules and actors (and the interactions between them) that affect institutional change (North, 1990). Thus, both *formal institutions* (e.g. national laws and international agreements), and *informal institutions* (e.g. systems of established and embedded social rules), serve to constrain and guide human behaviour (Hodgson, 2006). Consequently, CBC institutional change, which involves 'nested' multi-actor relations, requires acknowledging the existence and importance of both formal and informal institutions (Carlsson & Berkes, 2005; Carlsson & Sandström, 2008; Bodin et al., 2011; Clement, 2013; Guerrero et al., 2013; Ratner et al., 2013).

Therefore, strong local institutions that possess greater management autonomy, and which are accepted by communities, (i.e. possess *social institutional fit*), can promote institutional change for effective conservation management (Ostrom, 1990; Waylen et al., 2010; Calfucura, 2018). However, as discussed previously, devolving decision-making power to local communities in conservation management can be problematic due to elite capture (Ribot et al., 2006; Zulu, 2012; Galvin et al., 2018). However, the (mis)use of power affects *both* informal and formal institutional stability and the ability to change (North, 1993; Thompson, 1995; Redmond, 2005; Calfucura, 2018). Consequently, the design of institutions – especially when deeply entrenched in cultures of corruption, and conflicting or even contradictory governance mechanisms at different levels (local, regional, national and international) – can lead to failed institutional responses to conservation concerns (Rands et al., 2010; Haller et al., 2016; Calfucura, 2018). Therefore, CBC institutional change is complex, and functions amidst ongoing political, economic, and cultural change, and associated institutional turnover (Baral & Heinen, 2005; Schwitzer et al., 2014; Oldekop et al., 2016).

CBC literature emphasizes the need for 'appropriate' institutional arrangements. Yet a lack of consideration is given to enabling conditions and/or behavioural patterns required to facilitate change to a CBC mode of governance. Therefore, this necessitates improved understanding of *why* and under *what* conditions institutional change occurs at a global, national and local-level (Steinberg, 2009; Rands et al., 2010; Clement et al., 2015; Calfucura, 2018). Consequently, based on the above discussion, the proposed desired institutional result is simply identified here as *strengthened pro-CBC institutions*. *Figure 5.4.* depicts the three proposed desired outputs discussed above.

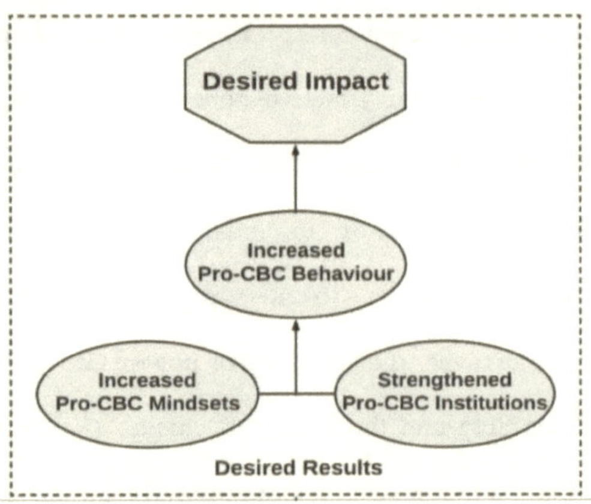

Figure 5.4.: The proposed desired results of Generic CBC ToC Pathway.

As introduced above *contextual change triggers* stimulate initiation and maintenance of interventions to bring about change. CBC contextual change triggers, inclusive of those within ecological or socio-economic, cultural and political domains, can be numerous and diverse, and may exist at different stages within an intervention's life span (Hagerman et al., 2010; Rodriguez-Izquierdo et al., 2010). An especially important consideration of CBC contextual change triggers is the *delivery mechanism*, in other words *how* the contextual change trigger is delivered, *who* delivers it, and *what* is the deliverer's relationship with the target beneficiaries (Wight et al., 2016). For example, if a government department (i.e. the 'deliverer') does not possess a productive working relationship with a local community (i.e. the target beneficiary), this will constrain the effectiveness of a contextual change trigger. Accordingly, such 'strained' government-community relations often result in external partners such as NGOs taking on the role of 'deliverer' (discussed further in *section 5.3.2.3.*).

Common CBC contextual change triggers include environmental degradation due to inappropriate development or environmental disasters, which in turn lead to deteriorating community livelihoods and well-being (Seixas & Davy, 2008; Shukla & Sinclair, 2010). For example, degradation of a specific (and often economically valuable) natural resource may 'trigger' a state-driven PA expansion strategy, which may lead to the development of conservation policy, that can then promote the implementation of CBC interventions (this expansion strategy itself is a potential CBC contextual change trigger) or may support local resource users in their demands that action be taken to protect local resources. However, contrast

Romero and Putz (2018) note how current national policy in Indonesia supports increased timber harvesting. Nevertheless, this contextual issue can trigger the need for changes in national policy and thus support a need for CBC-related interventions (*Table 5.1.*). An additional state-centric 'trigger' is state incapacity, which may 'trigger' support for a community-based mode of governance (Olsson et al. 2004; Seixas & Davy, 2008; Shukla & Sinclair, 2010).

Other notable common CBC contextual change triggers include high levels of poverty, a lack of alternative non-extractive resource-based livelihoods, and low institutional capacity (Roe et al., 2015; Biggs et al., 2017). Accordingly, the two examples introduced above by RARE (n.d.) and (Biggs et al., 2017) note the occurrence of limited alternative livelihoods and subsequently the need for interventions comprising actions in both cases linked to 'incentivizing' natural resource conserving practices (*Table 5.1.*). Consequently, these *contextual events or issues*, among others, may 'trigger' support for initiation and implementation of CBC interventions.

5.3.2.3. Identifying Common Actions to support a shift to CBC

As Wight et al. (2016: p522) suggest, most community interventions "exert their influence by changing relationships, displacing existing activities and redistributing and transforming resources." Accordingly, proposed *actions* should focus on addressing identified *initial contextual issues* which 'triggered' the need for the CBC intervention (see above), and subsequent 'issues arising' revealed by monitoring and evaluation (discussed subsequently in *section 5.3.2.5.*). This since both are constraining achievement of the *desired result*. Furthermore, in addition to 'reactive' actions, actions should strive to be 'proactive' in anticipating future potential 'issues arising' that could affect the ToC pathway.

I propose four common *action categories* that I consider to have emerged strongly from CBC literature in order to frame the design of common generic actions for this generic ToC pathway (Brooks, 2016; de Vente et al., 2016; Ernst, 2019; Dressel et al., 2020). These action categories refer to 'areas' in which action is commonly required within CBC interventions. Accordingly, these include the need to identify, and strengthen, increase or improve *actor-relations, socio-cultural alignment, knowledge dissemination,* and *institutional capacity*. However, it is important to acknowledge from the outset that the applicability of these action categories is highly context-specific, and these actions do not function in isolation, and are therefore, considered to interact with each other within the ToC pathway to achieve the desired result. Furthermore, it should be acknowledged that these may not be the only action categories, again required actions will be highly context-specific, and it is assumed that others would be able to identify other actions. Moreover, it assumed that further actions may emerge based on the empirical findings of this book, which will be discussed and lead to additions and/ or modifications in the final ToC pathway proposed in *Chapter 9*. Nevertheless, these action categories are deemed to serve as a valid baseline for further investigations into the three case studies.

I consider the proposed actions related to these action categories to positively affect, and be affected by, the extent to which the 14 key enablers identified in *Chapter 3: section 3.3.2.2.* are present, through iterative learning within the pathway. Consequently, the presence of these 14 enablers is hypothesised to increase the chance of achieving the desired result. A brief discussion introducing commonly proposed CBC actions emerging from the literature pertaining to each 'category' follows.

Action Category 1: Strengthen Actor-Relations

Social relations directly influence our beliefs, decisions, and behaviours (Ajzen, 1991). Whilst the initiation and implementation of CBC interventions may originate from either local resource-user' demands or outsiders' agendas, they nevertheless inevitably involve building 'nested' multi-actor collaborative partnerships (Seixas & Davy, 2008). However, as established above, the recent literature on CBC emphasizes the challenges facing these collaborative partnerships, leading to increasing calls for diverse, context-specific and transparent communication strategies to promote effective collaborative decision-making (Game et al., 2014; Kaplan-Hallam & Bennett, 2018; Baird et al., 2019a&b; Armitage et al., 2020).

Strengthening actor-relations essentially requires persuading all actors the *status quo* is no longer beneficial, and encouraging the collaborative and deliberative generation of new views to create new or transform existing institutions (Wijen & Ansari, 2007). This collaboration is most effective if relations of trust, respect and understanding have been built between actors for improved communication and coordination (Pomeroy et al., 2001; Redpath et al., 2013; Young et al., 2016; Baynham-Herd et al., 2018; Dressel et al., 2020). For example, RARE (n.d.) specifies the need for actions to *increase interpersonal communication among community members (Table 5.1.)*. Furthermore, Thompson (1995) proposes several key elements of institutional change relevant to CBC actor-relations, which include: a clear role for local people; ensuring accountability (in accordance with *enabler 12,*

Chapter 3: section 3.3.2.2.); accommodating diverse interests and perceptions; and working in collaboration with *external change agents*. These elements are showcased within the three CBC interventions discussed throughout. For example, these interventions refer to actions connected to the roles for the local community such as *train local workers in impact logging techniques* (Romero & Putz, 2018), *train local enforcement teams*, and *building a local management committee* (RARE, n.d.) (*Table 5.1.*). Furthermore, Biggs et al. (2017) note the need for actions to *support livelihoods not related to wildlife* which will require support of external change agents (*Table 5.1.*). It should be noted that many of these actions overlap with the categories below and will therefore contribute to for example *improved knowledge dissemination* and *strengthened institutional capacity*.

Consequently, actions proposed to strengthen actor-relations should be cognisant of the roles of CBC actors and networks, and the power dynamics contained within, to facilitate positive change in this regard (Barnes et al., 2017; Mbaru & Barnes, 2017). Numerous scholars emphasize it is necessary to identify both the individuals in power who may influence broader change, and those expected to embrace the change (Jentoft, 2007a, 2017; Raik et al., 2008; Castro & Mouro, 2011; Boonstra, 2016; Haller et al., 2016; Ojha et al., 2016; Kashwan et al., 2019). Therefore, notwithstanding potential 'issues arising', strong actor relationships are widely cited as crucial for promoting 'pro-conservation' mindsets, and behaviours of communities in both global (Bodin & Crona, 2009; Ruiz-Mallen et al., 2014; de Vente et al., 2016), and in particular African conservation contexts (e.g. Harris, 2007; Murphree, 2009; Horwich et al., 2012). Lastly, it should be acknowledged that the discussion above on strengthening actor-relations strongly links back to discussion on *enablers 4* (i.e. strong community ties) and *14* (i.e. nested support) in *Chapter 3: section 3.3.2.2.*

<u>*Action Category 2: Increase Socio-Cultural Alignment*</u>

Aligning governance rules with local socio-economic and -cultural priorities and conditions, is widely considered to positively affect CBC governance (e.g. Ehrlich & Levin, 2005; Harris, 2007; Govan, 2009; Waylen et al., 2010; Levine & Richmond, 2014; Brooks, 2016; Galvin et al., 2018). This action directly relates back to prior discussions pertaining to *social institutional fit (Chapter 3: section 3.2.2.*), and *enablers 2 (i.e. shared norms, values, interests and identities)* and *10 (i.e. strong alignment of rules with local priorities)* in *Chapter 3: section 3.3.2.2.*.

Conservation behaviours are commonly thought to be influenced by their perceived economic value to a community and the ability of conservation actions to outweigh the economic loss of any changes in behaviour (Nilsson et al., 2016). For example, Biggs et al. (2017) specifically note the need for actions that decrease the costs of living with wildlife (*Table 5.1.*). Accordingly, identifying and promoting *alternative livelihood strategies*, i.e. substitute or lower impact natural resource practices and or occupations, is a commonly proposed action (Roe et al., 2015; Wright et al., 2016). Promoting alternative livelihoods is also specifically noted by both Biggs et al. (2017) and RARE (n.d.) (*Table 5.1.*). The strategy of increasing local community involvement in CBC interventions by promoting economic incentives stems from the common belief that market forces will promote conservation (Hutton & Leader-Williams, 2003; Wright et al., 2015). For example, Biggs et al. (2017) suggest actions related to 'disincentivizing' illegal behaviour and 'incentivizing' wildlife stewardship, and in particular once again by supporting alternative livelihoods (*Table 5.1.*). Moreover, RARE (n.d.) suggest that allocating exclusive community fishing rights outside the MPA will incentivize not fishing within the MPA (*Table 5.1.*).

Notwithstanding the above, other 'non-economic' contextual factors may also be influential in promoting pro-conservation behaviours, such as their cultural acceptability, and local capacity to embrace the change (Nilsson et al., 2016). Accordingly, as introduced in *Chapter 3*, Galvin et al. (2018) found the alignment of a CBC intervention with a community's cultural worldview to be highly influential to the success of the intervention. Therefore, whilst 'incentivised' and 'neoliberal' conservation may be influential in promoting conservation behaviour, various scholars caution against 'commodifying' conservation (Child, M.F., 2009; Büscher & Dressler, 2012), since the significance of local natural resources goes beyond use and livelihoods, and includes religious, cultural, and recreational connections (Dudley et al., 2014; Golden et al., 2014b; Saunders, 2014; Walter & Hamilton, 2014). Numerous scholars specifically emphasize how local cultural contexts are highly influential to community perceptions, behaviours, and participation in conservation governance (e.g. Peterson et al., 2010; Waylen et al., 2010; Gurney et al., 2016; Infield et al., 2017; Baker et al., 2018). Yet, socio-cultural beliefs and practices (i.e. social norms) are complex, and may also clash with conventional conservation approaches, leading some to caution against 'cherry-picking' socio-cultural aspects with positive implications for conservation, while ignoring or trying to change those that are not well aligned (see Schultz et al., 2007; Schultz, 2011; Kinzig et al., 2013; Holmes et al., 2017; Infield et al., 2017). Consequently, many scholars propose CBC interventions should be strongly aligned with both a local community's socio-economic and -cultural context (e.g. Walter & Hamilton, 2014; Brooks, 2016; Gurney et al., 2016; Nilsson et al., 2016). Once again this connects back to *enabler 10* in *Chapter 3: section 3.3.2.2.*

<u>*Action Category 3: Improve Knowledge Dissemination*</u>
Conservation requires coupling an adequate knowledge base – which should include both 'Western Science' and 'Indigenous, Traditional and/or Local Ecological Knowledge' (LEK) – to decision-making and management actions and resource-user behaviour (Mascia et al., 2003; Rands et al., 2010; Aswani et al., 2018). However, diverse epistemic conceptualizations of 'nature' (i.e. forms of 'natural' knowledge), and attempts at organising and mapping it, have led to well-established and ongoing debates (Turnbull, 1997, Ingold, 2000; Aswani et al., 2018). A community's LEK represents "sensitivities, orientations, and skills that have developed over one's lifetime through actual engagement in and performance of practical activities" (Lauer & Aswani, 2009: p318). Therefore, it has been widely proposed that inclusion and dissemination of LEK within conservation interventions is required to counter the dominant effects of Western Science's 'reductionism', and the negative impact it is having on local communities and their perceptions of, and behaviours toward their natural resources (Goldman, 2003; Tengö et al., 2014; Golden et al., 2014a). Moreover, integration of LEK in decision-making is considered especially important to establishing an institutional setting, and identifying local change agents, social networks and power relations (Selman, 2004). The incorporation of LEK also links back to the previous action category of *increasing socio-cultural alignment* of CBC interventions. Greater recognition and incorporation of LEK would also contribute to efforts to *strengthen actor-relations.*

A commonly proposed action to improve CBC governance, involves developing a mutual understanding and agreement among CBC actors on common rules, conflict resolution strategies, and the sharing and building of common knowledge (e.g. Folke et al., 2005; Bodin & Crona, 2009; Ernst, 2019). As mentioned above, both Romero and Putz (2018) and RARE (n.d.) note the need for training local community members, and in the case of the latter specifically refer to increasing awareness of the potential negative effects of overfishing (*Table 5.1.*). Consequently, Wilson (2003: p265) states, "[i]f the knowledge needed for management ... is more accurate and easier to get, this helps management to be more rational...[and] the knowledge needed for management is contributed to, shared and controlled by more stakeholders, this helps management to be more equitable." Lastly, improved knowledge dissemination also directly influences the capacity of actors to actively engage in CBC institutions (Crona & Bodin, 2006; Barnes et al., 2019), which is discussed next.

Action Category 4: Strengthen Institutional Capacity
Institutional capacity refers to, "the collective ability of a group to combine various forms of capital within institutional and relational contexts to produce desired results or outcomes" (Beckley et al., 2008: p60). A systematic review of global CBC interventions revealed capacity building has been shown to better achieve 'win–win' CBC outcomes (Brooks, 2016). Accordingly, many scholars emphasize the need to build capacity – including individual skills development, and institutional capacity – to achieve positive social and ecological outcomes in CBC interventions (e.g. Pomeroy et al., 2001; Balint & Mashinya, 2006; Govan et al., 2009; Persha et al., 2011; Brooks et al., 2012; Levine & Richmond, 2014; Mountjoy et al., 2014; Brooks, 2016). This is also directly in accordance with the examples provided in

Table 5.1. Furthermore, capacity building has been shown particularly important when community capacity has been weakened by cultural erosion, and/ or development of dependency on centralized institutional structures (e.g. Brooks & Tshering, 2010; Akamani et al., 2015). Nonetheless, scholars have proposed that by building institutions on the "scaffolding" of existing institutional competencies (Clement et al., 2015: p465), including institutional networks and leadership, conservation management is more effective and legitimate than when institutional arrangements are 'enforced' on actors (see Cinner & Aswani, 2007; Clement et al., 2015).

In addition to local-level capacity building, capacity building of state and 'non-state' partners is also key to the ability of these actors to support CBC interventions (see O'Connell et al., 2019). This is also in accordance with *enabler 14* in *Chapter 3: section 3.3.2.2.*, i.e. the presence of nested support. Lastly, institutional capacity building (and *improved knowledge dissemination*) promotes *strengthened actor-relations*, which are important for multiple CBC outcomes (Brooks, 2016; Baird et al., 2019a; O'Connell et al., 2019).

5.3.2.4. Identifying Common Assumptions associated with a shift to CBC

A crucial element contributing to 'robust' ToC pathway development involves coupling a simple sequential ToC pathway with its underlying *assumptions* (Mayne, 2015, 2017). Furthermore, as introduced previously in *section 5.3.1.*, research on past experiences can better predict potential "at-risk assumptions" (Mayne, 2017: p157). Moreover, as introduced previously, assumptions are considered here to fall under rational and causal assumptions. In the present case, the generic *rationale assumption* made is that CBC interventions may offer a viable alternative approach to centralized and exclusionary conservation governance within appropriate contexts, able to deliver positive social and ecological outcomes. Furthermore, this book breaks down *causal assumptions* into 'causal pathway' assumptions and 'causal link' assumptions. The former relates to overarching causal assumptions made that positively influence the effectiveness of the ToC pathway as a whole. Accordingly, these specifically consider the presence (or the potential presence) of the 14 proposed CBC enablers (*Chapter 3: section 3.3.2.2.*) to positively influence achieving the desired result, and I therefore refer to these as the *causal pathway assumptions*. It is important to acknowledge that the presence of these enablers may in turn be strengthened by the ToC pathway actions in accordance with iterative systemic feedback. Accordingly, Biggs et al. (2017: p8) refer to "enabling actions" which can positively influence the presence of causal assumptions, and specifically emphasize enabling actions related to the ability to disincentivize certain 'anti-conservation' behaviours while incentivizing 'pro-conservation behaviours (*Table 5.1.*).

In addition to the 14 proposed CBC enablers, an overarching causal assumption for all actions proposed in the *Generic CBC ToC Pathway*, is the involvement and commitment of *change agents*. *Change agents* are self-motivated individuals possessing favourable personal, structural and relational characteristics, which deem them influential in motivating others, shaping and integrating new values into a social group's norms, and reinforcing and institutionalizing new patterns of behaviour through both formal and informal mechanisms within their specific context (Lippitt et al., 1958; Robbins & Judge, 2009; Crona et al., 2011; Moore & Westley, 2011; Scott & Davis, 2015; Englefield et al., 2019).

CBC change agents may emerge from local communities and/ or state and non-state partners. Accordingly, the concept of the assumed presence of *change agents* directly relates back to discussions on *enablers 3* (i.e. *strong local leadership*) and *enabler 14* (i.e. *'nested' support*) in *Chapter 3: section 3.3.2.2.*, the involvement of which has long been considered key for effective CBC governance (e.g. Agrawal & Gibson, 2001; Crona & Hubacek, 2010; Haller et al., 2016; Lyons & Cavaye, 2016; Crona et al., 2017). Moreover, both Biggs et al. (2017) and RARE (n.d.) note that presence of *strong local leadership* referring to assumed *equitable distribution of benefits*, and therefore no elite capture, and the assumed *respect for local leaders*, respectively (*Table 5.1.*). Additionally, in relation to the latter enabler of *'nested' support*, the Romero and Putz (2018) example notes the assumption that *resources be available to adopt management practices,* and *the presence of regulatory frameworks to support such activities (Table 5.1.).* Further resource-specific causal pathway assumptions commonly include a *small resource system with low mobility* (i.e. fisheries are more challenging to manage than forestry resources since many fish species are highly mobile across vast areas), the possibility of *storing benefits from the resource* (i.e. again this is easier within forestry than fishery contexts) and the presence of alternative livelihoods for resource users exists (Ostrom et al., 1999; Agrawal, 2002; Roe et al., 2015). Lastly, an additional causal pathway assumption is that all interventions make the assumption that participants will be willing to participate, as is specifically noted by both Romero & Putz (2018), and RARE (n.d.) (*Table 5.1.*).

In addition to the causal pathway assumptions, several key *causal link assumptions* are identified for the generic pathway. *Causal link assumptions* concern the effectiveness of a *specific* action to achieve the desired result. Whilst, overlap exists within these two 'groups' of causal assumptions, I propose notable causal link assumptions that emerge from the literature in relation to each action category.

The ability to *strengthen actor relations* assumes, as noted above, that all actors are willing to engage and collaborate with other actors, and that actors possess a shared vision and objectives of or the potential to achieve this. This latter assumptions links back to *enabler 2* (i.e. *shared norms, values, interests & identities*). In addition, this relates to the *ease of enforcement and conflict resolution* (i.e. *enabler 11*), and the *accountability* of those actors involved in management activities to other actors (i.e. *enabler 12*).

Increasing the socio-cultural-alignment of CBC initiatives primarily assumes the presence of customary institutions and practices linked to conservation (as introduced in *enabler 8* in *Chapter 3: section 3.3.2.2.*), which can positively influence achieving the desired result, or if in an eroded state have the potential to be revitalized (Cinner & Aswani, 2007; Cinner et al., 2012a). This also links back to the RARE (n.d.) example which specifies the need to *build local management committees (Table 5.1.)*. Furthermore, it assumes the presence of (potentially) suitable and socially acceptable alternative livelihoods that do not constrain achieving ecological conservation outcomes (Nilsson et al., 2016). Moreover, a related key assumption is a willingness for local resource users to abandon or modify present livelihoods, and participate in and build capacity related to the newly introduced livelihood. These assumptions are strongly emphasized by Biggs et al. (2017), and RARE (n.d.) *(Table 5.1.)*.

A central causal link assumption regarding the ability to *improve knowledge dissemination* amongst CBC actors is the acceptance of and willingness to incorporate LEK. Furthermore, it assumes the ability to reconcile both relevant LEK and scientific knowledge (Tengö et al., 2014). Moreover, it assumes the willingness of those that possess the knowledge to employ these skills for the benefit of the intervention, and to share this knowledge, where these knowledge holders are inclusive of community members, local leaders, and external partners (see *Table 5.1.*).

Practical causal link assumptions associated with the ability to *strengthen institutional capacity* are that the presence of enabling legislation, and local regulations, which will increase the legitimacy required for CBC institutional capacity (Pomeroy et al., 2001; Cinner et al., 2009a). This is specifically emphasized by Romero and Putz (2018) (*Table 5.1.*). Furthermore, once again the presence of willing and motivated actors is assumed. Moreover, the political will and participation of state partners toward the CBC capacity building process is also assumed. Additionally, the availability of technical and financial resources of partners to facilitate the capacity building process is assumed (as discussed throughout - see also examples *Table 5.1.*).

Consequently, these causal assumptions ultimately consider, "the needs, interests and behaviour of stakeholders and other key actors" (van Es et al., 2015: p16), which is central to developing a 'robust' CBC ToC pathway, and achieving an intervention's desired result. However, equally crucial is the monitoring and evaluation of the presence of enabling factors and conditions, from which 'issues arising' will arise and require reformulated actions. Accordingly, common CBC 'issues arising' are discussed next.

5.3.2.5. Identifying Common 'Issues Arising' that affect a shift to CBC

To reiterate, 'issues arising' relate to those issues identified from monitoring and evaluation of a CBC initiative, and can include both newly arising issues and/ or persistent initial contextual issues constraining achievement of the desired results (*section 5.3.1.*). A common 'issue arising' is a lack of locally-perceived benefits from a CBC intervention, which is commonly linked to the occurrence of elite-capture, and is emphasized in all three of the CBC interventions in *Table 5.1*. Furthermore, this aligns with discussions in *Chapter 3: section 3.3.2.2.* related to *enablers 3* (i.e. *strong local leadership*) and *7* (i.e. *equitable distribution of benefits*). On a related note continued high opportunity costs of the intervention for the local community is also a common 'issue arising', and once again is emphasized by all three cases in *Table 5.1*.

Additional common 'issues arising' include a lack of alignment of CBC decision-making processes with local priorities (as discussed in relation to *enabler 10* in *Chapter 3: section 3.3.2.2.*). Furthermore, issues associated with local leadership capacity, a lack of shared values, norms and interests (and conflicts), and eroded customary institutions and practices, may often emerge from an intervention and can negatively affect CBC governance (Brooks et al., 2012; Brooks, 2016). This in accordance with *enablers 2, 3, 7, 9, 11,* and *12* (*Chapter: section 3.3.2.2.*), and the three examples depicted in *Table 5.1.*.

Additional common CBC 'issues arising' are strained/ ineffective actor-relations, which are often negatively exacerbated by power asymmetries. Social relational challenges within CBC interventions are commonly connected to institutional organizational network structures, cultures, histories and perceptions, and control and coordination, which may all create mistrust of actors and misunderstandings and conflict, requiring resolution (Bodin, 2017; Baird et al., 2019a&b). This may manifest in particular with a lack of knowledge dissemination, whether from local leaders or partners (Crona & Bodin, 2006; Barnes et al., 2019).

Issues commonly arise in CBC interventions related to intra-community relations, notably once again *elite-capture*. Accordingly, Biggs et al. (2017: p10) provide the example of how actions can cause a "breakdown in social cohesion" within a community, citing potential conflicts between trained community game guards, employed to counter illegal wildlife trade, and local poachers (*Table 5.1.*). This in accordance with *enabler 4* (i.e. strong community ties) discussed in *Chapter3: section 3.3.2.2*. Furthermore, RARE (n.d.) emphasize the 'issues arising' related to rules misaligned with established norms, providing the example of family traditions constraining compliance with a no-take fishing zone of a newly created MPA (*Table 5.1.*).

Extensive recent literature emphasizes challenges associated with community-partner relations for realizing positive social and ecological CBC outcomes (e.g. Alexander et al., 2016; Barnes et al., 2017; Bodin, 2017; Armitage et al., 2017; Baird et al., 2019a&b). With regards to community-state relations, Biggs et al. (2017) specifically note resistance by states to decentralization, as well as state corruption, commonly emerge as issues when attempting to decrease illegal wildlife trade (*Table 5.1.*). Furthermore, a lack of support from state actors associated with a CBC intervention, largely due to a lack of state capacity (i.e. both a lack of technical and financial capacity), can also commonly constrain CBC interventions and the ability of a state to collaborate with local communities, this perhaps most notably concerning issues related to monitoring and enforcement (Hayes & Persha, 2010; Kothari et al., 2013; O'Connell et al., 2019; Sanders et al., 2019). Consequently, weak local leadership, and a lack of external 'nested' support are two especially notable common 'issues arising' requiring attention within CBC interventions.

<u>5.3.2.6. Common External Influences that affect a shift to CBC</u>

As introduced above a ToC pathway is perhaps better considered "a model of the contribution to and not cause per se of the intended result" (Mayne, 2015: p128). Therefore, a desired result and may not be realized as a result of external influences (Mayne, 2015, 2017). This is also emphasized in the CBC intervention described by Romero & Putz (2018). Mayne (2015: p123) describes *external influences* as:

> "...events and conditions unrelated to the intervention that could contribute to or detract from the realization of the intended results. These could include other interventions with similar aims, and/or general economic or social trends. They are not part of the intervention theory of change per se."

Consequently, *external influences* consider the external context within which the intervention takes place – and which is out of the intervention's control – that can (potentially) affect the successful achievement of the desired result. Accordingly, Sanders et al. (2019: p3) recently identified important barriers to conservation action, as described by conservation practitioners, including what they refer to as "external issues", key examples of which include a lack of funding and a constraining environment characterized by a lack of political will, ineffective policy and weak and corrupt states. Likewise, the example from Romero & Putz (2018) identifies a lack of political will and inadequate legislation as constraining to natural forest management enterprises in Indonesia. Additional common potentially constraining CBC external influences can include population growth, and globalization (Sanders et al., 2019). In the present coastal context, a notable concern related to globalization is the negative influence of *Illegal Unregulated and Unreported fishing* (IUU). Moreover, cross-sector conflict, notably conflicting agendas of different state departments or development sectors, and/ or NGOs can also negatively influence CBC interventions. For example, a food security programme aimed at small-scale agriculture, could inadvertently 'incentivize' forest clearing, and thus negatively affect a community forestry intervention.

However, external influences can also enable achievement of an intervention's *desired result* (Mayne, 2015). For example, as introduced above in *section 5.3.1.*, international commitments, such as the *Aichi Targets* or proposed *Post-2020 Global Biodiversity Framework*, can be considered an enabling external influence for CBC interventions if they are able to promote CBC initiation, implementation and governance. Consequently, external influences are considered within this book to have the potential to either constrain or enable a ToC pathway to reach its desired results.

5.3.3. Conclusion: A proposed Generic CBC ToC Pathway
Mixed results of CBC interventions thus far are largely due to limited understanding of the complexity of the context, processes and iterative nature under which social and institutional change occurs. Accordingly, developing a ToC

pathway represents a flexible and useful approach for greater understanding of the how this CBC change process occurs. This involves identifying and considering key change elements such as *contextual change triggers, actions, assumptions,* and *issues arising* and *external influences* required to achieve the *desired results*. While it is acknowledged that others may produce different and/ or additional change elements, the change elements proposed here, based upon the literature, serve to provide a valuable 'baseline' for investigations into this dissertation's three case study sites. Accordingly, I propose a *Generic CBC ToC Pathway* in *Figure 5.4.* as a 'prior', 'literature-based' ToC pathway, which represents the conceptual framework for this book as it informs data collection, analysis and interpretation, and eventually the development of a 'field-based' *Empirical CBC ToC Pathway* in *Chapter 9*. To reiterate, in addition to the causal-link assumptions for each action, the 14 key enablers form the *causal pathway assumptions*, i.e. their presence is assumed to positively influence, and be influenced by movement through the entire pathway toward the desired results. Lastly, it is important to note that this pathway is not sequential, as actions can take place in any order and do interact with one another to produce these desired results. Moreover, these actions will influence and in turn be influenced by, the desired results through systemic feedback, which will lead to adaptation of the intervention, as depicted in *Figure 5.4.*

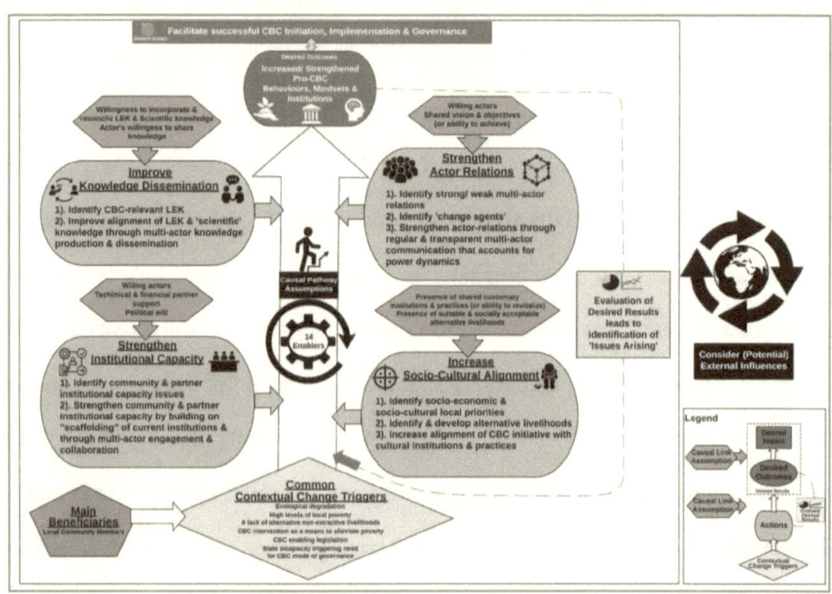

Figure 5.5.: A prior, literature-based *Generic CBC ToC Pathway*.

Chapter 6

An East African CBC Change Perspective:

The case of the Bay of Ranobe, Madagascar

6.1. Introduction and context

This chapter provides an East African CBC change perspective by presenting the findings of the investigation into the initiation and implementation of CBC governance in the first regional case study located in the Bay of Ranobe, Madagascar. Following a brief introduction of the national context and the country's CBC challenges and progress to date, the case study context is introduced. Thereafter, findings are presented in accordance with the *change elements* presented in *Chapter 5.* In doing so this chapter – together with the second regional case study presented in *Chapter 7* – addresses **objective 4** (*Box 6.1.*). These empirical findings, in conjunction with those in *Chapters 4* and *8*, are then revisited, and consolidated with other findings within the final discussion in *Chapter 9.*

Box 6.1.:

Objective 4: To investigate the factors, conditions and processes that enabled and constrained CBC in two regional case studies to learn lessons for South African CBC initiation, implementation and governance

6.1.1. National context

6.1.1.1. Madagascar: biodiversity, poverty and conservation

Located in the Western Indian Ocean, Madagascar represents the world's fourth largest island, with a distinct history and culture (Dewar & Wright, 1993; de Wit, 2003). It is perhaps most renowned for its natural capital, notably high levels of terrestrial biotic endemism (Goodman & Benstead, 2005; Ganzhorn et al., 2014), which has led to its 'high-priority' terrestrial conservation status (Brooks et al.,

2006). Furthermore, it has also recently been afforded high-priority marine conservation status (Selig et al., 2014; Jenkins & Van Houtan, 2016). Consequently, this global conservation status has influenced injections of conservation funding and resources into the country and supported its conservation and development agenda (Miller et al., 2013; Waeber et al., 2016; Corson, 2017). This 'agenda' is heavily linked to biodiversity-based tourism, which remains the primary attraction for foreign tourists and provides much needed employment and contributions to the country's gross domestic product (GDP) (Randriamboarison et al., 2013; WAVES, 2015). *Figure 6.1.* provides an overview of the country's key socio-economic statistics, in order to better contextualize the findings of this chapter.

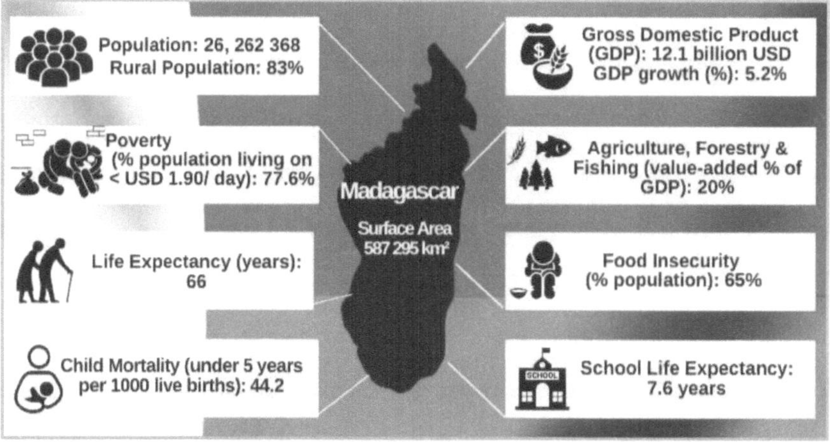

Figure 6.1.: A summary of key socio-economic statistics of Madagascar. **Source:** World Bank (2019); FAO (2019a).

Madagascar is characterised as a "Low-Income Food-Deficit Country" (FAO, 2019b). Not surprisingly, conflict exists between ecological and economic interests (see Lammers et al., 2017). Ambitious local rural economic development and conservation projects have faced numerous challenges within the context of an ineffective State and oversimplified private-sector interventions – particularly direct forest investment and privatization of rural customary land – with limited impact on improving local rural livelihoods (see Ferguson et al., 2014; Hänke & Barkmann, 2017; Hänke et al., 2017). The State's weak capacity and corruption has been illustrated by numerous global indices (e.g. World Justice Project, 2019; Transparency International, 2019). Most notable is the country's "severe political fragility" (OECD, 2018a). *Fragile* governments are characterised by a "lack of political will and/or capacity to provide the basic functions needed for poverty reduction, development and to safeguard the security and human rights of their populations" (OECD, 2007: p2). The influence of this fragility is evident within recent national assessments reflecting extremely high poverty rates, and a recent (and future projected) failure to improve the lives of the rural poor (UNDP, 2016; Collier, 2019). Food insecurity affects approximately 65% of the national population, with severe child malnutrition negatively impacting child development, particularly in rural areas (Le Manach et al., 2013; Rakotosamimanana et al., 2014; Borgerson et al., 2018). Furthermore, approximately 80% of the rural population are heavily dependent on natural resources but lack adaptive capacity in the face of environmental changes (Cinner et al., 2012b; Westerman et al., 2012; Osborne et al., 2016).

Coastal populations make up approximately 50% of the national population and are considered especially vulnerable and food insecure (Le Manach et al., 2013; Brenier & Vogel, 2017). The country possesses a coastline of about 5000 km, which comprises approximately 3,540 km of diverse reef systems boasting richly diverse and abundant marine ecosystems (Selig et al., 2014), particularly in the south-west (Gillbrand et al., 2007; Davies et al., 2009). These are home to coastal resources that provide small-scale fisheries (SSFs)[5] with an important and accessible source of protein and income (Davies et al., 2009; Barnes-Mauthe et al., 2013). Malagasy SSFs are estimated to account for approximately 102,000 fishers, catching an average of 135,000 tons of seafood annually, representing approximately 72% of the national catch (Le Manach et al., 2012; World Bank, 2015a).

However, local marine systems are experiencing changes due to cumulative human impacts (Halpern et al., 2015). Furthermore, large predicted future changes are of concern given ineffective safeguarding (Rabearivony et al., 2010; Molinos et al., 2016). More specifically, the country's coastal resources are considered under pressure from overharvesting by commercial and SSFs (Iida, 2005; Humber et al., 2011), environmental and climate change (Bruggemann et al., 2012; Cinner et al., 2012b), and increased coastal population growth and coastal in-migration of 'outsiders' (i.e. culturally-heterogeneous migrants), the latter driven by uncertain and declining inland agricultural productivity (Cripps & Gardner, 2016; Hänke & Barkmann, 2017). Moreover, some describe the problematic 'transplantation' and reconciliation of extant national conservation-related laws from French civil law into traditional customary law (Rakotoson & Tanner, 2006: p859), which are considered specifically within the coastal-marine realm to be rarely enforced, the

---

[5] The term small-scale fisheries (SSF) is used here in accordance with the FAO's (1999) definition specifying these fisheries as using small amounts of capital and energy mainly for local consumption. This encompasses those fisheries considered *artisanal*, *subsistence* and *small-scale commercial*. However, Madagascar (and Guinea-Bissau and South Africa) is characterized by the former two. *See further Glossary of key terms at beginning of this book.*

cause of governance conflict, and insufficient to protect marine species (Le Manach et al., 2012; Humber et al., 2015; Ratsimbazafy et al., 2019).

6.1.1.2. CBC in Madagascar

Early CBC efforts in Madagascar have been described as a "top-down", driven by "outsiders' agendas" and having developed within a "complex conservation bureaucracy" (Duffy, 2006; Hume, 2006; Dressler et al., 2010). The late 1980s observed the proliferation of various partnered-conservation programmes between big international non-governmental organizations (BINGOs), bilateral aid programmes, communities and the State (Duffy, 2006; Dressler et al., 2010). This period witnessed what Corson (2017: p2) refers to as a "critical historical conjuncture … in which agendas around biodiversity conservation and sustainable development emerged in the context of rising neoliberal policies—[which] then prompted a reconfiguration in power relations among public, private, and nonprofit actors". Consequently, a highly influential international donor community – as well as other private-sector industries such as mining (Scales, 2014) – emerged in relation to biodiversity conservation and community development (Duffy, 2006; Dressler et al., 2010; Corson, 2017). The above conservation partnerships enabled CBC to be implemented; firstly, in 1991 via the development of the World Bank assisted *National Environmental Action Plan* (NEAP) (Mercier, 2006), and secondly, in 1996 in accordance with newly developed national *Gestion Locale Sécurisée* legislation (GELOSE) (Antona et al., 2004).

The country launched the continents first NEAP in 1991, which promoted CBC through the transfer of limited management rights from the State to local CBOs (Gardner et al., 2018). NEAP aimed to restrain increasing environmental degradation by "reconciling [the] population with their environment" (Mercier, 2006: p50). More specifically, NEAP's second phase between 1996-2002 emphasized French- and Malagasy-scientist directed CBC implementation and governance (Dressler et al., 2010). However, weak governance and institutional capacity affected application of policies and regulations, with disproportionate centralized resource investment leading to limited local institutional capacity-building (Mercier, 2006).

The 1996 GELOSE legislation (*Decree No. 96-025*) aimed to provide secure local management of (initially) terrestrial natural resources (notably forests), by granting management rights over specific natural resources to local CBOs entering into a contract with the local municipality and a relevant State ministry (Antona et al., 2004). However, concern arose over whether implementation resulted in "a real transfer of rights" or simply represented "a policy by which local communities are made to manage resources in an externally defined way" (Antona et al., 2004: p839). Accordingly, GELOSE is criticized for not enabling adequate representation of community needs (Dressler et al., 2010), and recognition of customary legal systems (Rakotoson & Tanner, 2006; Pollini et al., 2014). The latter issue is pertinent since GELOSE required permanent closures, considered incompatible with temporary customary coastal resource seasonal closures (Cinner et al., 2009a). Moreover, insufficient regulations left key GELOSE principles lacking clear implementation tools (e.g. how to transfer property rights or legally recognize customary institutions) – which constrained implementation (Kull, 2002; Antona et al., 2004; World Bank, 2015a). Subsequent conservation management decrees also failed to address these implementation issues (Antona et al., 2004; Dressler et al., 2010).

GELOSE was first applied in a marine context in 1999 (Bruggemann et al., 2012). GELOSE's application to marine systems became problematic, largely because of a "jurisdictional mismatch" over marine resources under the authority of the *Ministry of Agriculture and Fisheries*, which is not permitted to sign a management contract with the community since biodiversity protection is the responsibility of the *Ministry of the Environment* (Cinner et al., 2009a: p492). Furthermore, local management of marine habitats lacked recognition with the *Ministry of Agriculture and Fisheries*, since the constitution of Madagascar provides for their centralized management (Cinner et al., 2009a). Consequently, Ratsimbazafy et al. (2019) note the complexity and subsequent impacts of inter- or even intra-institutional conflict for the effectiveness of any type of Marine Protected Areas (MPAs) in Madagascar, inclusive of Locally-Managed Marine Areas (LMMAs).

A key contextual change trigger for CBC emerged in 2003 when erstwhile President Ravalomanana, prompted by conservationist concerns, announced a "conservation emergency" (see Marie et al. 2009), and a five-year tripling of PAs, in accordance with the 'Durban Vision' (Gardner et al., 2018). This PA expansion was codified into law in 2001 (*Decree No. 848-05*), which led to the establishment of the *System of Protected Areas of Madagascar* (SAPM), and PA management expanding to focus on poverty alleviation, development and the conservation of cultural heritage through sustainable use of natural resources (Gardner et al., 2018). Consequently, this PA expansion involved designating IUCN category V and VI PAs – i.e. multiple-use PAs allowing sustainable extraction of certain natural resources by local communities as permitted by a zoning plan – to be co-managed via agreements between NGOs and CBOs (Gardner et al., 2013; Virah-Sawmy et al., 2014).

The above PA expansion decree endeavoured to simplify and redefine the legal PA establishment process, and provide a more flexible model permitting CBOs, NGOs and the private sector – in addition to the parastatal PA agency, i.e. Madagascar National Parks – to manage PAs (Rabearivony et al., 2010). However, concerns were raised over whether IUCN PA categories adequately reflect the country's new multi-use PAs agenda (see Gardner, 2011). Additional concerns for this rapid PA expansion were raised, largely stemming from initial terrestrial PA implementation attempts, including a lack of meaningful community engagement and the development of 'non-representative' CBOs, i.e. characterised by elite-capture (Raik & Decker, 2007; Toillier et al., 2008, 2011; Pollini et al., 2014). Moreover, concerns exist regarding who is the 'true' management authority of these PAs, with the PA system often considered to conceal NGO 'promoters' (i.e. partners) acting as the *de facto* managers (see Gardner et al., 2018). Nevertheless, some CBC initiatives, such as the *Anja Community Forest Reserve* located in the south, are considered to have set up successful institutional arrangements, and enabled positive outcomes in terms of biodiversity conservation, increased alternative livelihoods, and improved education and health services (Schwitzer et al., 2014; World Bank, 2015a).

Whilst customary institutions remain active in the country, levels of respect from both 'insiders' (i.e. long-term culturally-homogenous residents) and 'outsiders' towards these institutions have declined, especially when such institutions restrict access to necessary livelihood sources (Pollini et al., 2014; Cinner, 2012a; Cripps & Gardner, 2016). Nonetheless, the potential and realized benefits of CBC governance incorporating the use of *Dina* (i.e. a customary system of community-established rules based on socio-cultural norms ratified by village-presidents - Rakotoson & Tanner, 2006; Kaufman, 2014), has recently been acknowledged as a useful management approach for coastal conservation (Harris, 2007, 2011; Benbow et al., 2014). Many *Dina* have been implemented for generations through restricted sacred/ taboo fishing areas and days (Cinner, 2007; Cinner et al., 2009b). Furthermore, GELOSE legislation was the first legal mechanism that attempted to integrate *Dina* with state laws to enable CBC (Rakotoson & Tanner, 2006). However, integrating, strengthening and legally recognizing customary coastal governance in Madagascar has been problematic (Cinner et al., 2009b; Ratsimbazafy et al., 2019). Whilst some note the inefficiencies of Malagasy customary institutions – caused for example by a lack of trust, inflexible institutions and poverty (Cinner et al., 2009a&b) – many still believe these institutions hold greater hope for coastal-marine conservation governance than prior top-down approaches, as greater socio-cultural understanding of local institutions can inform self-compliance (Cinner et al., 2012c; Ratsimbazafy et al., 2019).

Dina has been used in the recent establishment of temporary fishery closures for octopus, which is a rapidly growing and economically valuable species (Harris, 2007; Benbow et al., 2014; Oliver et al., 2015). This CBC approach is thought locally to be more cost-effective, resilient and socially acceptable than top-down management (Harris, 2007). Furthermore, the success of these closures resulted in the State passing new fisheries legislation in 2005 for the seasonal closure of the big blue octopus (*Octopus cyanea*) fishery for six weeks from mid-December each year (Samolys et al., 2017). Moreover, the success of these closures has led to the rapid expansion of the number of LMMAs, recently estimated to account for more than 80 LMMAs covering 11,800 km^2 of Madagascar's coastal-marine habitat, and 17.7% of its continental shelf, with more than 148,920 beneficiaries (Samolys et al., 2017; Blue Ventures, 2018; Ratsimbazafy et al., 2019). This 'success' is predominantly driven by immediate tangible community benefits, acceptance by communities, peer-to-peer learning, and the support of BINGOs, (Mayol, 2013; Oliver et al., 2015). However, due to an onerous and expensive legal declaration process very few have received full legal PA status (Blue Ventures, 2018).

The Velondriake LMMA, located in the south-west of Madagascar, is partnered with the British NGO Blue Ventures, and was the first LMMA (IUCN category V PA) established in 2005, though only legally recognized in 2015 (Harris, 2011; MIHARI, 2015). Other LMMAs having received permanent protection status include: *Soariake* (partnered with the World Conservation Society (WCS) and the World Wide Fund for Nature (WWF)) and the cases study LMMAs of *Massif des Roses* and *Ankaranjelita* in the Bay of Ranobe (discussed in *section 6.1.2.*), all located in the south-west; as well as those located in the north-west, for example *Ankarea* and *Ankivonjy* (both partnered with WCS), and *Ambodivahibe* in the north-east (partnered with Conservation International) (MIHARI, 2015; Brenier & Vogel, 2017). However, whilst these (and other) LMMAs have delivered many social and ecological benefits, shortcomings exist in local-level governance, involvement of women, and the inability of CBOs to control fishing effort – especially 'free-riding' from 'outsiders' – upon the reopening of temporary closure sites (Westerman &

Benbow, 2013; Mayol, 2013; Benbow et al., 2014; Brenier & Vogel, 2017). Consequently, Blue Ventures (2018) recently stated that improvements needed for long-term community development and conservation include, among others: enhancing multi-actor participation; ensuring financial sustainability; enforcing rules; reducing the natural resource dependence of local communities through transformative livelihood change; and developing long-term visions to reconcile the differing objectives of conservation NGOs and other actors.

Therefore, whilst challenges persist, LMMAs are considered to have made much progress for community development and coastal-marine conservation in the country. Yet, as Gardner et al. (2018: p31) state "In all cases these structures remain works in progress, and will require years of further experimentation and evolution before they are optimized." Accordingly, the case study site of the Bay of Ranobe should be considered a work in progress, but does provide numerous insights into making the shift toward a community-based mode of governance, and is now introduced.

6.1.2. Local Context: The Bay of Ranobe
6.1.2.1. The Ecological System
The southern extreme of the Bay of Ranobe is situated approximately 27 km from Toliara city, in Toliara province, south-west Madagascar (refer to *Chapter 2: Figure 2.1.*). *Ranobe* translates to 'big water' which adequately describes this coastal lagoon stretching 8 km at its widest point and sheltered by a 32 km fringing reef (Belle et al., 2009). Rich marine biodiversity exists within diverse habitats comprising patch reef, seagrass beds, and estuarine and fragmented curtain mangroves (Davies et al., 2009).

The People

The population of the Bay of Ranobe is estimated at 20 000 and considered to have experienced substantial recent population growth (Bruggemann et al., 2012; Nordgren, 2014). The population is predominantly *Vezo, a* traditionally semi-nomadic, seafaring, and patriarchal people, dependent on the sea for their livelihoods and cultural identity (Astuti, 1995; Marikandia, 2001; Barnes-Mauthe et al., 2013). Spiritual traditions, such as ancestor worship and sacred places are central to *Vezo* beliefs (Langley, 2006; Cinner, 2007). Furthermore, *Fady* (i.e. taboos restricting daily activities, notably food prohibitions), and *Dina* (i.e. socio-cultural rules) govern daily life and have been specifically shown to restrict natural resource use in the south-west (Harris, 2007; Julia et al., 2008). Nonetheless, the *Vezo* belief of their right to use the sea and its ability to continually provide, can conflict with conservation governance objectives (Astuti, 1995; Iida, 2005).

Coastal-migration of other inland groups such as the *Masikoro*, *Mahafaly*, and *Antandroy* has increased due to decreased agricultural productivity (Jones, 2012). Whilst *Vezo* are known to live by the sea and fish, the *Masikoro* are known as "those who live in the interior... and raise cattle" (Astuti, 1995: p465). Nonetheless, Astuti (1995: p467) refers to the 'fluidity' of the *Vezo*, stating that "people with different ancestral customs can all be Vezo, because Vezo-ness is not determined by ancestry", as all "can be said to be Vezo if and when they learn the ways of the sea" (Astuti, 1995: p468). Therefore, her research concludes "people are Vezo if they behave Vezo-like" (Astuti, 1995: p473).

The Fishery

An estimated 8,000 fishers operate in the Bay of Ranobe, comprising 13 villages of varying size (Le Manach et al., 2013). Fishers operate nearshore, using pirogues (i.e. canoes) and a variety of traditional and modern fishing methods, including intertidal gleaning for octopus with various spears (predominantly by women and children), and various forms of nearshore line-, spear-, and net-fishing (principally by men) (Davies et al., 2009; Gough et al., 2009). A high diversity of capture species exists, yet coastal pelagic species dominate catches (Davies et al., 2009). Since the early 1990's the greater Toliara Bay has been considered overfished (Laroch & Ramananarivo, 1995; Brenier et al., 2011). More specifically, the Bay of Ranobe has been characterized as heavily exploited (Davies et al., 2009). Octopus represents the most commercially valuable marine resource in Toliara province, with as much as 93% of the local catch exported through international sea food companies (L'Haridon, 2006; Epps, 2007). Another export species is sea cucumber (Conand & Mara, 2000), which together with seaweed represent potentially profitable local aquaculture species (Darwin Initiative, 2017).

LMMA management

In October 2006, *Reef Doctor* (a locally-based British NGO working in the bay since 2002) and fishers from *Mangily* and *Ifaty* villages first met to discuss the establishment of a local fisher association and an LMMA (see *Figure 2.1.* previously for locations). Fishers agreed to form the CBO named *FIMIHARA* (derived from the Malagasy *FIkambanana MIaro sy HAnasoa ny RAnomasina* – translated to mean *Association to Protect and Enhance the Marine Environment*) to work with Reef Doctor to establish and manage an LMMA and develop other conservation and community development initiatives (Belle et al., 2009). In April 2007, *FIMIHARA* was legally recognized and incorporated into the local customary legal system (Belle et al., 2009), On 25 May 2007, the first LMMA called *Massif des Roses* (see *Figure 2.1.* previously) – covering an area of approximately 0.16 km^2 and located in the south of the bay – was established (Belle et al., 2009; Reef Doctor, 2012). In 2009, *Ankaranjelita,* a second larger LMMA (approximately 0.24 km^2) was established in the north of the bay (see *Figure 2.1.* previously). Unlike the temporary closures of other LMMAs in the south-west (see Harris, 2011), both LMMAs in the Bay of Ranobe are permanently closed to fishing but open to tourism, with entrance fees contributing to *FIMIHARA* running costs and community development projects (Belle et al., 2009). A *Dina* was established for the protection of the two present-day LMMAs, which are both now legally recognized by the State as part of the national PA system (pers. comm. 23 May 2018). Moreover, recently a third LMMA, *Vatosoa Marine Reserve,* was officially opened by the State on 22 January 2019, after the fieldwork for this case study was completed (Reef Doctor, 2019).

6.2. Findings

6.2.1. Contextual Change Triggers

A central contextual issue triggering support for a CBC mode of governance in the country emphasized by most respondents is the degradation of both terrestrial and marine resources, which has enabled the country's aforementioned conservation and development agenda (Cinner et al., 2012d; Waeber et al., 2016; Corson, 2017). However, the additional contextual issue of extreme poverty, especially in rural areas, results in conservation being a low priority for communities, as emphasized by all community and partner respondents. Poverty has long been closely linked to environmental degradation in Madagascar (e.g. Gezon, 1997; Sarrasin, 2013), where the poorest regions often correspond with the most acute environmental degradation (World Bank, 2015a). Furthermore, environmental degradation trends in Madagascar – most clearly illustrated by deforestation (Zinner et al., 2014) – as well as the State's inability to alleviate poverty and achieve sustainable development goals have raised concerns (see Waeber et al., 2016). Moreover, increased articulation with external markets in the past two decades – notably the introduction of an international export-market for fresh seafood in the south-west – is thought to have increased the value and local exploitation levels of marine resources, notably octopus (see L'Haridon, 2006), and raised concerns about marine resource degradation (Harris, 2007; Mayol, 2013). Consequently, the value of both terrestrial and marine resources, and ongoing degradation of these systems, as well as poverty are key contextual issues triggering change to CBC in the country.

As introduced previously, the State promoted a national PA expansion strategy, notably the proliferation of multi-use PAs. Furthermore, this led to the generation of PA legislation, which although characterized as onerous and requiring streamlining by numerous respondents, assisted in 'triggering' the proliferation of LMMAs. However, State incapacity is a key contextual issue to the CBC change process in the country. Accordingly, most partner respondents argued this enabled a shift to 'NGO-partnered' CBC initiatives since they possess both the technical and financial resources that the State does not. Furthermore, all State respondents supported BINGO presence and involvement. More specifically, the State's increasing support through legal recognition of LMMAs, has allowed BINGOs to become highly influential in triggering community involvement in these

CBC initiatives (Ferguson & Gardner, 2010; Scales, 2014), and thus is considered an additional contextual change trigger.

6.2.2. Review of Enablers

This section analyses the extent to which the 14 enablers identified in *Chapter 3* are present in the Bay of Ranobe case study. Accordingly, a summary of the *absence, partial presence, partial absence,* or *presence* of these enablers can be found in *Figure 6.2.* Furthermore, *Table 6.1.* provides a brief overview of the findings that influenced these enabler presence ratings. These findings are discussed further below. Lastly, reference is also made within the following discussion to relevant responses from national Malagasy CBC partner respondents to contextualize these findings.

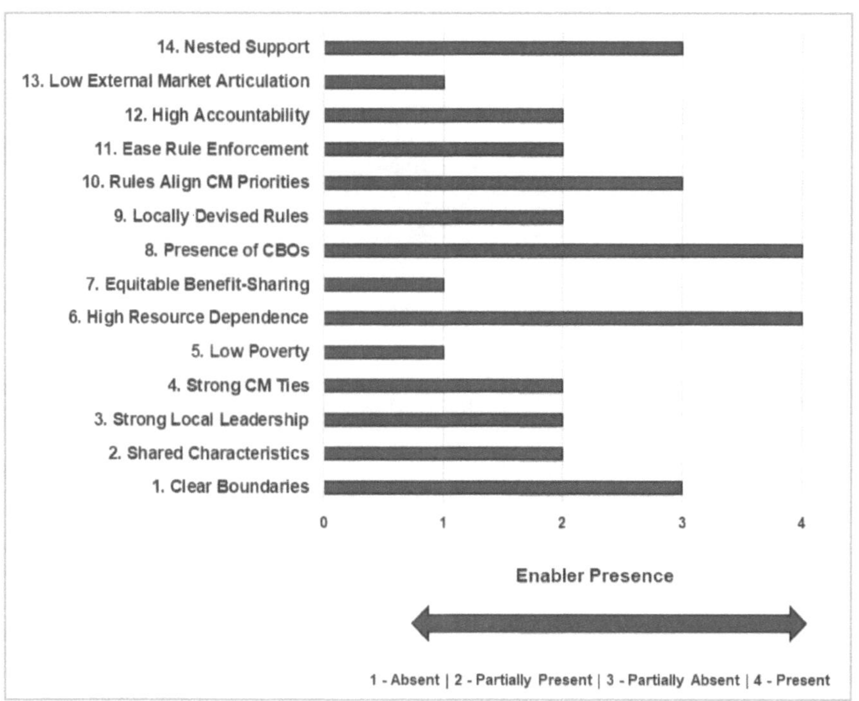

Figure 6.2.: A graphical summary of the presence of the 14 enablers in the Bay of Ranobe. **Note:** CM – Local Community.

Table 6.1.: A summary of the review of the 14 enablers in the Bay of Ranobe.

	Enablers	Presence	Explanation
Resource System & Users	1. Clearly-defined resource system & user boundaries	**Partially Absent**	Community members confirmed clearly-defined boundaries of the two LMMAs. Furthermore, the community can be considered clearly geographically defined. However, due to coastal migration of inland tribes, social boundaries are less well defined.
	2. Shared norms, values, interests & identities	**Partially Present**	Coastal in-migration and erosion of customary practices confirmed especially by elders means cultural norms, values, interests and identities within the community only partially shared.
	3. Strong local leadership	**Partially Present**	*FIMIHARA* has taken ownership of managing LMMAs and especially finances. However, elite-capture and poor representation notably characterized by a lack of feedback were identified by most community members. In contrast, traditional leaders, including village presidents and elders, were perceived positively and considered influential.
	4. Strong community ties	**Partially Present**	Strong family and overall intra-village bonds, however, weaker inter-village bonds were identified by most community members. Furthermore, aforementioned weak community representation due to perceived elite-capture of knowledge and benefits affects community ties.
	5. Low levels of poverty	**Absent**	High levels of poverty reported by all community and partner respondents.
	6. High levels of dependence on resource	**Present**	High levels of natural resource dependence reported by all community and partner respondents. Accordingly, most community members identified no alternative livelihood outside fishing (inclusive of boat- & net-building & fish-selling).
	7. Equitable distribution of benefits from common property resources	**Absent**	High levels of community-perceived elite-capture of benefits by *FIMIHARA* village-representatives and their families.
Institutional Structure & Externalities	8. The presence of community institutions	**Present**	Well-established and legally recognized CBOs present.
	9. Locally devised access and management rules	**Partially Present**	Rules established at *FIMIHARA* meetings, but process is largely perceived by community members to be captured by representatives. Furthermore, poor involvement of women in decision-making identified by many female community members.
	10. Rules strongly align with local priorities/ needs	**Partially Absent**	Rules strongly align with customary practices and institutions. However, poverty alleviation is the main local priority which community members do not perceive is being accomplished. Accordingly, many community members commonly suggested closed LMMAs should be temporarily opened to improve livelihoods.
	11. Ease in enforcement of rules, and conflict resolution	**Partially Present**	Rules largely understood but community-perceived compliance is low and with repeat offenders. This considered due to poverty, with many stating they need to feed their families. Conflict resolution handled by *FIMIHARA* and local village-level Dina committees, though in particular community members perceive the former to be less, and the latter to be more effective. That said partial cultural erosion acknowledged by respondents.
	12. High levels of accountability	**Partially Present**	Guardians are not always active and often absent. Guardians are accountable to *FIMIHARA*, however, due to family ties (one is the son of *FIMIHARA* president) voicing community concerns largely absent. Low community-wide perceived accountability of *FIMIHARA* to the community.
	13. Low levels of articulation with external markets	**Absent**	Presence of international export seafood market.
	14. The presence of nested governance with high levels of initial external support	**Partially Absent**	High levels of initial NGO support for implementation and capacity-building and ongoing NGO financial and technical support. National and local State recognition. Private sector support. Local-level governance by *FIMIHARA* although elite-capture is perceived locally.

<u>6.2.2.1. Clearly-defined resource system & user boundaries:</u>
Based on all data collected both the resource and resource-user systems can be considered geographically clearly-defined. However, many respondents emphasized less well-defined socio-cultural boundaries as a result of coastal in-migration by inland ethnic groups, notably the *Masikoro* (discussed further below). Consequently, due to social boundary 'issues', this enabler is considered *partially absent*. Moreover, this finding reinforces those of common scholars such as Ostrom (1990), Cox et al. (2010), and African CBC studies (e.g. Cinner et al., 2009a; Child, 2019)

<u>6.2.2.2. Shared norms, values, interests & identities</u>
All respondents considered linking conservation to customary practices a key enabler in promoting community participation and motivation in managing the LMMAs. However, whilst the *Vezo* have a long tradition of fishing, most respondents, notably village elders, noted partial erosion of customary institutions and practices. Accordingly, numerous village elders confirmed a decline in respect for customary laws – especially within the younger generation – but attributed this to the demands of extreme poverty.

Most *Vezo* respondents considered coastal migration of non-*Vezo* groups such as the *Masikoro*, and the presence of international-export seafood traders as reasons for 'diluted' *shared norms, values, interests and identities*. While many *Masikoro* dwell within all villages, levels of integration, notably through intermarriage or child-birth, were beyond the research focus. Nonetheless, whilst *Masikoro* can and do access marine resources, many *Masikoro* respondents repeatedly emphasized their perceived exclusion. Their perceived exclusion emerged strongest in those living in the two northern villages, who highlighted prejudice based on their 'tribal-identity' specifically regarding their representation on *FIMIHARA*, as well as their limited ability to derive benefits. More specifically, some of these respondents noted an inability to meaningfully participate at meetings, and a lack of consideration of their opinions. In turn, many *Vezo* respondents emphasized the negative impact of 'non-traditional' fishing tribes like the *Masikoro* on local marine resources, due to a lack of traditional ecological knowledge and the use of destructive fishing methods, as observed elsewhere in the south-west (Cripps & Gardner, 2016). Consequently, it can be inferred from respondents that levels of shared norms, values, identities and interests are only *partially present*.

6.2.2.3. Strong local leadership

The majority of community respondents perceived *FIMIHARA* as the most central actor associated with *Ultimate Decision-Making Power*, as depicted by the social network map in *Figure 6.3* representing the responses to SRNA conducted. As established in *Chapter 2*, this designation refers to all topics concerning decision-making power over the LMMAs. However, whilst *FIMIHARA* has taken ownership of managing the LMMAs including their finances, elite-capture of benefits and knowledge (i.e. characterized by poor representative feedback) was identified as a concern by most non-local-elite community respondents (discussed further in *section 6.2.2.7.* related to 'equitable benefit distribution'). Nevertheless, most respondents emphasized this power should reside with 'all actors' (i.e. *All STK* [Stakeholders] - *Figure 6.3.*).

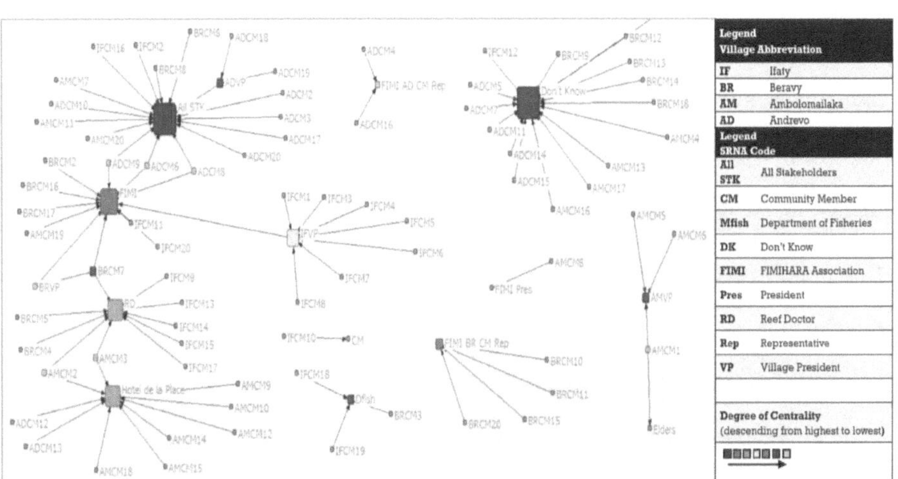

Figure 6.3.: A social network map depicting 'actors' community members perceive to have the *Ultimate Decision-Making Power* related to the LMMAs in the Bay of Ranobe.
Note: Degree of centrality is indicated by size (the bigger the square icon the higher the centrality – i.e. most ties to other actors and therefore deemed the most powerful 'actor' regarding decision-making) and colour (see legend). See legends for village abbreviations and SRNA code.

220

NGO respondents in both the present case and nationally specifically noted many past development project failures could frequently be attributed to not only weak local management capacity but 'culturally-misaligned' appointments to positions of power. Accordingly, many of these respondents emphasized CBO leadership is often in conflict with customary leadership structures. Accordingly, Razafindrakoto et al. (2015) emphasize Madagascar's development is hindered by social fragmentation within Malagasy society, which leads to elite-capture and a society torn between a *democratic-* and *meritocratic-* (i.e. biased toward 'educated' individuals) system, and *customary-system* (i.e. aligned with values inherited from the past). The above notions were identified by most community respondents (including some elected leaders) to be applicable to *FIMIHARA*.

Community respondents most notably emphasized the highly influential nature of their respective *village presidents* – most notably in the southern villages of *Ifaty* and *Beravy* – which they characterized as playing key roles as 'communicative middlemen' between community members, the State, other partner organizations, and *FIMIHARA* village-representatives. As one *Ifaty* community respondent stated, "if [the village president] says somethings needs to be done then the community will follow" (IF5). Additionally, *village elders* were consistently identified by community respondents as playing significant roles specifically in monitoring compliance with, and participating in ceremonial events used to formalize *Dina*. Such ceremonies have been shown in both the present case, and elsewhere in the south-west, to have legitimized CBC initiatives (Westerman & Gardner, 2013). Notwithstanding many positive responses from community members, they had mixed perceptions on their relations with their village presidents, as was the case with their *FIMIHARA* local representatives (*Figure 6.4.*). The exception being the positive responses from approximately 77% of *Ifaty* respondents (*Figure 6.4.*). Furthermore, a village-based trend emerged with responses from the northern villages of *Ambolomailaka* and Andrevo particularly negative (*Figure 6.4.*). It could be inferred from responses that negative perceptions are largely due to a perceived lack of involvement and poor representation of other villages in the initial stages (refer to *section 6.2.2.9.*). Consequently, *strong local leadership* was only considered *partially present*. Accordingly, most community respondents specifically noted the pressing need to develop more democratic local leadership, and emphasized the need for greater democratic selection and rotation of *FIMIHARA* village-representatives. Therefore, these findings reinforce the importance of strong leadership to enable CBC interventions, as noted by previous studies (Pomeroy et al., 2001; Agrawal, 2002; Galvin et al., 2018; Biggs et al., 2019).

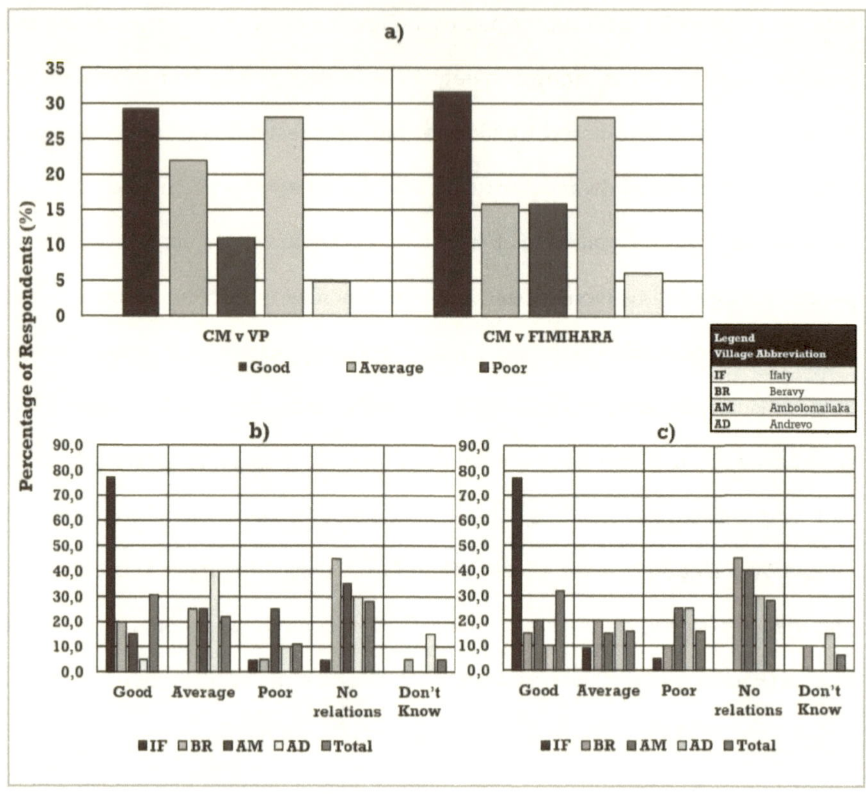

Figure 6.4.: Community-perceived relations with local leaders including village presidents (VP) and FIMIHARA-representatives by percentage of respondents **a)** overall, **b)** village presidents by village and **c)** FIMIHARA-representatives by village. **Note:** *CM* = Community-Member. *No Relations* indicates a community-member had limited to no interaction with these 'actors'. See legend for village abbreviations.

6.2.2.4. Strong community ties

In accordance with the above discussion regarding the strength of local leadership, community members perceived local traditional authorities, in particular the village presidents of *Ifaty* and *Beravy,* as central actors providing *Interactional Support*, i.e. those actors deemed most approachable regarding community members natural resource concerns. This relationship is depicted by the social network map in *Figure 6.5.* Furthermore, family-ties and intra-village ties were characterized by community respondents as strong in all four villages, and crucial to knowledge acquisition and diffusion. This is consistent with other studies in the south-west (e.g. Fritz-Vietta et al., 2017). However, gender-based issues in knowledge-diffusion also emerged and should be acknowledged. As a female community respondent stated, "I don't know the community role as only men go to the [*FIMIHARA*] meeting" (IF12). A lack of involvement of women in conservation decision-making has been revealed previously in the area (Westerman & Benbow, 2014), and therefore, this finding continues to emphasize the issue of gender-exclusion in Malagasy CBC governance. Lastly, whilst strong intra-village ties emerged, a lack of trust could be inferred from the high centrality of the *No Trust* designation in *Figure 6.5*. This perhaps notably regarding concerns with local representation on *FIMIHARA*, as discussed above. Accordingly, one community respondent simply stated, "*Gasy* [i.e. Malagasy] people don't trust one another" (AM10). Consequently, the presence of strong community ties is considered only *partially present*.

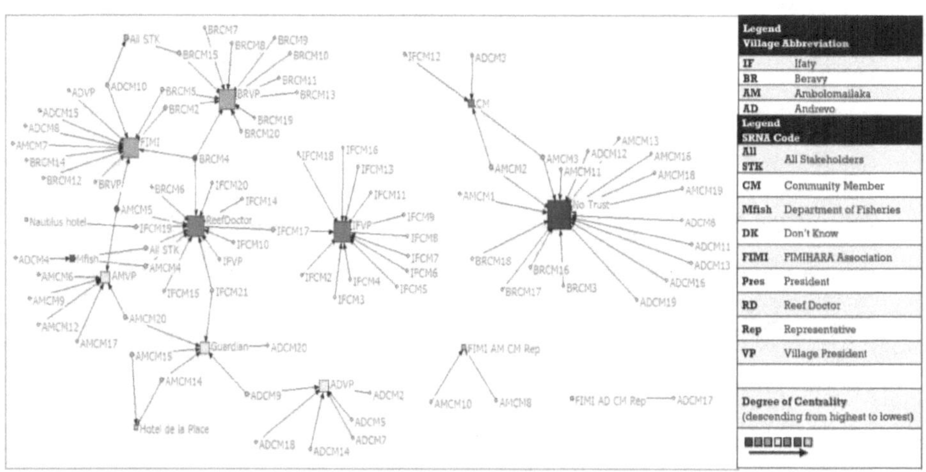

Figure 6.5.: A social network map depicting 'actors' community members perceive as the most approachable for their natural resource access and use concerns in the Bay of Ranobe (i.e. *Interactional Support*). **Note:** Degree of centrality is indicated by size (the bigger the square icon the higher the centrality – i.e. most ties to other actors and therefore deemed the most approachable) and colour (see legend). See legends for village abbreviations and SRNA code.

As discussed throughout, the country is plagued by *high levels of poverty*, which was acknowledged by all respondents. Furthermore, all respondents confirmed that poverty was the primary motivating factor for non-compliance with LMMA regulations (i.e. their permanent closure), and the primary constraining factor for governance in the Bay of Ranobe in general. Moreover, partner organizations noted local elite-capture is largely driven by extreme poverty and a need to care for one's own family. Consequently, all respondents identified poverty alleviation as a key enabler for facilitating CBC initiation, implementation and governance, especially since marine resources are their primary livelihood source (refer to *section 6.2.4.6.).* These findings are consistent with other studies in the south-west (e.g. Barnes-Mauthe et al., 2013). Accordingly, national CBC studies depict how poverty can affect a local community's long-term view of conservation (e.g. Lammers et al., 2017). Consequently, this enabler is considered *absent*, and requires urgent poverty alleviating action.

6.2.2.6. High levels of resource-dependence

Approximately 66% of community respondents eat and/ or sell marine resources, with 62% confirming they lacked an alternative livelihood. These levels of natural resource-dependence are most notable in the northern most village of *Ambolomailaka* with 85% of respondents relying on marine resources as a sole source of livelihood. However, all community respondents confirmed these natural resources are not meeting their basic requirements, a common finding in rural Malagasy communities (e.g. Borgerson et al., 2018). Whilst all community respondents expressed concern for the state of natural resources, they acknowledged that high levels of poverty dictate harvesting activities in the

LMMAs persist, regardless of closures. Therefore, it can confidently be inferred that in the present context a high level of dependence on the local natural resources is *present*. Consequently, while this enabler cannot be viewed in isolation of other enablers, these findings question – at least within this context – whether *high resource-dependence* is an enabler for CBC. This since as described above most community respondents confirmed their high resource-dependence, and lack of alternative livelihoods, as the primary reason for non-compliance with LMMA regulations, and acknowledged its contribution to the declining status of the resources, and therefore, its constraining influence on management of the LMMAs.

6.2.2.7. Equitable distribution of benefits from common property resources

National partner respondents, as well as all Bay of Ranobe community respondents, identified the importance of obtaining tangible benefits for enabling CBC. In reference to this, one community respondent stated, "otherwise LMMAs are just decorations in the sea" (AD7). In fact, all respondents identified benefits such as an improved standard of living and increased food-security, as crucial to increasing community motivation required to enable CBC governance.

The SRNA confirmed that most community respondents acknowledge *FIMIHARA* as the most central actor regarding the distribution of benefits from the LMMAs. However, as previously stated, community-perceived local elite-capture is a key concern. More specifically, an overwhelming lack of *trust* for *FIMIHARA's* ability to equitably distribute benefits (and knowledge) emerged. This most notably stated by northern village respondents, i.e. 70 and 80% of *Ambolomailaka* and Andrevo respondents respectively. The exception once again being *Ifaty* members (i.e. approximately 91% of *Ifaty* respondents perceiving relations of *trust*). Therefore, this village-based pattern of community-perceived inequality requires further attention. Elite-capture of benefits was also identified by all national partner organization respondents as a core national CBC constraining factor. Consequently, this enabler is considered *absent*, requires improvement, and reinforces the findings of established scholars reviewed in *Chapter 3: section 3.3.2.1.* (e.g. Agrawal, 2002; Galvin et al., 2018; Biggs et al., 2019).

6.2.2.8. The presence of community institutions

The community is represented by village councils which are headed by village presidents. Furthermore, villages possess *Dina* committees to deal with conflict resolution regarding the breaking of established *Dina*. Moreover, as introduced in *section 6.1.2.2.*, *FIMIHARA* represents a legally recognized and functional CBO responsible for managing the LMMAs. Consequently, this enabler is considered *present*, though these CBO require improvements to be more effective.

6.2.2.9. Locally devised access and management rules

Most community respondents perceive *low* levels of involvement in management activities associated with *FIMIHARA* and the LMMAs (*Figure 6.6.*). They reasoned that while their *FIMIHARA*-representatives are highly involved, their representation within *FIMIHARA* by these representatives is poor. Once again this is most notable in the northern villages of *Ambolomailaka* and *Andrevo*, and *Ifaty* was again the exception with most responses characterizing their representation as *Good*. Consequently, *Ifaty* responses brought the above overall percentage of *low* responses down in *Figure 6.6.* As established in *section 6.1.2.2.* a longstanding relationship between *Ifaty* and Reef Doctor, stemming from their original involvement in establishing *FIMIHARA* and the LMMAs, could be inferred as influential to positive responses from *Ifaty* respondents throughout. In turn, once again the two northern villages perceived less involvement, which again may be due to a lack of original involvement, but also a result of a greater distance to attend meetings.

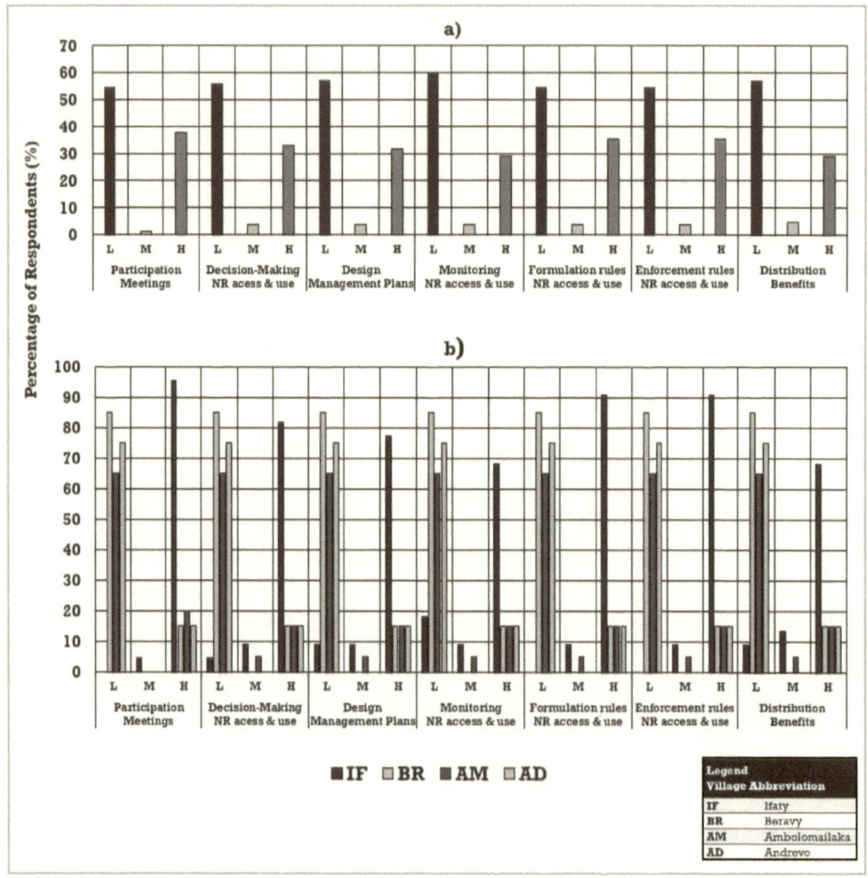

Figure 6.6.: Community-perceived involvement across diverse management activities associated with LMMAs in the Bay of Ranobe by percentage of respondents a) overall, and b) by village. **Note: L** = Low; **M** = Medium; and **H** = High; and **NR** = Natural Resource.

At a national-level, partner organizations consistently emphasize *passive* community involvement in their partnered CBC initiatives as *moderate* to *high* (i.e. for meeting attendance). However, described *active* involvement as commonly restricted solely to more educated community-members (i.e. in accordance with aforementioned *meritocratic* local elite-capture). Accordingly, partner organizations often stated incorporation of diverse community opinions and interests in conservation decision-making is problematic, and especially vulnerable to socio-cultural hierarchies that are subject to local power asymmetries. Therefore, this enabler is considered only *partially present*, and requires improvements for strengthened CBC governance. Consequently, these findings mirror those of several scholars as identified in *Chapter 3: section 3.2.2.1.* (e.g. Ostrom, 1990; Agrawal, 2002; Cinner et al., 2009a; Cox et al., 2010).

6.2.2.10. Rules strongly align with local priorities/ needs

While rules were perceived by respondents to be strongly aligned with cultural institutions such as Dina, as discussed throughout persistently high levels of poverty dictates that community members widely perceive failure to address their socio-economic needs. That said, aquaculture projects are increasingly addressing these concerns (discussed further in *section 6.2.2.13.*). Partner respondents emphasized that communities possess a wealth of knowledge about, and realize the need to conserve their natural resources for the future. However, as established previously, they noted high levels of poverty dictate local communities do not prioritise conservation. Furthermore, all community respondents confirmed this. Moreover, partner organizations confirmed this dilemma often results in intra-community conflict and weak local leadership over natural resources. Therefore, while strongly aligned with cultural priorities, although efforts exist to alleviate poverty, since *high levels of poverty* persistent this enabler is considered *partially absent*, and requires further action. Consequently, these findings reinforce the need for greater both socio-economic and cultural alignment of rules within a CBC intervention, as introduced in *Chapter 3: section 3.2.2.1.*, and most notably found in other African CBC studies (e.g. Cinner et al., 2009a; Galvin et al., 2018).

6.2.2.11. Ease of enforcement of rules, and conflict resolution

Dina committees and *FIMIHARA guardians* are involved in enforcement of the regulations of the LMMAs (refer to *section 6.2.2.11.*). However, a noteworthy cultural construct in the present case, emphasized by both community and partner respondents, is the non-confrontational nature of the *Vezo*. Most *Vezo* respondents (and some partners) identified this as a constraint to the ability of community members to express themselves. Accordingly, community respondents often noted an inability to confront their village-representatives, due to this established 'cultural-politeness'. Furthermore, this may be exacerbated by local-level power asymmetries. Moreover, respondents confirmed this politeness, and a shared understanding of *need* (i.e. high levels of poverty and limited alternative livelihoods), can constrain enforcement and conflict resolution, especially concerning confronting fellow fishers. However, many community respondents emphasized the responsibility of the fisher is "to do what is right", and for others to confront them if they are not following the rules (although most stated they did not confront others). Lastly, community respondents perceived State-enforcement as corrupt, with most describing the ease with which fines for infringements could potentially be overcome with bribes to State officials. However, most respondents also acknowledged a lack of finances to employ this strategy. Consequently, this enabler is considered *partially present* and mechanisms for improved enforcement and conflict resolution are required, most notably from the State.

<u>6.2.2.12. High levels of accountability</u>

FIMIHARA guardians are responsible for patrolling the LMMAs and monitoring access. However, most respondents confirmed they are not active all day, and often absent during shifts. While guardians are accountable to *FIMIHARA*, community respondents were reluctant to voice concerns due to the aforementioned 'cultural politeness'. Furthermore, community respondents specifically perceived *low* levels of accountability regarding *FIMIHARA*, which is not surprising considering community-perceived relations with their *FIMIHARA representatives* (as discussed throughout). Moreover, they highlighted that family ties can potentially constrain accountability. For example, respondents pointed out that one guardian is the son of the current *FIMIHARA* president, which affects their openness to confront accountability concerns. *Reef Doctor* and the *Institute of Fisheries and Marine Sciences* (IHSM) carry out ecological monitoring in the bay and attempt to report findings back to community members through *FIMIHARA* meetings. However, as mentioned ecological knowledge is often perceived to be 'captured' at the *FIMIHARA* village-representative level. Therefore, this enabler is deemed only *partially present,* and improved levels of local representative and State (see *enabler 11* above) accountability are required. Consequently, these findings reemphasize the importance of high levels of accountability as was introduced in *Chapter 3: section 3.3.2.1.* (e.g. Pomeroy et al., 2001; Agrawal, 2002).

<u>6.2.2.13. Low levels of articulation with external markets</u>

As mentioned previously (in *section 6.2.1.*), the south-west has undergone increased articulation with international seafood export-market, which both respondents and previous studies suggest has increased the value and local exploitation levels of target species in the area, most notably octopus (see L'Haridon, 2006). In 1995, two companies started to collect seafood in the Toliara region (i.e. *COPEFRITO SA* and *MUREX INTERNATIONAL*), and their impact was quickly visible with an increase from 8.5 t in 1994 to 114.3 t in 1995 in frozen octopus production (L'Haridon, 2006). Furthermore, provincial fisheries landings data indicates an increase in octopus landings in Toliara from 49.1 t/ year (1994) to 264.19 t/ year (1996) (L'Haridon, 2006). However, national octopus production showed a decrease from 1189 t to 528 t between 2004 and 2005 (L'Haridon, 2006). Consequently, while the reliability of the above landings data may be questioned, this data indicated the introduction of an international seafood export-market resulted in increased harvesting activities, and subsequent depletion of stocks. Unfortunately, no more recent national data could be located to make a more conclusive analysis.

Many community respondents noted if they could get more money for their catch, this could stem overfishing. However, this was not a unanimous view, as some fishers acknowledged this may attract greater fishing effort. Furthermore, some community respondents, and Reef Doctor staff emphasized that aquaculture activities in the bay, which also have access to export-markets, are producing increasing contributions to livelihoods. For example, recent figures from Reef Doctor show local aquaculture activities generated a total income of USD$ 27,385 in the final year of a recent three-year Darwin Initiative project (between 2014 and 2017), and benefitted 247 households (Darwin Initiative, 2017).

Therefore, when this enabler is considered together with *high levels of poverty* and *natural resource dependence*, it can be inferred that these external markets may be exacerbating increasing harvesting activity, and therefore, may be negatively affecting the effectiveness of management of the LMMAs. That said, some community and Reef Doctor respondents suggested the aforementioned local aquaculture projects that derive income from local markets could potentially curb pressure on wild-capture marine resources. Therefore, due to the aforementioned presence of international seafood export-market I consider low levels of external market-articulation as *absent*. Consequently, while the directionality and nonlinearity of this enabler is in perhaps still in question (as introduced in *Chapter 3: section 3.2.2.2.*), pending further research in the Bay of Ranobe, this finding does reinforce that of Agrawal's (2002), since a higher level of market-articulation appears to be negatively influencing the success of this intervention.

6.2.2.14. The presence of 'nested' governance institutions with high levels of initial external support

As introduced previously in *section 6.1.2.2.,* a highly 'nested' governance structure comprising community representatives, the State, private sector (i.e. hoteliers and dive operators), and Reef Doctor, led to the formation of *FIMIHARA*, and subsequently the two LMMAs. However, as with community-perceived local representative relations, community-perceived relations with partner organizations were mixed, and many community respondents described a general lack of strong partner relationships (*Figure 6.7.*). Not surprisingly, once again the village of Ifaty was the exception to this (i.e. with approximately 77% of respondents perceiving positive partner relations), which again can be inferred as resultant of its initial involvement, and in particular the village's well-established relations with Reef Doctor.

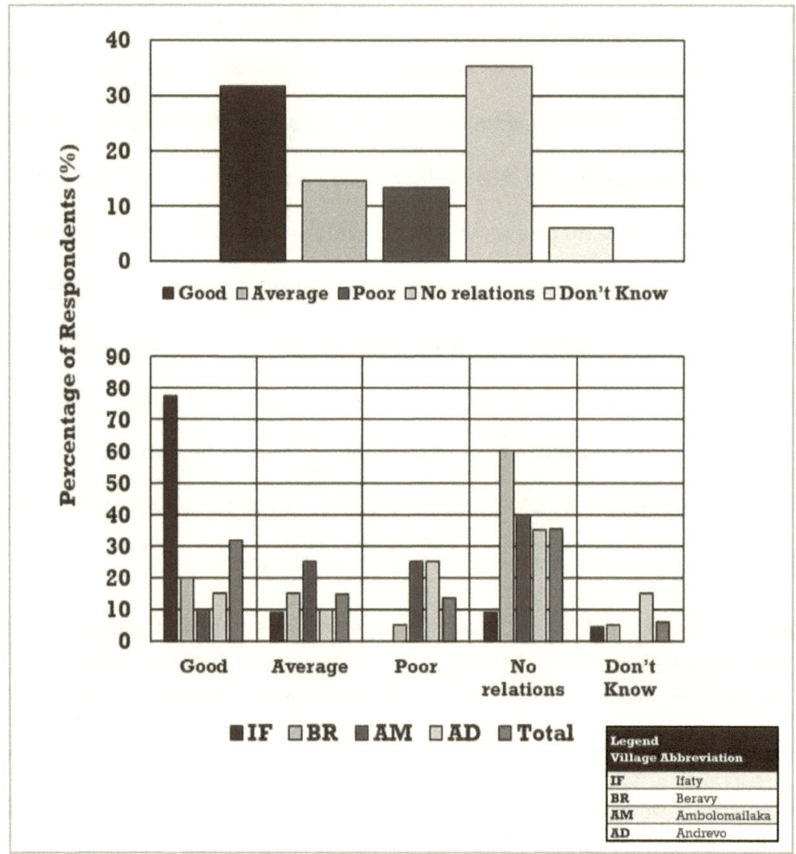

Figure 6.7.: Community-perceived relations with partner organizations by percentage of respondents: **a)** overall and **b)** by village.

Both knowledge-diffusion and benefit-sharing were perceived by all respondents both at a national- and local-level to be problematic, perpetuated by local power asymmetries, and resulting in strained and ineffective social relations and networks. Nevertheless, all respondents also acknowledged that a lack of local capacity, most notably concerning weak community-perceived *FIMIHARA* representation (as discussed throughout), deemed external technical and financial support of partner organizations necessary. This is also consistent with findings of other CBC initiatives in the area (e.g. Harris, 2007; Long et al., 2017). More specifically, community respondents noted the required external support is largely due to concerns over internal corruption within *FIMIHARA* (as discussed throughout). Accordingly, some community respondents stated it is "more productive with *vazaha* [i.e. foreigners]" (AM10), and "we want *vazaha* to oversee finances in particular... to avoid corruption" (IF20). Nonetheless, the desire for *vazaha* involvement was not unanimous, for example, as one *Ambolomailaka* respondent stated, "we don't want foreign conservation because it is not going as promised" (AM3).

Partner organizations described their role in CBC initiatives as, "facilitators of conversation between different stakeholders" (PO5), and "knowledge, skills, and fund providers" (PO6). Accordingly, all community respondents perceived Reef Doctor as central to CBC governance efforts (i.e. active in numerous activities including being a source of *Interactional Support* and *Knowledge Diffusion*). Furthermore, Reef Doctor, and governmental organization SAGE, assisted the community with the legal declaration process resulting in the LMMAs obtaining national PA status. Moreover, Reef Doctor and IHSM continue to collaborate to provide necessary ecological monitoring data, upon which FIMIHARA bases its decisions.

At a state-level, both *GELOSE* and *NEAP*, and recent LMMA legislation, promote increased State recognition of local community's management role in conservation. However, greater State support in 'streamlining' enabling legislation, was identified by partner organizations as key to enabling positive CBC outcomes. Furthermore, these respondents particularly emphasized a lack of legal CBC recognition has negatively influenced legitimacy, and subsequently constrained the formulation and enforcement of rules by local communities (i.e. *enabler 11*). Accordingly, partner organizations confirmed giving communities' greater responsibility, with both State and NGO support (especially with enforcing laws), as a positive strategy for CBC in the country. However, community respondents commonly argued that *Dina* has existed for a long time, and therefore communities should establish conservation activities and rules, and only thereafter should partner organizations come into support, emphasizing it must first come from the community. Therefore, the need to establish harmony between customary and conventional conservation approaches, with the support of external agents (notably State legislation and technical and financial BINGO support), was deemed key by all respondents to improving CBC initiatives in the country. Consequently, this enabler is considered *partially absent* in the Bay of Ranobe, since the presence of 'nested' governance systems requires improvements. Moreover, these findings strongly correlate with those of established commons scholars (e.g. Ostrom, 1990; Pomeroy et al., 2001; Agrawal, 2002), and more recent African CBC literature (e.g. Cinner et al., 2009a; Galvin et al., 2018), as introduced in *Chapter 3: section 3.3.2.1*.

6.2.3. *Actions*

This section reports on the key actions taken to facilitate change toward a CBC mode of governance in the Bay of Ranobe. A preliminary action to support the implementation and governance of CBC was the aforementioned State promoted national PA expansion strategy. More recently, State action emerged in the form of legal recognition of LMMAs, and CBOs as their management authority. Accordingly, *Article 6* of *Law No. 2015–005* now recognizes four PA governance types: public governance; shared governance (co-management); private governance; and community-based governance (Ratsimbazafy et al., 2019).

The first step involved in declaring the LMMAs was the formation and legal recognition of a CBO, in the present case *FIMIHARA*. The second step involved the implementation of a *Dina* for the two LMMAs and its legal recognition by the State. This involved establishing a *Dina,* during several community meetings, regarding the no-take LMMA boundaries. This *Dina* was subsequently formalized through customary ceremonies by village elders. Whilst all respondents identified the establishment of *Dina* as a key action toward successfully facilitating implementation and governance of the LMMAs, and subsequently promoting community support and recognition, it can be inferred from community respondents that both *FIMIHARA* and the *Dina* lack full support from the community. Once again, this emerged from most respondents as being largely due to high levels of poverty, and dependence on marine resources for food and livelihood, with limited alternatives.

NGO-CBO enabled partnerships require building relations of trust among partners, the CBO and the greater community. Partner organization respondents confirmed a key action in this regard involves combining community development and conservation initiatives, which is consistent with findings elsewhere in the country (e.g. Freudenberger, 2010; Kiefer et al., 2010), and specifically within the south-west (e.g. Harris, 2012; Robson & Rakotozafy, 2015). Reef Doctor has carried out numerous community development and ecological projects since 2002 in the Bay of Ranobe, including ecological monitoring, school restoration, education programmes raising local environmental awareness, and alternative livelihoods initiatives, notably aquaculture (Belle et al., 2009; Darwin Initiative, 2017). The implementation of these projects, and the time invested in the community were perceived by both community and Reef Doctor respondents to have assisted in strengthening actor relations, specifically by increasing the support and trust of the local community for their presence. This also aided in the action of increasing alignment with socio-cultural priorities. All partner organizations and community respondents emphasized not to force conservation approaches onto communities, and therefore the necessity for greater flexibility allowing conservation efforts to better harmonise with established customary institutions and practices.

FIMIHARA and other partners have strengthened relations through numerous meetings, especially during the planning phases of these projects. Reef Doctor continues to support *FIMIHARA* financially, notably through sourcing funds from donors for alternative livelihood projects, for example through the *Global Environment Facility's (GEF) Small Grants Programme* (e.g. mangrove restoration since 2009-2012 - GEF, 2012), and recent *Darwin Initiative* projects (e.g. local aquaculture since 2014 – Darwin Initiative, 2017)). These projects were considered by most respondents as key to aligning CBC initiatives with local livelihood priorities. Moreover, Reef Doctor continue to provide technical support regarding management of these projects, and, together with IHSM, through ongoing ecological monitoring data.

Additional case-specific noted actions include local management capacity-building, and increased community involvement through improved local representation characterized by increased knowledge diffusion supported by the various partner-implemented community development projects introduced above. Furthermore, it emerged from Reef Doctor and other *FIMIHARA* partner respondents, such as hoteliers, that whilst partners attempted to take a 'back seat' and allow local representatives to take the initiative at meetings, they did provide support and advice to assist in guiding management by *FIMIHARA*. Consequently, the aforementioned actions sought to strengthen the enablers to support the CBC change process. However, all respondents confirmed much work remains to reach local CBC social (i.e. improved standard of living) and ecological (restored fish and coral reef systems) objectives.

6.2.4. External Influences

While there is clearly state support for involving local communities more in the governance of their resources, and NGOs facilitating the formation of CBC initiatives through various programmes, projects and activities (i.e. *actions*), there were *external influences* that contributed to, or detracted from CBC implementation and governance in the country. Firstly, State fragility and instability represents an external influence. Whilst usually constraining to CBC initiatives this State incapacity has enabled BINGOs to take on the State's role in CBC initiatives. However, numerous respondents stated State corruption, which extends to interactions between the State and commercial interests, undermine CBC efforts. For example, in the Bay of Ranobe LMMAs some respondents raised concerns that Chinese commercial fishing in the vicinity is negatively affecting fish stocks. Furthermore, both community and non-State partner respondents emphasized the State often ignores community concerns regarding this matter. This is consistent with other national studies (e.g. Gardner et al., 2018). Regarding State instability, BINGO respondents in particular highlighted that this has constrained their ability to source funding for CBC initiatives, and the speed at which implementation and legal recognition of all types of CCAs occurs. Nevertheless, many respondents noted that the global conservation prioritization of the country has promoted international donor funding that enables conservation efforts including CBC initiatives.

Secondly, as with other case studies within this book, international institutional commitments, in this case notably associated with the CBD (e.g. The *Aichi Targets* – CBCD, 2011; and *Post-2020 Global Biodiversity Framework* - CBD, 2020), *FAO Voluntary SSF Guidelines*, and regional commitments including those of the African Union, are also external influences, which can be considered to have potentially enabled implementation of CBC interventions. Furthermore, more specific to the present coastal context, in 2014 the State promised to triple the number of MPAs in the country, and planned to establish a legal framework for community-based management of fishing grounds (IUCN, 2014b).

Lastly, an additional enabling external influence was the formation of the conservation award-winning *MIHARI* network, a national network of LMMA partners and communities focused on knowledge exchange (Blue Ventures,

2019b). Most partner respondents acknowledged the positive influence of *MIHARI* in promoting a CBC mode of governance in the country.

6.2.5. Issues Arising

A key overarching 'issue arising' in the present context is a lack of consideration and alignment of CBC initiatives with local needs or priorities, which is well-recognized in Madagascar (e.g. Antona, 2004; Raik et al., 2008). This appeared to encompass both local socio-economic and -cultural elements. Respondents most notably emphasized the negative influence of not only poverty on conservation efforts, but also additional socio-economic issues. These included low-levels of education, and poor health service delivery. Approximately a third of Madagascar's population is deprived of access to education (World Bank, 2014), and all partner organization respondents identified this as a major negative influence on the effectiveness of CBC initiatives, in accordance with past national studies (e.g. Dolins et al., 2010). Furthermore, respondents confirmed local health service delivery remains problematic, mirroring a country-wide concern (Marks et al., 2016). In response to these community 'issues arising', many partner respondents suggested cross-sector community development – encompassing health, education and the environment – may better garner community support for conservation, as previously noted in national (e.g. Freudenberger, 2010; Kiefer et al., 2010), and specifically research conducted in the south-west (e.g. Harris, 2012; Robson & Rakotozafy, 2015). Therefore, a key 'issues arising' and requiring reformulated actions is the ability of the State and partners to address local socio-economic concerns such as high levels of poverty through identifying alternative livelihood strategies. Whilst this has occurred already through local community development projects, partners suggested more can be done in this regard.

In relation to socio-cultural issues, as established above all respondents considered linking conservation efforts to customary practices as a key enabler for CBC implementation and governance. However, most respondents acknowledged that endeavours to accomplish this have proven problematic, as has been noted by others in the country (e.g. Fernández-Llamazares et al., 2018; Osterhoudt, 2018). Erosion of customary norms, practices and institutions can be especially problematic for CBC, as noted in the present case (refer to *section 6.2.2.2.*), and by other national CBC studies (e.g. Cinner, 2007; Desbureaux & Brimont, 2015). Nonetheless, these remain an important component defining individual and social-group identity within Malagasy society (see Walsh, 2002), and may promote greater social cohesion required to develop a greater sense of community ownership to enable CBC initiatives.

Partner organizations also emphasized that the 'failures' of many past developmental projects to deliver social benefits, can frequently be attributed to community members being appointed to positions of power in conflict with established *traditional authorities*. Furthermore, many partner respondents emphasized that representatives were often elected without the necessary management skills. Therefore, a key 'issues arising' and requiring reformulated actions, which is highly connected to the need to address local socio-economic concerns, is the ability to better harmonize CBC initiatives with local customary institutions and practices.

An additional 'issues arising', which most respondents noted, was the need to strengthen institutional capacity. Accordingly, whilst most national partner organization respondents noted a high willingness of community members to engage in national conservation initiatives, and believed communities aspire to independently manage their CBC initiatives in the future, they suggested both limited local institutional capacity and financial resources available for interventions still prevent this at present. However, in contrast to partner respondents, most community respondents stated they did not aspire to manage the LMMAs independently in the near future, consistently citing a need for continued external support. Consequently, an additional 'issues arising' that can be inferred from these community responses is a need to firstly strengthen local institutional capacity to reduce persistently high levels of 'aid-dependency'. That said, financial support would still be required from partners even if the local capacity allows the community to manage their own CBC intervention.

On a related note, institutional power dynamics emerged as a key 'issue arising', as highlighted throughout, and by previous national studies (e.g. Raik & Decker, 2007; Corson, 2012, 2017). Accordingly, as discussed, many respondents identified elite-capture as a core 'issue arising' that constrains CBC interventions in the country. This related to both local and partner relations in the Bay of Ranobe. Local power dynamics were shown to manifest most notably through the perceived control of knowledge, and benefits by local representatives. Consequently, poor local representation appears a key 'issues arising'. Furthermore, some respondents perceived a 'capture' of decision-making power by external partners such as Reef Doctor. Moreover, in reference to the State, non-State partner respondents emphasized institutionalized corruption forms a part of everyday life in Madagascar. As one of these respondents noted "often State officials need a 'gift' to make it happen [i.e. establish CBC initiatives]" (PR3). Therefore, whilst most partner respondents acknowledged a lack of State capacity has enabled a shift to CBC initiatives partnered by NGOs, and the State desires and requires external partner support, it continues to meddle. Accordingly, one non-State partner respondent stated, "Marine natural resources represent wealth, so [the State] don't want to let go" and consequently the "State still has the final say over the community's voice" (PR4).

Consequently, these findings emphasize the importance of not only bonding ties (i.e. connecting individuals of similar social status e.g. *intra-community* and *family ties*), but also bridging ties (i.e. connecting different subgroups e.g. *community–village-representatives* and *village-representatives–partner-organization* ties), for positive CBC outcomes. More specifically, some respondents noted improved *community–village-representative* relations in particular present a potentially valuable opportunity to garner greater community support for the LMMAs. However, this requires current powerful actors to share their power with marginalised actors, which respondents considered problematic. Nevertheless, as Gore et al. (2013) suggest, informal institutions in Madagascar, such as CBOs found in CBC initiatives, may increase accountability and reduce corruption, and may operate more efficiently than State governance. However, this is contextually-specific and influenced by strong and effective intra-community and community–partner relations. Consequently, all these 'issues arising' require specific actions if CBC progress is to be made in the future.

6.3. Conclusion

Many past Malagasy CBC studies describe inadequate community involvement, a lack of improvement of the standard of living, ill-defined resource rights and inadequate CBC legislation, elite-capture, and a lack of community and State capacity (e.g. Pollini & Lassoie, 2011; Raik & Decker, 2007; Pollini et al., 2014). The present case study appears to mirror these concerns. A critical exploration of CBC in Madagascar at both a national-level (through national partner organization interviews) and at the local-level through the Bay of Ranobe case study, has identified key *contextual change triggers, enablers, actions,* and *external influences* and 'issues arising' to (potentially) facilitating CBC initiation, implementation and governance. Perhaps the most notable contextual change trigger has been the presence, capacity and willingness of BINGOs to facilitate CBC implementation and governance within the context of State incapacity. Whilst CBC initiatives have received State support in the form of enabling legislation since the 1990s beginning with NEAP and GELOSE, and more recently through LMMA legislation; 'issues arising' related to State and local governance capacity, characterized by elite-capture, mean implementation has been slow and/ or problematic. Thus, in accordance with other Malagasy studies (e.g. Gore et al., 2013; Jones et al., 2019b), findings indicate ongoing external support and reduced state corruption are still required for CBC moving forward.

A key finding here is that facilitating change toward a CBC mode of governance requires aligning with local priorities, notably in the present case the need to alleviate high levels of poverty. Furthermore, weak local governance capacity and alignment with customary institutions requires improvement. Consequently, if enablers currently not present can be strengthened by future actions this may stimulate positive social and ecological CBC outcomes, and generate greater support for CBC in Madagascar. Accordingly, this requires acknowledging and addressing 'issues arising' that concern key CBC enablers, including perhaps most notably the need to develop *strong local leadership* and greater *locally devised access and management rules* for more equitable local CBC governance. Lastly, the above findings emphasize the need for high levels of not only initial but ongoing 'nested' external support. An infographic summary of the key findings of this chapter can be represented by *Figure 6.8.*

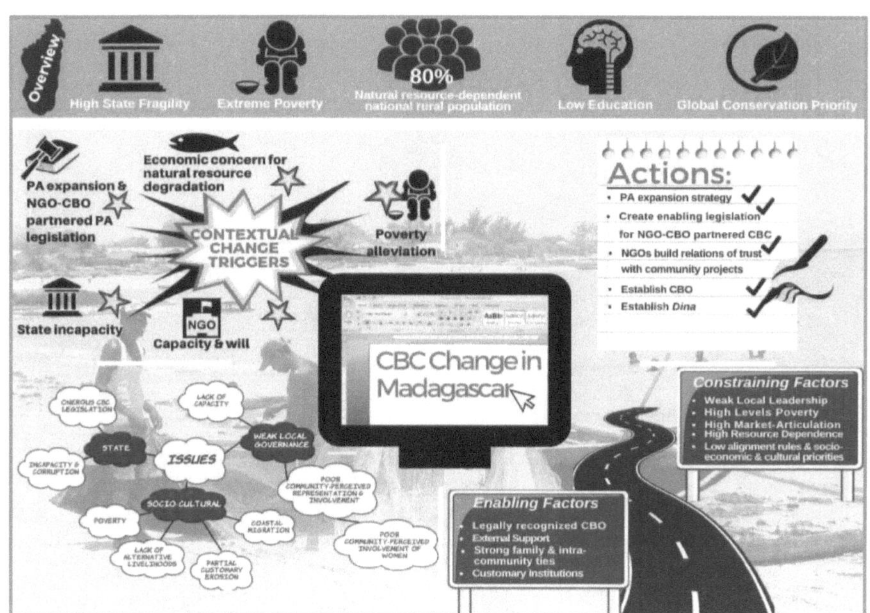

Figure 6.8.: An infographic summary of key CBC change process findings in the Bay of Ranobe, Madagascar.

248

Chapter 7

A West African CBC Change Perspective:

The case of the Urok Islands, Guinea-Bissau

7.1. Introduction and context

This chapter provides a West African CBC change perspective by presenting the second regional case study located in the Urok Islands, Guinea-Bissau. Following a brief introduction to the national context and the country's CBC challenges and progress to date, the case study context is introduced. Thereafter, findings are presented in accordance with the change elements presented in *Chapter 5*. Once again, this chapter, together with the first regional case study presented in *Chapter 6*, addresses **objective 4** (*Box 7.1.*). These findings are revisited and consolidated with other findings within the final discussion in *Chapter 9*.

Box 7.1.:

Objective 4: To investigate the factors, conditions and processes that enabled and constrained CBC in two regional case studies to learn lessons for South African CBC initiation, implementation and governance

7.1.1. National context:

7.1.1.1. Guinea-Bissau: instability, poverty & conservation

Guinea-Bissau is a small coastal West African country, and like Madagascar is considered one of the world's poorest, most corrupt and most fragile (UNDP, 2015; OECD, 2018a; Transparency International, 2019). Its recent designation of "severe" *political and environmental fragility* (OECD, 2018a), arguably stems from a prolonged history of institutional instability (Abreu, 2012; Bruneau, 2017). This includes a decade-long 'bloody' liberation struggle eventually resulting in independence from Portugal in 1974 (Davidson, 2017). Furthermore, some describe a colonial legacy of an "underdeveloped" bureaucracy (Abreu, 2012), with persistent institutional struggles to reform, among other sectors, health, education and human rights (Sangreman et al., 2018). Consequently, the country has been described as possessing "permanent chaotic institutional instability" (Sangreman et al., 2018: p67). *Figure 7.1.* provides an overview of the country's key socio-economic statistics, in order to better contextualize the findings of this chapter.

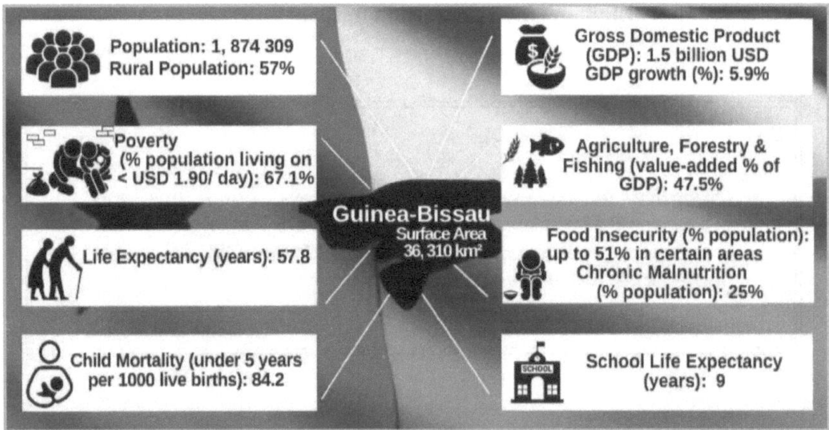

Figure 7.1.: A summary of key socio-economic statistics of Guinea-Bissau. **Source:** FAO, (2019a); UNESCO (2019); WFP (2019); World Bank (2019).

Approximately 67.1% of the national population and 88% of the rural population survive on less than USD 1.90, and USD 1 per day respectively (Cockayne & Williams, 2009; World Bank, 2019). Extreme poverty is further emphasized by a recent *UNDP Human Development Index* ranking of 178 out of 188 countries (UNDP, 2015), as the country continues to struggle to emerge from its 'poverty trap' (Só et al., 2018). Up to 51% of families are affected by food insecurity in certain rural areas

(WFP, 2019), and child malnutrition remains a serious national health issue, evidenced by one of the world's highest under-five mortality rates (UNICEF et al., 2016; World Bank, 2017). Nevertheless, some have noted more recent poverty alleviation (Nissanke & Ndulo, 2017).

The country possesses West Africa's longest coastline (making up approximately 70% of the country), and largest continuous mangrove area (Feka & Morrison, 2017). Furthermore, it possesses an exclusive economic zone rich in high quality marine resources, because of coastal upwelling, ocean currents and nutrient river discharges (RGB, 2014; Belhabib et al., 2016). Consequently, fisheries represent a main economic activity, with marine fisheries contributing about 59% of national employment and the country's main protein source (Campredon & Catry, 2018). However, Guinea-Bissau's approximately 80% coastal population is highly vulnerable to climate change, especially since its fisheries are characterised by very low adaptive capacity (Allison et al., 2009; Vasconcelos et al., 2015; Feka & Morrison, 2017).

Small-scale fisheries (SSFs)[6] are of particular local socio-economic importance, and employ some 5,600 coastal fishers, with an average national annual marine fisheries landings estimated at 14,311 t in 2015, though acknowledged to be in decline (Belhabib et al., 2015a; GB-UEMOA, 2016; Intchama et al., 2018). The *National Artisanal Fisheries Services* department issue two types of licenses under law (*Decree No. 24/2011*) to either national boat owners or fishers with foreign vessels, with the latter's catches mostly landed in neighbouring countries and thus lost to the local economy (Intchama et al., 2018). Furthermore, some note how

---

[6] See definition previously in *Chapter 6: section 6.1.1.1.* and the *Glossary of key terms.*

unqualified and at times corrupt officials negatively affect conflict-resolution, and have perpetuated local fisher dissatisfaction with the State (e.g. Fernandes, 2012). Moreover, fisheries monitoring is sporadic, landings data is scarce, and Illegal, Unreported and Unregulated (IUU) fishing plagues the country (Intchama et al., 2018; Denton & Harris, 2019; Okafor-Yarwood, 2019). More specifically, the impact of West African 'in-migrant' small-scale fishers (Cross, 2016; Campredon & Catry, 2018) – as well as foreign commercial fleets (Okafor-Yarwood, 2019; Virdin et al., 2019) – on transboundary regional fish stocks is well established.

Notwithstanding the importance of fishing more than 70% of the population rely on subsistence agriculture, with cashews (a locally favoured cash crop), and to a lesser extent rice, the most important crops (Temudo & Abrantes, 2014; Monteiro et al., 2017). However, whilst cashew-inspired land cover changes reflect local socio-political agendas, they are widely considered to have negative implications for both local poverty alleviation and biodiversity conservation (Lundy, 2012; Temudo & Abrantes, 2014; Cabral & Costa, 2017).

7.1.1.2. CBC in Guinea-Bissau

Since 1990, the State and its partners have strived to conserve and manage the country's unique coastal zone, upon which the country's biodiversity depends (World Bank, 2016). One such collaborative conservation initiative was *The Coastal and Biodiversity Management Project* (2004-2010), financed by the Global Environment Facility (GEF), European Union (EU), IUCN and the World Bank (World Bank, 2016). This initiative resulted in the legal establishment of five national parks, and the national parastatal conservation agency *Institute for Biodiversity and Protected Areas* (IBAP) in 2005 (*Decree No. 2/2005*). Furthermore, the IUCN – operating in Guinea-Bissau since 1989 – is a central conservation role player, and was catalyst to the formation of several additional key conservation organizations, including the national *Coastal Planning Office* (CPO) and several now IUCN member NGOs (e.g. *Action for Development, Palmeirinha* and *Tiniguena*).

More recent collaborative conservation initiatives include *GEF's Small Grants Programme* (SGP), which specifically stipulates local community participation in sustainable use and management of the country's natural resources (GEF, 2018). Additionally, recent *EU Fisheries Partnership Agreements* between EU countries and Guinea-Bissau provide substantial financial support for the creation and management of MPAs, and emphasize the contribution of local communities to sustainable management (OECD, 2018b).

IBAP's mandate is "to propose, coordinate and implement the policy and actions pertaining to biodiversity and areas protected throughout the national territory" (RGB, 2014: p51). More specifically, IBAP's focus is on improving PA management and the promotion of sustainable development in local communities through: participatory management; the creation of both strict conservation and sustainable use zones; and supporting NGOs and local communities to develop sustainable development initiatives (RGB, 2014: p51-52). IBAP played an important role in the 2010 revision of the *1997 Framework Law on Protected Areas*, officially decreed in 2011 (*Decree No. 5-A/2011* - RGB, 2014). This legislation serves national biodiversity conservation management, and officially recognizes and defines IBAP's role and clarifies its relations with other relevant natural resource management institutions (RGB, 2014). More specifically, this legislation defines and categorises PAs in accordance with the IUCN system, including sustainable use PAs. Furthermore, more recently IBAP has been responsible for developing the *National Biodiversity Strategy and Action Plan* (RGB, 2014).

Other relevant national CBC legislation of particular relevance to coastal conservation in general and the establishment of MPAs in particular, include three laws decreed in 2011, namely; the *Forest Act (Decree No.5/2011)*, the *General Law on Fisheries (Decree No. 10/2011)* and the *Artisanal Fisheries Regulation Framework Law (Decree No. 24/2011)*. More specifically, these, among other, laws determine fishing zonation, prohibitions of hunting and the use of mangroves (RGB, 2014). In addition, the country has committed to numerous regional and international conservation-related institutions and conventions (see RGB, 2014; FAO, 2019a). In terms of its *Aichi Target* obligations, the country's PA expansion – accounting for approximately 16.7 and 10% of the terrestrial and marine national territory respectively (UNEP-WCMC, 2019a) – has resulted in it being one of the few West African countries to comply with the CBD's *COP 10* and *2020 Aichi* agreements (GEF, 2018).

CBC is enabled by an IBAP developed zoning system dictating PA establishment where central zones are left (at least in principle) completely untouched (and often overlapping with sacred areas), zones allowing limited sustainable (but conservation compatible) activities, and lastly zones for resident villagers (World Bank, 2015c). However, uncertain funding means PA management increasingly requires generating local revenues, hence recent interest in tourism-incentive programs (e.g. EU's External Action Service – see Benzinho & Rosa, 2015), and financial mechanisms (e.g. REDD+ - see Skutsch & Ba, 2010; Vasconcelos et al.,

2015). Four priority coastal conservation areas have been identified, with PAs embracing the above zoning plan subsequently established in: the *Cufada Lagoons;* the *Cacheu Mangroves;* the *Cantanhez Forest;* and the *Bijagós Archipelago.*

The *Cufada Lagoons National Park* (CLNP) – located in the south and created in 1997 – was the country's first recognized Ramsar site in 1990, and represents an important habitat for threatened chimpanzee populations, resulting in continued calls for landscape-scale conservation and improved community-involvement to mitigate human-chimpanzee conflict (Vasconcelos et al., 2002; Sousa et al., 2005; Carvalho et al., 2013). The estuarine *Cacheu Mangrove National Park* (CMNP) – located in the north-west and legally recognized in 2000 – possesses one of the richest fishing regions, and is key to the economically valuable national shrimp industry (Vasconcelos et al., 2002, 2015). CMNP also represents the nation's greatest concentration of protected mangroves, with the lowest (and recently considered stabilised) deforestation rate (Fernandes, 2012; Vasconcelos et al., 2015). Decreased deforestation rates in the CMNP – also shown to have occurred in adjacent 'unprotected' areas – are attributed to conservation management activities since 2002, including community-monitoring, increased community involvement in management, and the introduction of socially acceptable alternative livelihoods (Vasconcelos et al., 2015). However, population migration, certain agricultural practices, poaching and uncontrolled logging remain challenges in both CLNP and CMNP (Vasconcelos et al., 2002; Carvalho et al., 2013; Amador et al., 2015). Furthermore, the State's recent announcement of a proposed thermoelectric plant within CLNP attests to the ongoing 'balancing act' between ecological and economic development facing IBAPs mandate (UNIOGBIS, 2017).

A highly relevant and illustrative example of national CBC progress and challenges (in addition to the present case study) can be found in the well-established south-west coastal biodiversity hotspot of the *Cantanhez Forest National Park* (CFNP - Temudo, 2012). CFNP comprises a 'patchwork' landscape of settlements, agricultural fields, savannah, and sub-humid, secondary and mangrove forest, and is home to several rare and endangered species (Temudo, 2011; Casanova et al., 2014). Recognised as a Portuguese colonial hunting reserve in 1941, and a PA in 1980, it was only declared a national park in 2008 and legally recognized in 2011 (Sousa et al., 2017). However, CFNP's establishment was specifically initiated in 1992 by three NGOs newly formed by former public servants (Temudo, 2012). In 2002, NGOs, traditional authorities and IBAP signed an agreement allowing limited forest use by local people for shifting agriculture and hunting (Sousa et al., 2017). However, limited investment in development resulted in a locally perceived development failure (Temudo, 2012; Sousa et al., 2017). Negative local conservation perceptions also stem from the presence of conflicting customary and Western conservation beliefs, wildlife crop-raiding, and the banning – through a fines and fences approach and even forced relocations – of specific forested areas for farming (Temudo, 2012; Sousa, 2014; Casanova et al., 2014; Sousa et al., 2017).

Integration of local communities into conservation in the country remains highly complex, notably due to an extremely ethnically diverse population, which comprises 30 different ethnic groups (Spinola et al., 2008; Gomes et al., 2017), all with diverse associated cultural rules, customs, and beliefs regarding natural resource use (Costa et al., 2013; Casanova et al., 2014; Amador et al., 2015). A notable example being differing beliefs over hunting, and the trade and consumption of bush meat (Temudo, 2012; Casanova et al., 2014). Hunting represents a well-publicised national conservation concern since many communities perceive wild animals necessary to daily survival, and consider domestic animals safety nets for times of war, famine or important celebrations (Hockings & Sousa, 2013; Casanova et al., 2014; Amador et al., 2015). In addition to hunting practices, sacred spaces are integral to both conservation and customary beliefs in Guinea-Bissau (Casanova et al., 2014; Cross, 2014, 2016). Nevertheless, whilst local customary practices are influential to conservation in the country, they have been shown insufficient to alleviate pressure on natural resources degradation.

PA establishment in the country often results in changes to customary institutions. Whilst *de jure* decision-making power may reside with local customary authorities, *de facto* control involves co-operation between customary authorities and NGOs (e.g. Casanova et al., 2014). Furthermore, a general decline in power and recognition of customary authorities is considered to be negatively affecting local conservation management in the country (see Temudo, 2011; Costa et al., 2013; Casanova et al., 2014). Local studies reveal colonial and neoliberal land-reform and conservation agendas, and NGO's, have influenced customary decision-making processes and constrained effective CBC governance (e.g. Temudo, 2012; Casanova et al., 2014). Moreover, a lack of engagement with women has been shown to incite negative perceptions of conservation actions (e.g. Costa et al., 2017).

Notwithstanding the challenges above the country's participatory PA management approach at least in policy and rhetoric promotes community participation, which is thought to have increased local awareness and enabled national conservation 'successes', including reaching the landmark of protecting approximately 26% of its national territory (World Bank, 2015c; UNEP-WCMC, 2019a). Accordingly, the *Urok Islands Community Marine Protected Area* (CMPA), is considered to have embraced a participatory management approach with some 'success' (Brenier et al., 2009; Equator Initiative, 2019), and is now introduced.

7.1.2. Local Context: The Urok Islands

<u>7.1.2.1. The Ecological System</u>

The Urok Islands form an integral component of the Bijagós Archipelago, located south-west of the capital city of Bissau (refer to *Chapter 2: Figure 2.3.*). The archipelago comprises 88 islands, and is the only active deltaic archipelago on the African Atlantic coast (Campredon & Catry, 2018). The ecological system comprises mudflats, mangroves, savannah grasslands, palm groves and dry forested areas, and represents key reproductive and nursery areas to marine turtles, and numerous species of fish, shark, crustaceans and molluscs (Campredon & Catry, 2018; Simier et al., 2019). The area is also recognized as highly important to large concentrations of migratory birds (Delany et al., 2009; Lourenço et al., 2017).

<u>7.1.2.2. The Social System</u>

The People

Approximately 30,000 people inhabit 185 *tabancas* (i.e. villages) across some 20 permanently inhabited islands in the archipelago (Brenier et al., 2009; Catry et al., 2017). More specifically, about 3000 people inhabit the Urok Islands, 68% of which are between 0 and 30 years of age (IMVP, 2017; Tiniguena, 2019). Moreover, Formosa – the largest of the Urok islands – is home to approximately 1873 inhabitants (Indjai, 2017 – refer to *Figure 2.3.* for location). The traditional hunter-gatherer, matriarchal and matrilineal *Bijagó* make up approximately 90% of the archipelago's population (Cross, 2014). Local livelihoods comprise rice farming, palm plantations, and predominately subsistence fishing (Brenier et al., 2009). Whilst *Bijagó* origins are disputed, many consider these heavily influenced by Portuguese 'pacification', even though this was met with strong local resistance (Abreu, 2012; Cross, 2014). However, Portuguese oppression, including a policy of complete segregation of indigenous peoples from colonials, allowed what some describe as an 'isolated' existence from colonial influence which promoted a greater 'homogenous ethnicity' when compared to mainland inhabitants (Klute & Fernandes, 2014).

Bijagó society is organized into *camadas* (i.e. age-classes or cohorts), which requiring *fanado* (i.e. initiation ceremonies) to progress, and the youth possessing obligations to *garandi* (i.e. elders) (Cross, 2014). The *Bijagó* are animist, revering spirits infused within natural phenomena, which some scholars suggest depicts a deeply knowledgeable and respectful cultural connection with *their* environment (Cross, 2014, 2016). Sacred sites – which strongly correlate with areas of high biodiversity (Brenier et al., 2009) – are monitored by family clans and reserved for initiation rituals, with customary guidelines restricting access and governing rituals and other behaviour (Cross, 2014). Whilst customary *Bijagó* governance varies between islands, village areas are still customarily held by an *oronho* (i.e. land chief descendant of the original settler clan), who is advised by a *garandesa* (i.e. Council of Elders). The latter represent the primary customary management and regulation body, including regulating access to local natural resources

(Brenier et al., 2009; Cross, 2014). Other population groups have recently settled from the mainland and other regions (Tiniguena, 2019). Yet despite this in-migration of 'outsiders', and the growing pressure from globalization, the cultural and natural values of the Bijagó are considered by many scholars to have largely endured (Klute & Fernandes, 2014; Belhabib et al., 2015b; Campredon & Catry, 2018).

The Fishery

Most of Guinea-Bissau's fishing occurs in the Bijagós Archipelago (Correia et al., 2018). Frequently caught species belong to the *Ariidae*, *Lutjanidae*, *Clupeidae*, *Cichlidae*, *Sparidae*, *Sphyraenidae* and *Mugilidae* families (Brenier et al., 2009; Simier et al., 2019). Whilst the local climate is influenced by the Canary and Guinean Currents – during the dry and rainy season respectively – dominant fish species occur throughout the year (Lafrance, 1994; Belhabib et al., 2016; Simier et al., 2019). *Bijagó* men predominantly practice subsistence fishing, most commonly using nets, lines, and harpoons (Tvedten, 1990; Brenier et al., 2009). Furthermore, women traditionally gather shellfish – the primary local source of protein and a key element of cultural ceremonies – on mudflats close to their village (Brenier et al., 2009).

The main threats facing these islands are overfishing, tourism, and ongoing offshore oil exploration and bauxite mining in neighbouring regions (Campredon & Catry, 2018; Tiniguena, 2019). These activities, and ever-increasing drug trafficking, further convolute the social system, placing immense pressures on long prevailing subsistence fishing practices and customary institutions (O'Regan & Thompson, 2013; Klute & Fernandes, 2014; Belhabib et al., 2015b; Campredon &

Catry, 2018). Whilst the archipelago is generally free of industrial fishing – though more recently attracting growing European and Asian interest (Intchama et al., 2018; Virdin et al., 2019) – this is not the case for regional neighbours, which negatively impacts the archipelago's highly mobile fish stocks (Kaczynski & Fluharty, 2002; Campredon & Catry, 2018; Correia et al., 2018). Furthermore, as established in *section 7.1.1.1.* the negative impact of large numbers of West African 'in-migrant' small-scale fisher operations within the archipelago is well-recognized.

CMPA Management

The Urok CMPA, the country's first, was recognized under legal decree on 12 July 2005 (*Decree no. 8/2005*). The CMPA spans approximately 619 km^2 and encompasses the three inhabited islands of *Formosa*, *Nago* and *Chediã* (UNEP-WCMC, 2019b – refer to *Figure 2.3.*). The CMPA forms part of the national PA network, and the *UNESCO Bolama and Bijagós Archipelago Biosphere Reserve* (BBABR), the latter established in 1996. The BBABR includes both the *Orango*, and *João Vieira and Poilão* Marine *National Parks* (Brenier et al., 2009 – refer to *Figure 2.3.*). Furthermore, the archipelago has been a Ramsar Site since 2014 (RAMSAR, 2014). The CMPA is also part of the *West Africa Protected Areas Network* (RAMPAO), and *The Regional Programme for Coastal Conservation in West Africa* (PRCM) (Wabnitz et al., 2008; PRCM, 2012). In addition, the State and its conservation partners are currently working toward its classification as a *UNESCO Natural and Cultural World Heritage Site* (UNESCO, 2013; Tiniguena, 2019).

The CMPA is managed by the *Urok Management Committee* (UMC), which includes a range of actors (*Figure 7.2.*). *UMC* management mechanisms are grounded in participatory management, which strives to ensure a bottom-up consultation-negotiation decision-making process through participation of the local community and/ or their representatives. The *UMC* comprises 13 local representatives from the three islands (6 – Formosa; 3 – Nago; 3 – Cheida). Community participation occurs through local *Village Management Committees* (VMCs), annual island assemblies, and the *UMC* annual general assembly (Brenier et al., 2009). IBAP represents the State, and provided technical assistance in the creation of the CMPA Additional technical assistance was provided by the International Foundation of the Banc d'Arguin (FIBA), as part of the *West African Regional Strategy for Marine Conservation Areas* under the framework of the PRCM (Renard & Touré, 2012; PRCM, 2012). Tiniguena (a national NGO) facilitates participatory governance and the collaboration of all governance actors including *VMC*s. In addition to the aforementioned actors the *UMC* also includes the national *Commission on Fisheries Surveillance* (FISCAP - charged with monitoring local compliance), the national *Centre for Applied Fishing Research* (CIPA), the national *Institute of Studies and Research* (INEP), and the national *Coastal Planning Office* (CPO - see *Figure 7.2.*). Furthermore, the MPAs' zoning requires the issuing of fishing licences which falls to the national *Ministry of Fisheries* (Brenier et al., 2009).

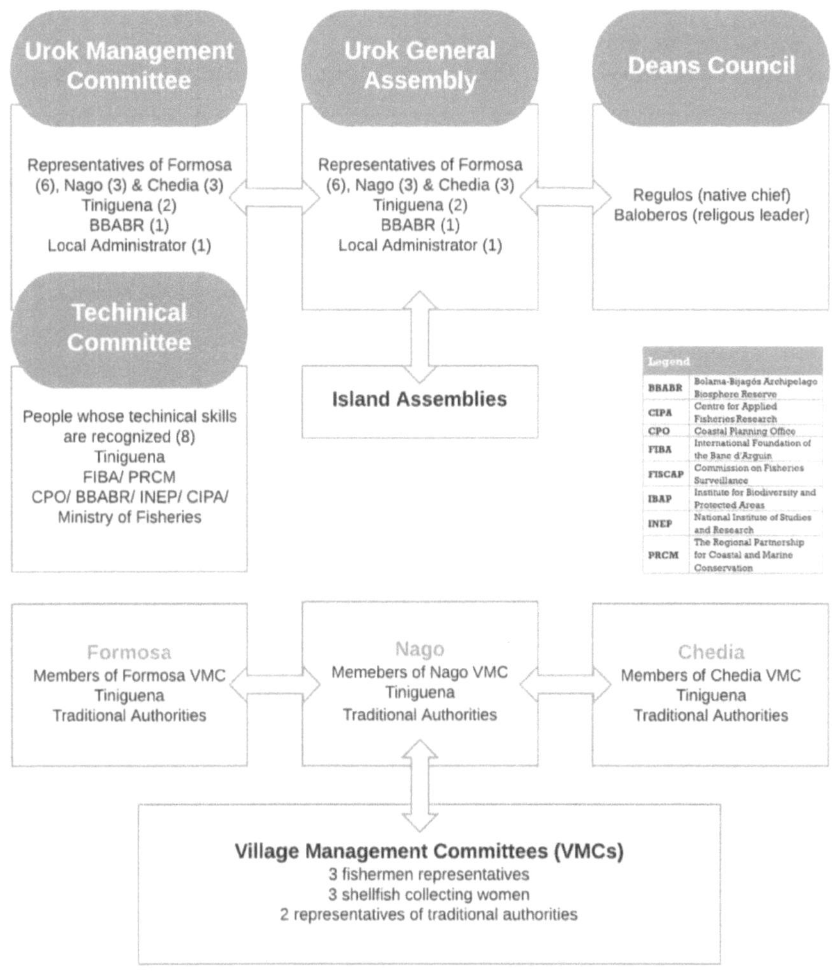

Figure 7.2.: Urok CMPA Institutional Organization Map. **Source:** Translated and Adapted from Biai et al. (2003). Note: see legend for organization abbreviations.

The *UMC*, in accordance with the aforementioned national PA zonation, developed a CMPA management plan identifying three distinct zones: 1) a central zone comprising the mangroves and backwaters, reserved solely for subsistence and ceremonial fishing by residents; 2) a peripheral zone extending to the limits of the first channels, reserved for traditional commercial fishing solely by residents (i.e. over and above subsistence), with the aim of providing local economic security; and 3) an exterior national solidarity zone, extending up to the boundary of traditional territory, open to both residents and non-residents on condition they respect rules governing equipment and techniques (Biai et al., 2003 – see *Figure 7.3*). Nevertheless, few community members access resources beyond zone 1 due to a lack of appropriate boats.

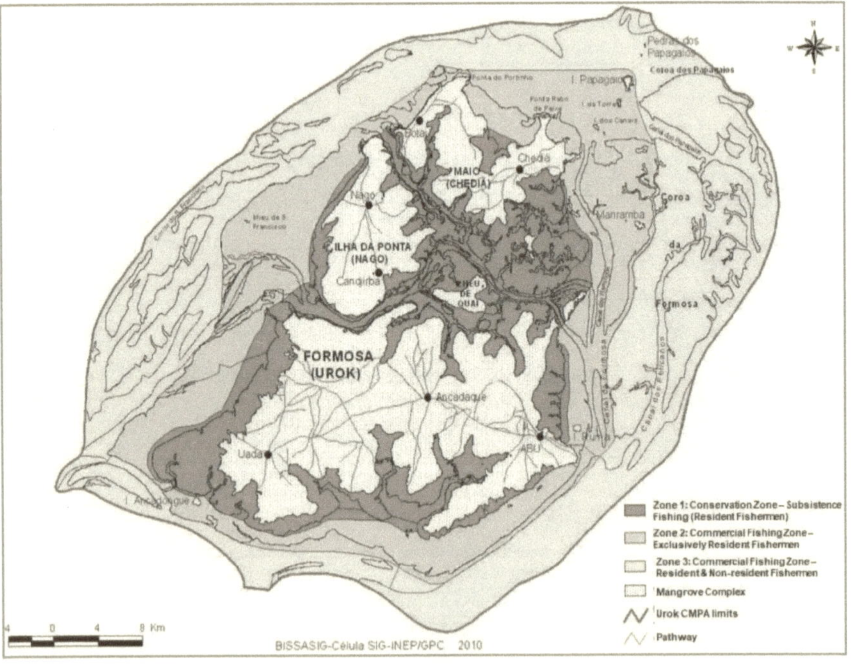

Figure 7.3.: Urok Islands CMPA zonation map. **Source:** Tinguena (2019).

Actors monitoring the management plan include the local *UMC* representatives, a committee of elders, and a partners making up the technical committee (Brenier et al., 2009 – refer *Figure 7.2.*). Management rules adopted cover: the most vulnerable areas and environments; threatened species and strategic resources in terms of conservation, culture and the local economy; and fishing vessels and methods used to exploit other resources (e.g. a ban on monofilament nets and the closure of backwaters - Brenier et al., 2009). Past evaluations deem the CMPA participatory management approach inclusive of multiple actors, and concluded work undertaken to be positive and to have enabled development of mutual trust by combining conventional conservation actions with customary institutions (Brenier et al., 2009).

7.2. Findings

7.2.1. Contextual Change Triggers

As with findings in Madagascar, all partner respondents identified State incapacity and the presence and will of other partner organizations to take on the role of State, as contextual issues that have triggered implementation of CBC initiatives in the country. This role is notably taken up by IBAP and various NGOs, which in the case of the CMPA is Tiniguena. Additional contextual issues considered to have 'triggered' national CBC initiatives include concerns regarding the economic value of marine resources to the country, and overharvesting by SSFs and commercial IUU, which resulted in a national MPA expansion agenda (Cross, 2016). This 'agenda', and the ecological significance of the archipelago, emerged from partner respondents as a key contextual change trigger that promoted political will towards CBC initiatives, and willing partners and donors to support these initiatives.

Local contextual change triggers included the Urok Islands' immense natural and cultural wealth, and the interest and commitment of the resident community to adopt and implement 'good' governance, emerged as a key contextual change trigger (Tiniguena, 2019). More specifically, this 'triggered' the State to officially approve the formal creation of the CMPA (Tiniguena, 2019). Furthermore, many respondents acknowledged that the establishment of the aforementioned nearby State MPAs of *Orango,* and *João Viera and Polião* Marine *National Parks*, 'triggered' the community to seek support from Tiniguena to establish the CMPA. Moreover, a major community-identified contextual change trigger for establishing the CMPA – as depicted in the Malagasy case study – was poverty, and concern for a perceived reduction in natural resources. In particular, community responses cited the need for, and an inability to develop basic service delivery such as schools, clinics, water and transport as key contextual issues 'triggering' community support of local integrated conservation and development projects, such as the CMPA. This has also been noted elsewhere in the country (e.g. Amador et al., 2015).

7.2.2. Review of Enablers

This section analyses the extent to which the 14 enablers identified in *Chapter 3* are present in the Urok Islands case study. Accordingly, a summary of the *absence, partial presence, partial absence,* or *presence* of these enablers can be found in *Figure 7.4.* Furthermore, *Table 7.1.* provides a brief overview of the findings that influenced these enabler presence ratings. These findings are discussed further below. Lastly, reference is also made within the following discussion to relevant responses from national CBC partner respondents in Guinea-Bissau to contextualize these findings.

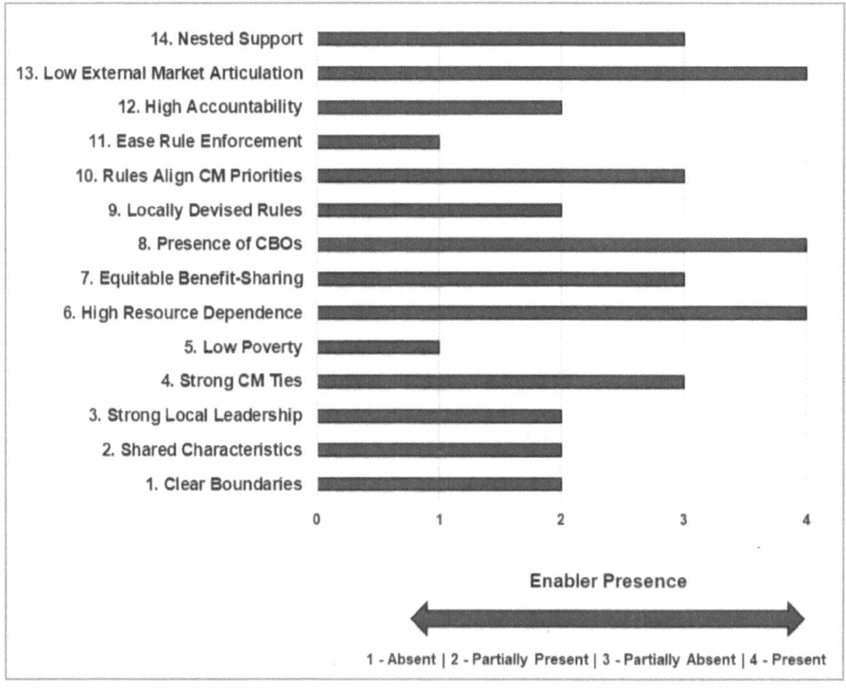

Figure 7.4.: A graphical summary of the presence of the 14 enablers in the Urok Islands. **Note:** CM – Local Community.

Table 7.1.: A summary of the review of the 14 enablers in the Urok Islands.

	Enabler	Presence	Explanation
Resource System & Users	1. Clearly-defined resource system & user boundaries	**Partially Present**	The three CMPA zones are considered clearly-defined, as is the geographical community. However, due to in-migration of 'outsiders' socio-cultural resource-user boundaries are less well defined.
	2. Shared norms, values, interests & identities	**Partially Present**	Community respondents confirmed coastal migration and partial erosion of customary practices. Nevertheless, forest-based taboos and shellfish remain culturally important to most community members.
	3. Strong local leadership	**Partially Absent**	Strong and respected traditional leaders. However, *UMC* representation was often perceived by community members to be poor, and characterized by perceived *elite-capture* of knowledge and benefits.
	4. Strong community ties	**Partially Present**	Strong family and intra-village bonds, however, weaker inter-village bonds. Furthermore, whilst poor community-perceived representation exists, these community-representative relations were generally positively perceived.
	5. Low levels of poverty	**Absent**	High levels of poverty confirmed by all community and partner respondents.
	6. High levels of dependence on resource	**Present**	Most community members emphasized the heavy dependence upon forests and small-scale agriculture. However, a confirmed lack of alternative protein sources dictates high levels of dependence on fish and shellfish. The latter is also of particular cultural importance.
	7. Equitable distribution of benefits from common property resources	**Partially Absent**	High levels community-perceived elite-capture of benefits by *UMC* and Abu as a village, and even at times Tiniguena. Nonetheless, community members still have access and benefit from both forest and coastal resources.
Institutional Structure & Externalities	8. Presence of community institutions	**Present**	*UMC* and *VMC*s are well-established CBOs present. *UMC* legally recognized. VMCs well respected by community members.
	9. Locally devised access and management rules	**Partially Present**	Rules established at *UMC* meetings (notably the annual general assembly), but process is largely perceived by community members to be captured by representatives. Furthermore, low levels of attendance from the greater community at these meetings.
	10. Rules strongly align with local priorities/ needs	**Partially Absent**	Poverty alleviation and access to basic services represent the main local priorities. However, since many community members perceived a lack of benefit-distribution, socio-economic alignment is not strong. However, community possesses access to marine natural resources, which allows for access to culturally important shellfish. Therefore, cultural alignment is strong.
	11. Ease in enforcement of rules, and conflict resolution	**Absent**	Community emphasized a lack of boats for monitoring, and absent/ ineffective State monitoring. Rules largely understood including prohibition from selling catch. Conflict resolution by *UMC* is perceived by community to be ineffective.
	12. High levels of accountability	**Partially Present**	Community members may voice concerns through representatives at *UMC* meetings, but low community perceived *UMC* representative accountability. Additionally low perceived State accountability to the community, especially regarding monitoring and enforcement.
	13. Low levels of articulation with external markets	**Present**	Prohibited from selling catch, and limited to no access to Bissau markets.
	14. Presence of 'nested' governance with high levels of initial external support	**Partially Absent**	High levels of initial for CMPA implementation and on-going capacity-building support from Tiniguena. State legal recognition. Ongoing Tiniguena financial and technical support. Lack of State support, especially with monitoring and enforcement. Local governance by *UMC* (with partner members), including establishment and enforcement of rules (although weak).

<u>7.2.2.1. Clearly-defined resource-system and resource-user boundaries:</u>
Community respondents acknowledged clearly-defined CMPA zones as stipulated in the *UMC* management plan (as introduced in *section 7.1.2.2.*). Nevertheless, respondents acknowledged that few community members access beyond zone 1 (inner zone) due to a lack of appropriate boats. Furthermore, whilst the community may be considered geographically clearly-defined, as it is isolated to the three islands, many community respondents emphasized 'in-migration' of 'outsiders' has resulted in a less clearly-defined socio-cultural resource-user boundaries. This finding in accordance with the *Bay of Ranobe*, and previous studies (Cox et al. (2010; Cinner et al., 2009a; Child, 2019). Findings related to socio-cultural aspects of the community in the Urok Islands are discussed further in the subsequent enabler. Consequently, this enabler is considered *partially present*.

<u>7.2.2.2. Shared norms, values, interests & identities</u>
All community respondents confirmed the presence of long-standing customary practices – such as protection of sacred areas and the use of *malto malgos* (i.e. 'curses') – continue to promote and encourage community participation and motivation in conserving natural resources. However, many of these respondents also described the partial-erosion of customary institutions, which may have negative implications for resource-user compliance with established natural resource rules, as noted elsewhere in the archipelago (e.g. Bordonaro, 2009), and across the country (e.g. Temudo, 2012; Casanova et al., 2014; Sousa et al., 2017). Whilst, coastal-marine related customary institutions and practices are being challenged by 'outsiders' and local customary erosion, community respondents emphasized customs within local coastal forests remain effective. For example, one community respondent emphatically stated in relation to forest resources, "if [the

curse] says don't touch and you touch, you will die!" (AM3). This continued ability of customary practices to control natural resource use has also been shown elsewhere in the archipelago (see Madeira, 2016). It could also be inferred from community responses that customary matriarchal power and recognition are in decline, as is perhaps best exemplified by a lack of female representation on *VMCs* and the *UMC*. However, notwithstanding this lack of involvement, female respondents confirmed they were still widely respected. Consequently, the impact of gender-based roles on natural resource in the area requires further research.

Poverty was also emphasized by all respondents to be constraining the effectiveness of customary institutions for CBC. Consequently, most community and partner respondents called for a 'hybrization' of customary and conventional conservation management. Accordingly, one community respondent stated, "we need to change the mindset, we need to mix what's useful from traditions with the modern" (AM9). Whilst the community-perceived value of the CMPA varied, a general pattern of increased negative perceptions in villages located further from the project hub of *Abu* emerged, i.e. where Tiniguena is located (*Figure 7.5.*). This is perhaps best depicted by negative perceptions of the CMPA emerging from all respondents from *Uada* and *Kuian* villages (*Figure 7.5.*). Furthermore, this 'village-based' pattern of negative perceptions (as was similarly depicted in the Madagascar case) emerged across a wide range of factors discussed below. Therefore, such division potentially eludes to a community with diverse interests. However, notwithstanding the aforementioned issues, numerous community responses emphasized the need to protect their resources, and a continued sense of pride in the CMPA. This was perhaps best exemplified by many referring to "our" protected area when being interviewed. As one *Abu* elder, *VMC* and *UMC* representative proudly stated, "we have enough to preserve our natural resources, we have the moral power to do it" (AB2). Consequently, based on the above discussion this enabler is considered *partially present*.

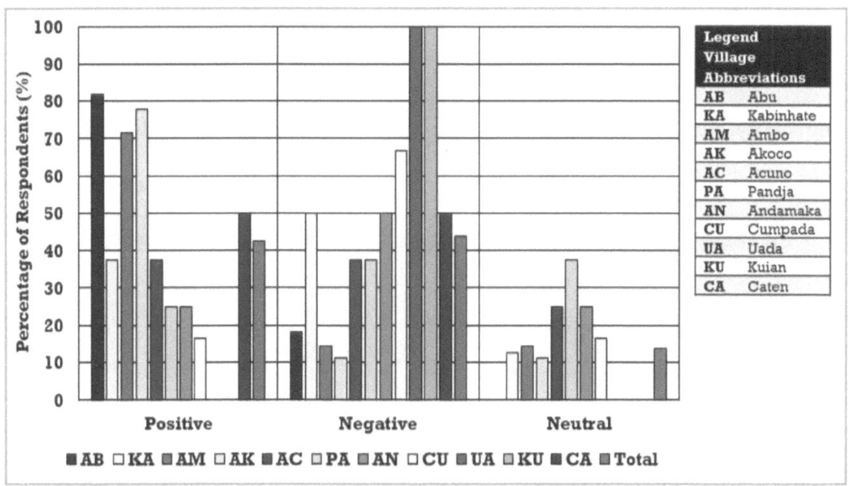

Figure 7.5.: Community perceived value of CMPA by percentage of respondents for each village. Note: Villages are ordered by distance from *Abu* on the left to *Caten* on the right (i.e. Caten is located the furthest from Abu). See legend for village abbreviations.

<u>7.2.2.3. Strong local leadership</u>

All community respondents noted elders, traditional authorities (TAs - e.g. local village chiefs), and local *UMC* and *VMC* representatives as highly influential 'middlemen', notably as a link between the community and Tiniguena. Accordingly, both partner and community respondents emphasized the link between both *VMC* and *UMC* representatives and community members as crucial to the representation of community interests and concerns. Community responses generally described their relations with local leaders as *Good* (*Figure 7.6.*). This was particularly noted regarding TAs and *VMC* representatives, most notably in villages located further away from *Abu*. These village respondents stating for example, "we respect them like our own father" (CU4). All community respondents considered the TA's role in organising the community and conserving cultural traditions as especially important, specifically within ceremonial events. This is of relevance to management of the CMPA as the consumption of shellfish occurs at all ceremonial events. In addition to TAs, the power and influence of elders emerged especially noteworthy. For example, a *Kuian* respondent stated, "If elders don't agree, it won't happen!" (KU5). The importance of elders and local representatives in management decisions was also echoed by numerous partner respondents.

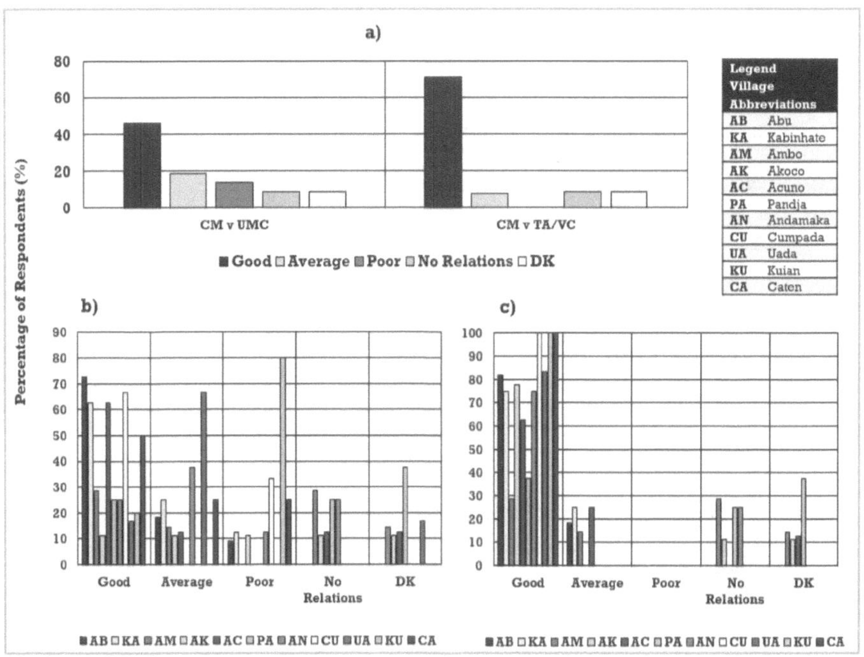

Figure 7.6.: Community perceived relations with local leaders including Urok Management Committee representatives (UMC), and Traditional Authorities (TA) and Village Management Committee (VMC) representatives by percentage of respondents **a)** overall, and by village for **b)** local UMC representatives; and **c)** TAs and VMCs. Note: **CM** = Community-Member. 'No Relations' indicates community-member had limited to no interaction with this 'actor'. In **b)** and **c)** villages are ordered by distance from Abu (AB) on the left to Caten (CA) on the right (i.e. Caten is located the furthest from Abu). See legend for village abbreviations.

Notwithstanding the enduring power and recognition of TAs and elders, community respondents emphasized weak local CMPA governance capacity by their *UMC* representatives. This concern emerged three-fold. Firstly, regarding a community-perceived inability to adequately disseminate knowledge and to provide feedback from meetings. Secondly, community respondents perceived inequitable distribution of benefits arising from the CMPA (discussed further in *section 7.2.2.7.*). Thirdly, these community respondents strongly emphasized the inability of UMC representatives to effectively monitor, and especially enforce,

CMPA rules. Accordingly, the first two concerns emphasize perceived local elite-capture, which community members consistently identified as negatively influencing community mindsets, and subsequently behaviour pertaining to the CMPA. Whilst, community respondents identified knowledge-acquisition occurring predominantly through their local *UMC* representatives, many described not knowing how well the community was being represented. This was emphasized by two community respondents who simply stated, "we didn't seem them yet!" (CU4). Therefore, whilst community-perceived relations with local leaders, including both their *VMC* and *UMC* representatives were largely positive, and many community respondents stated their representatives were good at hearing their concerns, community-perceived representation was largely considered *poor* due to a lack of perceived feedback and delivery of tangible benefits. Nevertheless, the power and influence of TAs and VMCs appears to have endured and thus, this enabler was considered *partially absent*. Consequently, as with the findings in the *Bay of Ranobe*, these findings reinforce the importance of the need for strong local leadership to enable CBC interventions, which also confirms those of previous studies (Pomeroy et al., 2001; Agrawal, 2002; Galvin et al., 2018; Biggs et al., 2019).

7.2.2.4. Strong community ties

Community respondents perceived *other community members* (i.e. CM) the most central regarding *Interactional Support* as depicted in *Figure 7.7*. To reiterate this refers to those actor's community members deemed the most approachable regarding their natural resource concerns. Accordingly, as an *Acuno* respondent stated, "there is good unity within the village" (AC4). Therefore, strong family and intra-community ties were confirmed by respondents (as was noted in the *Bay of*

Ranobe). However, as discussed above community-perceived inter-village divisions exist. Furthermore, the negative influence of in-migration of 'outsiders' was commonly noted by community members and partners, which is also consistent with *the Bay of Ranobe*. Moreover, as established community-perceived challenges exist with local leadership. Consequently, ties within the greater community are considered to have weakened, and therefore, this enabler is considered partially *present*.

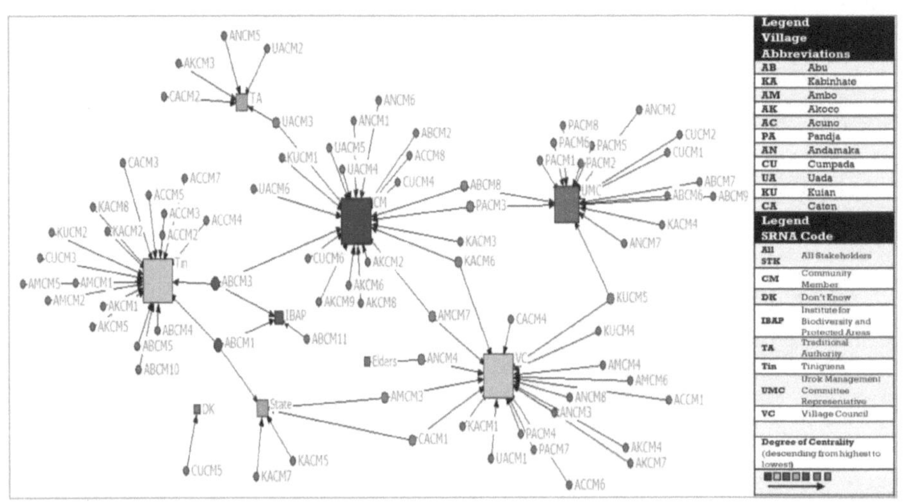

Figure 7.7.: A social network map indicating those CMPA actors community members perceived most approachable with their natural resource concerns (i.e. 'Interactional Support').
Note: Degree of centrality is indicated by the size of icon (the bigger the higher the centrality, i.e. the actor most approachable) and colour (see legend). See legends for village abbreviations and SRNA code.

7.2.2.5. Low levels of poverty

As established previously high levels of poverty were confirmed by all respondents, and emerged as a major constraint for implementation and governance of the CMPA. Respondents also emphasized that a lack of basic services, market-access and alternative livelihoods exacerbates local poverty levels. Moreover, perceived elite capture was considered largely a result of high poverty levels (as was the case in the *Bay of Ranobe*). Consequently, this enabler is considered *absent*.

7.2.2.6. High levels of dependence on resource

All respondents confirmed high levels of natural resource dependence as a result of high levels of poverty, and a lack of alternative livelihoods outside of natural resource harvesting. Community respondents specifically emphasized their reliance upon forest-based resources. Accordingly, an elder from *Ambo* stated, "when the wood ends, we take a seat, then we starve" (AM7). Furthermore, whilst cashew has long been dubbed a local 'panacea' for poverty alleviation, and many community members pursue this livelihood, cashew production is known to have increased deforestation in the archipelago (Madeira, 2016). Moreover, community respondents consistently described the irony of how harvesting palm-based products and cashew has become a common method of trade for rice (i.e. a former staple traditional crop), as shown elsewhere on in the archipelago (see Madeira, 2016).

In addition to forest-based resources, community respondents also confirmed the fishery – inclusive of net-fishing (principally by men) and shellfish harvesting (principally by women) – contributes the major component of their protein. Whilst, limited domestic livestock are kept (largely among the wealthier members), these are usually reserved for special ceremonial occasions such as weddings. At times bushmeat is also consumed as was observed first-hand. Lastly, whilst *ad-hoc* alternative livelihoods such as brick-making were also observed, this was largely to trade with neighbours. Therefore, while this enabler is considered *present*, the extent to which this influences the management of the CMPA is debatable. Nevertheless, since marine resources like shellfish are central to cultural ceremonies I propose that this resource-dependence is enabling CMPA management, although the nonlinearity is acknowledged, and further research is required. Consequently, this resonates with Agrawal's (2002) critical factors for CPRM success, namely "High levels of dependence on the resource system". Nevertheless, this enabler is discussed extensively subsequently in the final discussion in *Chapter 9*.

7.2.2.7. Equitable benefit-distribution of from common property resources

A common finding amongst community respondents was a perceived lack of benefits. Accordingly, as one community respondent stated, "The CMPA needs to improve, it is too slow to show benefits" (AB7). This perception was once again especially notable for respondents in villages located further from *Abu*. Accordingly, respondents from *Caten*, *Andamaka*, and *Uada* stated respectively, "we go to work, but everything goes to *Abu*, we receive nothing!" (CA5); "Abu have a lot and we don't!" (AN4); and "they never bring anything for us" (UA2). Consequently, numerous 'non-*Abu*' respondents emphasized elite-capture by the village of *Abu*.

At the time of the fieldwork visit the *UMC* had managed to save a substantial amount of money spanning the years of its operation (*Figure 7.8.*). Therefore, why the *UMC* was not allocating these funds to highly desired community development projects (i.e. such as schools, water, health and sanitation), was often queried by community members. In contrast, partner respondents described the completion of numerous projects including the building of a clinic, water wells in most villages and a school. However, partner respondents also emphasized the high costs of such community development projects as constraining. Nevertheless, community respondents consistently noted a lack of perceived benefits has 'fuelled' negative community perceptions of, and disillusionment for the CMPA and the *UMC* (as discussed previously). Accordingly, many Urok partner respondents (as well as other national partners) emphasized that community perceptions were heavily dependent upon obtaining tangible benefits. This emerged particularly true in relation to community-acknowledged changes to the Bijagó cultural age-structure. For example, it was noted that the youth are now seeking positions of power – formerly reserved for elders – as well as economic gain. Many community respondents suggested this was a result of increasing influences from the 'outside world'. Consequently, while community members have access to resources, a lack of perceived benefits, especially in relation to projects associated with desired basic service delivery, this enabler is considered *partially absent, and* requires improvements. Lastly, these findings mirror those in the *Bay of Ranobe*, and reinforce findings of established scholars reviewed in *Chapter 3: section 3.3.2.1.* on the importance of equitably distributed benefits for CBC success (e.g. Agrawal, 2002; Galvin et al., 2018; Biggs et al., 2019).

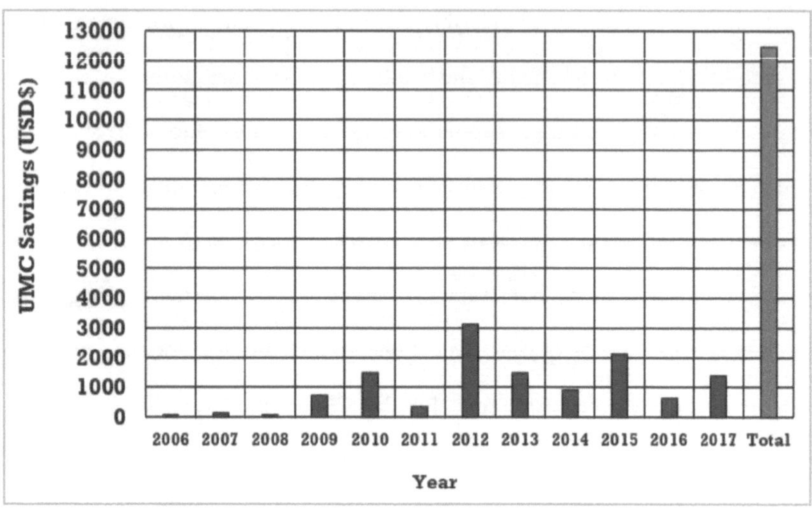

Figure 7.8.: UMC total and annual savings between 2006 and 2017. **Source:** Tiniguena.

7.2.2.8. Presence of community institutions

The community is represented at a village-level by their respective *TAs*, and *VMC*s. As established, community respondents consistently emphasized these representatives as highly influential to decision-making and managing village behaviour (*section 7.2.2.3.*). Furthermore, the *UMC* is a well-established and legally recognized community-based institution, which comprises local representatives supported by partner organizations (refer to *Figure 7.2.* previously). Consequently, this enabler is considered *present*, although local representation requires improvements.

7.2.2.9. Locally devised management rules

Community responses regarding this enabler may be interpreted with regards to representative and community-wide involvement. High community-perceived involvement through *VMC* and *UMC* representatives emerged, with active participation of these representatives across *UMC* management activities (*Figure 7.9.*). However, community respondents perceived *low* active community-level participation in management decisions and activities. This was particularly emphasized once again by villages located further from Abu. More specifically, these 'further-village' respondents frequently described the capture of decision-making power by Abu, which they suggested was due to its size (i.e. as the biggest village), and most notably it being the location of the *UMC* and *Tiniguena* headquarters. Furthermore, some community respondents (notably again from 'further' villages) perceived elite-capture of decision-making by *Tiniguena*, and described the NGO's power to influence the decisions of local representatives.

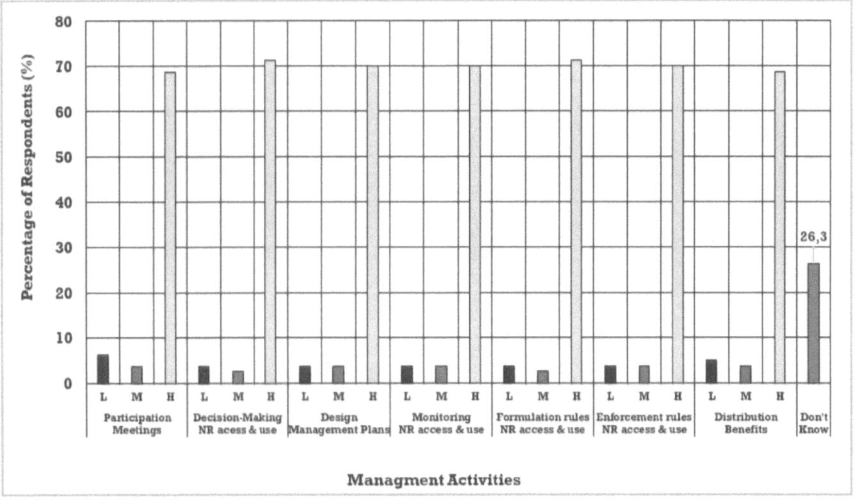

Figure 7.9.: Community-perceived involvement across diverse management activities associated with the CMPA in the Urok Islands by percentage of respondents. **Note: L** = Low; **M** = Medium; and **H** = High; and **NR** = Natural Resource.

Notwithstanding the aforementioned issues raised by community respondents, it should be acknowledged that *low* community-perceived involvement may stem from a lack of attendance at the *UMC* annual general assembly, which has witnessed declines in recent years (*Figure 7.10.*). Both partner organizations and community members confirmed whilst the community were originally supportive of establishing the CMPA, interest has declined, most notably citing a lack of effective representative feedback and perceived tangible benefits. This pattern was confirmed by national partner respondents regarding their respective community projects, and is consistent with findings from other community initiatives implemented within the archipelago (Madeira, 2015). Consequently, this enabler is considered *partially present*, since active community-involvement requires improvements. Moroever, these findings echo those in the *Bay of Ranobe*, and confirm the importance of this enabler as emphasized by extensive literature (as discussed in *Chapter 3*).

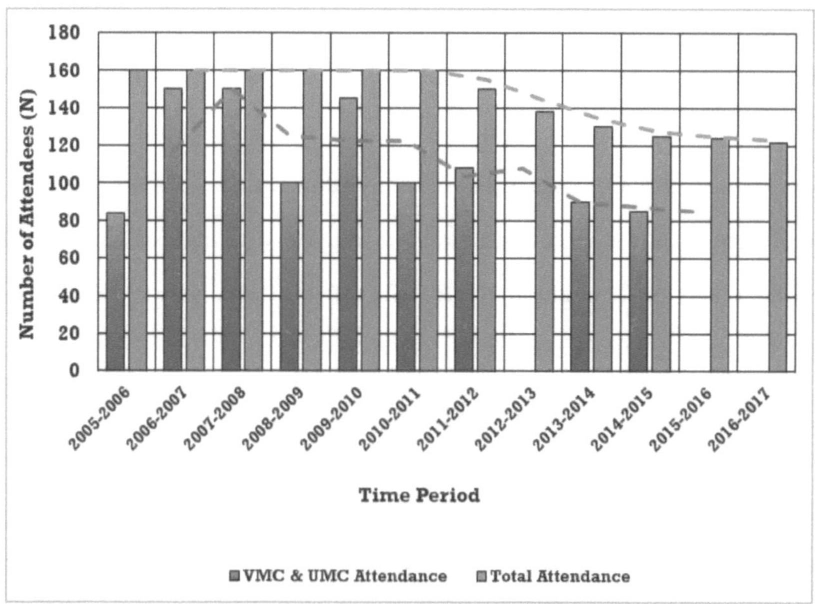

Figure 7.10: Total and *Village Management Committee* (VMC) and the *Urok Management Committee* (UMC) representative attendance at the UMC annual general assembly between 2005 and 2017 by number of attendees.

7.2.2.10. Rules strongly align with local priorities/ needs

Partner respondents confirmed *UMC* (and State) rules currently prohibit the sale of marine resources and any tourism activities. However, this was acknowledged by respondents to be having negative implications for a community possessing high levels of poverty and resource-dependence, and lacking alternative livelihoods or access to external markets (the latter discussed in *section 7.2.2.13.*). Community respondents consistently emphasized the need for improved access to basic services, however, while financial resources appear to be available for this (as discussed previously), community members perceive a lack of alignment with these local priorities. Furthermore, attendance at a *UMC* local representative meeting confirmed proposed future *UMC* fiscal development efforts, which may

284

address these concerns. Therefore, while the community does possess access to natural resources, delivery of basic services deemed locally important is lacking, and therefore, this enabler is considered *partially absent* and requires attention. Consequently, these findings serve to reinforce those in the Bay of Ranobe, and the importance of CBC alignment with local conditions, as is well-established in the commons literature (Ostrom, 1990, Cox et al., 2010), and notably by other African CBC studies (e.g. Cinner et al., 2009a; Galvin et al., 2018).

7.2.2.11. Ease in enforcement of rules, and conflict resolution

As one community respondent stated, "We can't control those that don't respect the rules" (AB12). Furthermore, all community respondents noted difficulty in independently enforcing CMPA rules, particularly regarding 'outsiders', which is consistent throughout the archipelago (Madeira, 2015). Accordingly, community respondents specifically noted 'outsiders' (commonly citing migrant Senegalese small-scale fishing crews) are often observed fishing on the boundary of, and at times within prohibited CMPA zones. Although *FISCAP* is the designated state representative for monitoring and surveillance, all Urok community and partner respondents confirmed their incapacity and ineffectiveness, as do recent studies (e.g. Okafor-Yarwood, 2019). Current local surveillance efforts involve *UMC* local representatives and the local police. However, UMC representatives and UMC partners emphasized logistical challenges to their monitoring and enforcement efforts, most notably a lack of boats and a second team to monitor the CMPA.

In addition to the above, the community perceived the *UMC* to lack the ability to resolve conflicts. Whilst *UMC* representatives stated infractions are often penalized through graduated sanctions, community members perceived a lack of *UMC* follow through in this regard. However, like the Malagasy case, most community respondents confirmed this was due to concern for their fellow community member's need to eat (i.e. poverty). Based on attendance at a recent *UMC* local representative meeting numerous strategies are in place to increase the effectiveness of enforcement, including purchasing an additional boat. Furthermore, *UMC* representatives acknowledged fines provide an important source of much needed income to cover necessary management expenses such as monitoring. Consequently, due to the lack of State technical support, and the aforementioned UMC logistical challenges, this enabler is considered *absent*.

7.2.2.12. High levels of accountability of monitors and other officials to resource users

Community members acknowledged the opportunity to voice concerns at *VMC* meetings and through representatives at *UMC* meetings, but perceived *low* representative accountability. These perceptions of accountability largely stem from a lack of perceived feedback. Furthermore, all respondents confirmed a lack of active involvement, institutional capacity, and financial resources has meant State institutions responsible for research, and monitoring and enforcement, such as *CIPA* and *FISCAP,* are inactive and their perceived accountability to the community is *low*. Additionally, all respondents specifically confirmed a lack of attendance by these two State institutions at *UMC* annual general assemblies. Interviews with *CIPA* members revealed they only entered the project in 2011 and were forced to cease their involvement in 2013 due to a lack of funding. Consequently, considering the above *downward accountability* challenges, this enabler is considered only *partially present*, and requires substantial improvements most notably from State partners *CIPA* and *FISCAP*.

7.2.2.13. Low levels of articulation with external markets

A lack of access to external markets emerged as highly influential in maintaining high natural resource dependence. The prohibition of fish sales and highly limited transportation to the capital Bissau, mean low levels of articulation with external markets exists. Consequently, all respondents acknowledged limited local economic opportunities, which is cited as a major constraint to economic development throughout the archipelago (Madeira, 2015, 2016). Accordingly, community respondents emphasized a desire for greater access to markets in Bissau. Whilst, low levels of market articulation at present may be restricting natural resource harvesting activity, planned fiscal development by the *UMC* may reverse the present situation. Consequently, while this enabler is considered *present*, as discussed throughout this book, the issue of directionality in enablers is problematic, and in this specific case further research is necessary to see what the influence of increased market-access will.

7.2.2.14. Presence of 'nested' governance with high levels of initial external support

As established previously the *UMC* governance structure is theoretically horizontal in structure and incorporates numerous external supporting actors (see *Figure 7.2.* in section 7.1.2.2.). Community respondents acknowledged high levels of initial and ongoing external support from Tiniguena, which stems from a relationship developed through Tiniguena's original involvement with community development projects on Formosa Island since the 1990s. Furthermore, in reference to the levels of community acceptance of 'non-State' partners, a community respondent stated, "it is easy for the community to accept them as the State is absent, we need them!" (KA3). Accordingly, most community respondents specifically acknowledged their inability to control 'outsiders' means partners are necessary.

Some 'non-State' partner organizations suggested corruption within a political system characterised by a 'hunger' for power and status inhibits effective State involvement. Nevertheless, State support comes in the form of parastatal conservation agency IBAP, which provides CCA legal recognition and national PA status. However, as mentioned previously, monitoring and enforcement issues persist and require improved support from *FISCAP*, and fisheries research partner *CIPA*. Moreover, some national partner respondents noted many State officials lack sufficient knowledge of their sector or its laws, and often act in conflict with these laws. However, other partners suggested whilst the State did possess many foreign- and well-educated officials, there was a lack of practical (hands-on) capacity, due to their lack of 'field-experience' (PO14). Nevertheless, all respondents confirmed ongoing external support necessary for the CMPA. Consequently, while Tiniguena continues to support this community, due to a lack of State support, which requires urgent attention, this enabler is considered *partially absent*.

7.2.3. Actions

A brief discussion follows on key actions taken in the present case to facilitate change toward a CBC mode of governance, though reference is made to other partnered national CBC initiatives. As observed in Madagascar (*Chapter 6*), and introduced above, a key preliminary action for CBC implementation and governance in Guinea-Bissau was political will, and State support in the form of the creation of enabling legislation. Furthermore, CMPA implementation and governance, in accordance with other community development and conservation initiatives in the country (and the case in Madagascar), notably involved the formation of an *NGO-CBO* partnership. Accordingly, Tiniguena respondents noted the first action involved building relations of trust. As observed in Madagascar (*Chapter 6*), building relations of trust commenced with Tiniguena's involvement in community development projects on *Formosa Island* since the 1990s, and subsequent work on the other two Urok Islands. These projects include the construction of schools and a clinic, development of vegetable gardens, and more recently paid bird and shellfish community-monitors. Other national partner organizations also confirmed using this approach in their respective community initiatives, which is also consistent with both past national (e.g. Costa et al., 2017), and Bijagós CBC studies (e.g. Madeira, 2016).

The NGO-CBO partnership in the CMPA subsequently entailed the formation and recognition (by the State through IBAP) of a CBO (i.e. the *UMC*), in accordance with national mandates for improved conservation through participatory governance. Community respondents, notably *UMC* and *VMC* representatives, acknowledged this was made possible by the existence of established and functioning *VMC*s (especially since many current *VMC* representatives serve on the *UMC*). Nevertheless, Tiniguena continues to support the *UMC* both financially and technically, through which capacity has been built. Financial support has been accomplished for example by funding from a *GEF Small Grants Programme* since 2016, which supported activities such as mapping forest cover and land use; creation of plant nurseries; training on agroecology techniques in family farming; and the production and broadcasting of radio programmes to promote agroecological practices and awareness (GEF, 2019). Lastly, Tiniguena's local capacity building actions extend to specific skill development through the employment of its local staff and the empowerment of community members through specific projects (e.g. community-monitoring and vegetable gardens).

7.2.4. External Influences

The State's ability to enable CBC hinges predominantly on the presence of enabling legislation and political will, as well as financial and technical resources. The State via IBAP provides CBC legal recognition and national PA status, and have promoted MPA expansion. However, all partner respondents identified State instability as negatively influencing CBC. Furthermore, this has influenced the role of non-State partners in CBC interventions. Accordingly, Baldursdóttir et al. (2018: pS27) refer to the increased "'NGO-ization' of aid" in the country due to the State's lack of capacity and stability, and describe how civil society is considered more active than the State in numerous sectors including education, human rights, and the environment. Respondents also acknowledged the aforementioned in-migration of 'outsiders', particularly regional small-scale fishers, as a key external influence on positive CBC social and ecological outcomes (Cross, 2016; Campredon & Catry, 2018). Moreover, issues related to commercial and IUU fishing are well-established in the region (Intchama et al., 2018; Denton & Harris, 2019; Okafor-Yarwood, 2019; Virdin et al., 2019). Lastly, as with other case studies, international institutional commitments, in this case notably associated with the CBD (e.g. The Aichi Targets – CBCD, 2011; and *Post-2020 Global Biodiversity Framework* - CBD, 2020), *FAO Voluntary SSF Guidelines*, and regional commitments including those of the African Union, are external influences on this case study.

7.2.5. Issues Arising

A brief supplementary and consolidatory discussion follows of the key 'issues arising' related to Guinea-Bissau, and the Urok Islands CMPA in particular. Community respondents emphasized low education, high levels of poverty, a lack of alternative livelihoods and basic services, and weak local governance institutions (inclusive of partially eroded customary institutions), as key 'issue arising'. Accordingly, a lack of education was commonly cited by both community and partner respondents – and by other research in the archipelago (see Madeira, 2015, 2016) – as a common and persistent issue requiring future actions so as to promote poverty alleviation, decrease 'aid-dependency', and ultimately improve the chances of successful CBC outcomes. While it is safe to assume these issues have existed in the community (and country as a whole) for some time, they are considered to have 'emerged' once again as key to the CMPA (and other community projects), and require further action.

All respondents strongly emphasized that high levels of poverty, a lack of alternative income generating activities, and low external market-access were causal to the community remaining heavily resource-dependent. Accordingly, both partner and community respondents noted effective management of the CMPA is highly linked to alignment with local priorities, most notably the ability of the community to derive benefits in the form of basic service delivery. Whilst Tiniguena respondents confirmed basic service delivery forms a major focus of their actions in the Urok Islands, limited community-perceived progress emerged, which has led to pessimism among many for the CMPA, a finding consistent with other national CBC studies (Madeira, 2015). Once again these negative perceptions were in keeping with the village-based pattern established throughout (i.e. greater negative responses from respondents within villages located further from *Abu*, notably *Uada* and *Kuian*). As one *Kuian* respondent stated, "we talk, we work, but nothing happens" (KU5). In contrast, *Abu* respondents possessed more positive perceptions than any other village regarding the value of the CMPA. Accordingly, one *Abu* respondent stated, "the CMPA comes when we are sleeping, but now we are awake!" (AB13). More specifically, many *Abu* respondents noted the CMPA has delivered benefits in the form of improved fish and shellfish stocks. Consequently, responses elude to a key 'issue arising' being a community potentially divided on the CMPA's value (refer to *section 7.2.2.2.*). Since national studies have extensively shown how influential local perceptions are for conservation outcomes in the country, especially those pertaining to the local governing power and recognition of customary institutions (e.g. Temudo, 2011, 2012; Costa et al., 2013; Casanova et al., 2014; Sousa et al., 2017), this requires future actions.

Whilst, all partner respondents suggested they believed community members did wish to eventually manage the CMPA independently, most community respondents stated they did not desire this. Furthermore, national partner organizations confirmed this 'trend' in their respective community-partnered initiatives. Accordingly, both community and partner respondents cited a lack of community capacity and high levels of dependency on external financial and technical resources as constraints to independent CMPA management by the community. Therefore, as was the case in the Bay of Ranobe, high and persistent levels of 'aid-dependency' were emphasized as a key and persistent issue emerging from community responses. This is perhaps best expressed by two community respondents who stated that, "if we didn't need [partner organizations] then other projects would not come" (KA12), and "we need more projects, then we can survive here" (KA3). Consequently, the aforementioned notion of 'NGO-ization of aid' observed in the country (*Cf.* Baldursdóttir et al., 2018), appears relevant within the Urok Islands, and requires attention.

In addition to concerns over local leadership of *UMC* representatives, and community involvement in CMPA decision-making, established above, an additional 'issue arising' is the perceived 'condition' of community-partner relations. All partner organizations described their role as facilitators in the CBC implementation and governance process, and in the case of the CMPA consistently stipulated decision-making power resided with all CMPA actors through *UMC* participatory and deliberative processes, notably the *UMC* annual general assembly. However, community perceptions of relations with partner organizations and their projects were mixed, with negative perceptions – largely depicting the village-based pattern described throughout – confirmed to originate largely from a perceived failure to deliver community-prioritised basic services (*Figure 7.11.*). Many community respondents specifically commented on a lack of local job creation through conservation and development projects, as one stated, "Tiniguena brings people with knowledge to work, not to teach and employ locals" (KA7). Nevertheless, whilst divided, community-perceived relations with partners emerged largely positive (*Figure 7.11.*). This positive relation perhaps best exemplified by one *Kabinhate* respondent who stated, "Tiniguena are from the same house" (KA6). Moreover, all national partner organizations suggested community relations within their partnered-projects were generally good, however, again reiterated that this was almost exclusively subject to delivering tangible benefits. Accordingly, one of these partner respondents suggested that many partner organizations focus on conservation and *not* development, and suggested specifically within the context of the Bijagós Islands, that without the latter you will never achieve the former.

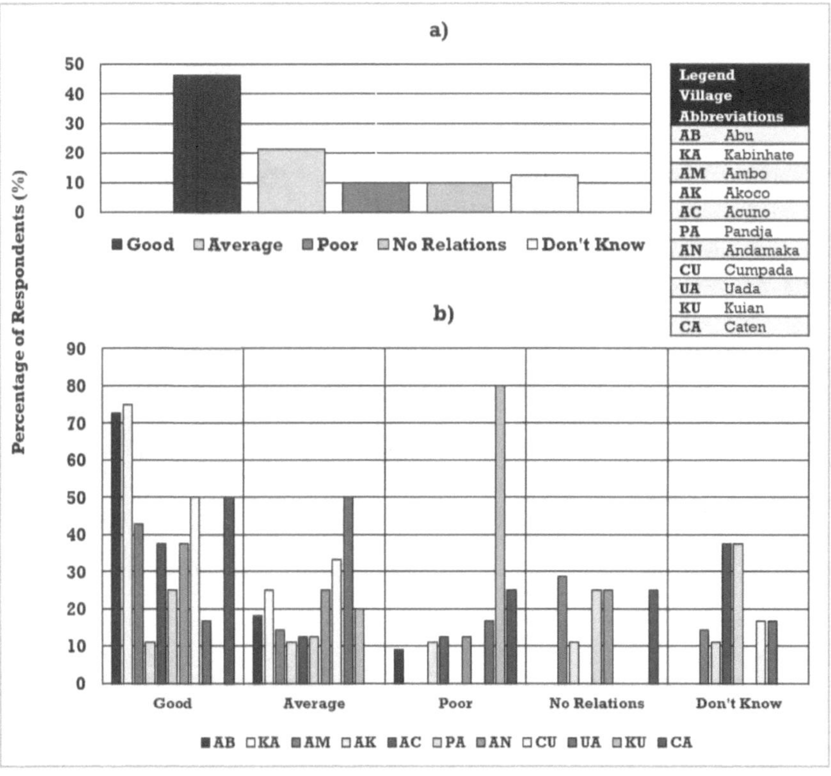

Figure 7.11.: Community perceived relations with CMPA partner organizations by percentage of respondents **a)** overall, and **b)** by village. Note: Villages are ranked by distance from Abu on the left to Caten the village located the furthest from Abu on the right. See legend for village abbreviations.

Further to community-partner relations, the community-perceived decision-making power over the CMPA (i.e. the *Ultimate Decision-Making Power*), was commonly cited as Tiniguena, as depicted by *Figure 7.12*. Furthermore, it is worth noting that few community responses perceived *Ultimate Decision-Making Power* to reside with local leaders, inclusive of *VMC* and *UMC* representatives. Numerous community responses even emphasized community-perceived project-capture by *Tiniguena*, and a lack of community-perceived ownership of local conservation and development projects. Once again these responses were largely consistent with the aforementioned village-based trend. However, some community respondents (notably again from 'non-Abu' villages) could not (or would not) answer this question (i.e. the *Don't Know* responses in *Figure 7.12.*). Lastly, whilst most community respondents noted *all community members* (CM) should ideally be the decision-making power across all management activities, especially noting the distribution of benefits, most confirmed this was not the present reality (*Figure 7.12.*). Yet, local incapacity was commonly noted as an 'issue arising' that is causal to this.

Therefore, like Madagascar, the key 'issues arising' identified here require reformulated actions focused on alleviating poverty, and strengthening actor relations and the institutional capacity of local representatives. Lastly, it is noteworthy that while the *Bay of Ranobe* case (in *Chapter 6*) noted high levels of market-articulation may be exacerbating natural resource degradation, it emerged from respondents that a lack of market-articulation in the Urok Islands appears to be exacerbating natural resource harvesting patterns, and decreasing community support for the CMPA. This aspect requires further research attention.

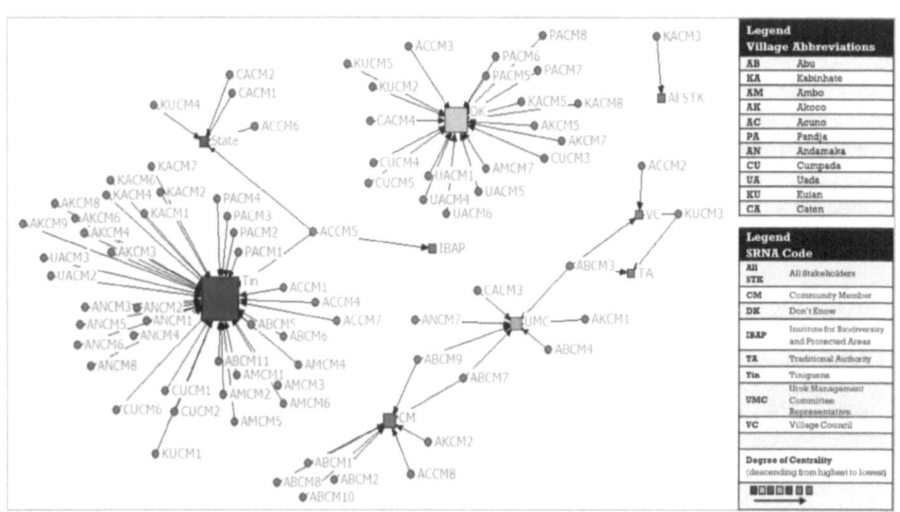

Figure 7.12.: A social network map depicting actors perceived by community respondents to have the 'Ultimate Decision-Making Power' related to the Urok CMPA. **Note:** Degree of centrality is indicated by size (the bigger the square icon the higher the centrality, i.e. the more powerful perceived actor) and colour (see legend). See legends for village abbreviations and SRNA code.

296

7.3. Conclusion

A critical exploration of CBC in Guinea-Bissau at both a national-level (through national partner organization interviews) and at the local-level through the Urok Islands case study, has identified key *contextual change triggers,* and the *enablers, actions,* and *external influences* and 'issues arising' regarding facilitating CBC initiation, implementation and governance. Perhaps the most notable contextual change trigger – as observed in Madagascar (*Chapter 6*) – has been the presence, capacity and will of partners to facilitate CBC implementation and governance within the context of a weak State, characterized by instability and incapacity, most notably pertaining to monitoring and enforcement. However, whilst a weak State together, with a lack of State funding, has proved problematic for CBC implementation and governance in the country, it has also enabled NGOs and donors to positively influence CBC initiatives. Further to State incapacity, local governance issues emerged, notably community-perceived elite-capture of knowledge-diffusion and benefit-distribution by local representatives. Furthermore, whilst potentially beneficial customary institutions are still present (especially with regards to coastal forest resources in the Urok Islands) these may be considered partially eroded due to poverty, in-migration of outsiders, and the introduction of external partners, the latter having resulted in local 'aid-dependency'.

Key actions in the present case include the creation of IBAP and an MPA expansion agenda, which has promoted participatory governance legally recognized by enabling CBC legislation. This is perhaps largely a result of concern for valuable natural resources. Moreover, Tiniguena's presence and relations with the community have been built through long-tern presence and many community development projects, which also emerged as an enabling action.

Facilitating a 'shift' to a CBC mode of governance was shown in the present context to require strong alignment with local priorities. Accordingly, key CBC enablers requiring attention include perhaps most notably the need to: address levels of poverty, natural resource dependence and external market-articulation; develop strong local leadership; and increase locally devised access and management rules, for more equitable local governance. Moreover, findings emphasize the continued need for high levels of not only initial but also ongoing 'nested' external support that is better able to harmonise local community and conservation practices and objectives. *Figure 7.13.* presents an infographic summary of the key findings of this chapter.

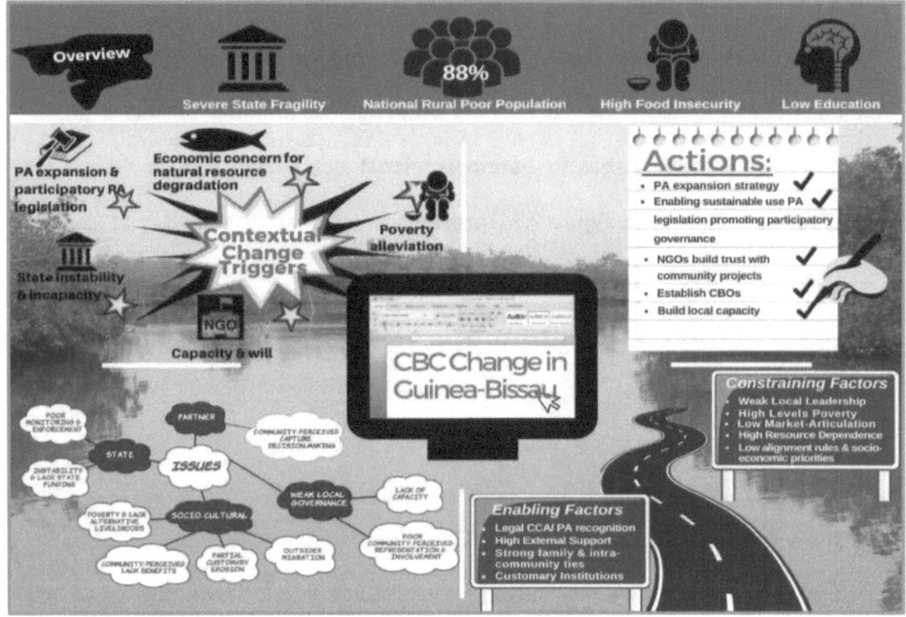

Figure 7.13.: An infographic summary of key CBC change process findings in Guinea-Bissau.

299

Chapter 8

A South African coastal-marine CCA implementation 'case-in-progress':

The case of the Olifants Estuary

8.1. Introduction and Context

This chapter presents a South Africa coastal CCA implementation 'case-in-progress'. Since it is of specific relevance to the present case study, a brief introduction to estuarine management in the country is provided. This is followed by a brief description of the case study context. Thereafter, the findings are presented in accordance with the change elements established in *Chapter 5*, in order to address **objective 5** (*Box 8.1.*). As with the previous two case-study chapters, these findings are revisited in the final discussion (*Chapter 9*).

Box 8.1.:

> **Objective 5:** To explore a South African coastal CBC case-in-progress, to better understand the factors, conditions and processes enabling and constraining its implementation

8.1.1. Introduction

South Africa possesses the continents second longest coastline, and 291 functional estuaries that provide important nursery areas for many fish and invertebrate species (Whitfield, 2016; Paterson, 2018b; Wepener & Degger, 2019). The country's estuaries were recently estimated to be worth R803 million per year[7], including an estimated R35.7 million per year[8] for subsistence estuarine and coastal harvesting (Turpie et al., 2017). However, this is considered merely 58% of the potential value of these estuaries were they in their "natural condition"

(Turpie et al., 2017). Accordingly, merely 28% of the country's estuaries remain in a completely "natural state", with approximately 43% considered *threatened* (Veldkornet et al., 2015; Paterson, 2018b). Recent national estuarine research specifically reveals the negative impact of increasing urban encroachment (e.g. Veldkornet et al., 2015), and emphasizes that estuaries found in 'more' rural areas are often in a more "natural state" (e.g. van Niekerk et al., 2019). Notwithstanding the above estimates, about 59% of the country's estuaries remain subject to no formal protection (Paterson, 2018b).

Over the past two decades many laws have been introduced that either directly or indirectly influence estuarine protection, however, implementation has faced several challenges (see Paterson, 2018b). This is because estuarine management is highly complex and subjected to marine, riverine and terrestrial influences and legislation, which requires collaboration amongst a myriad of State departments and other actors. Historically the *Department of Water Affairs* managed estuaries under the *National Water Act of 1998*, however, this legislation does not address the management of the water resource once fresh and ocean water mix (de Villers, 2016). Furthermore, whilst the erstwhile *Department of Fisheries* under the *Marine Living Resource Act 18 of 1998* managed marine species, this legislation did not include provisions for the management of estuarine conditions. Additional complexity stems from biodiversity conservation legislation and mandates as managed by the *Department of Environmental Affairs* (DEA), under the *National Environmental Management Act of 1998* (NEMA), the *National Environmental Management: Biodiversity Act of 2004* (NEMBA) and the *National Environmental*

Management: Protected Areas Act of 2003 (NEMPAA), as discussed previously in *Chapter 4: section 4.1.3.3.*

Therefore, whilst an increasingly recognized steady decline of national estuary conditions necessitated improved and integrated estuarine management, no clear legislation or lead State department for estuarine management existed in the past (de Villers, 2016). Consequently, efforts to achieve effective integrated estuarine management led to provisions in the *National Environmental Management: Integrated Coastal Management Act of 2008* (NEM: ICMA - introduced in *Chapter 4: section 4.1.3.3*). More recently, the *National Estuarine Management Protocol* (NEMP) was published under NEM: ICMA in 2013. NEMP states, "Human impact activities need to be regulated and managed for estuaries to be adequately conserved and sustainably utilised" (*Supra* section 1). This is achieved in principle through the development of *Estuarine Management Plans* (EMPs), which are managed by either local or district municipalities or other State departments, dependent on the estuary's spatial distribution or presence within a PA (*Supra* section 5). Furthermore, local government is required under the *Local Government: Municipal Systems Act of 2000* to develop and promote *Integrated Development Plans* (IDPs), inclusive of EMP considerations. Moreover, local government is specifically required to promote community participation and liaise with other spheres of government (i.e. national, provincial and local) concerning estuarine management (NEMP: section 5.). Nevertheless, much confusion persists over which level of government possesses authority within estuarine management (see Paterson, 2018b).

NEMP stipulates the requirements of an EMP and proposes a variety of options for implementation, addressing at minimum (among others): the conservation and utilization of living resources; social issues; management of water quality and quantity; land use and infrastructure planning and development; and compliance and enforcement (*Supra* section 7.3.). In reference to the latter, EMPs must specify intended spatial zonation of activities in the estuary, and which organs of state and relevant laws are required to implement proposed zonation (*Supra* section 6.5.). In the present case study, notable examples include the possible implementation of a no-fishing zone by the *Department of Agriculture, Forestry and Fisheries* (DAFF), or declaration of a PA under NEMPAA by the *DEA*.

In addition to the above legislative and management issues, land-use zonation is of particular relevance to the present case study, and is closely linked to the land reform process (introduced previously in *Chapter 4: section 4.1.3.*). This process requires formation of a community-based organization (CBO), usually but not exclusively in the form of a local *Communal Property Association* (CPA) (Paterson, 2015). The CPA is responsible for designing and implementing a land-use plan. In the present case, this land-use plan would include land surrounding the *Olifants Estuary,* a section of which is to be included within a proposed *Community-Conserved Area* (CCA). Consequently, the land claim process adds further complexity to the EMP development process, and subsequently the declaration of a CCA.

Therefore, EMPs require high levels of collaboration across the three spheres of government (i.e. national, provincial and local), which to date has proven problematic (e.g. Paterson, 2018b). Furthermore, development, implementation and co-ordination of EMPs should actively engage all relevant stakeholders (NEMP: section 8), and (in accordance with NEM: ICMA) once implemented contribute to co-operative coastal governance (NEMP: section 11). Consequently, estuarine management is highly complex and requires collaboration amongst a myriad of State departments and other actors. Notwithstanding this complexity, concerns for degrading estuary conditions and a lack of formal protection to date suggest that PA status for estuaries of high biodiversity value is required, and that the appropriate legal mechanism including declaration of CCAs. An introduction to the social-ecological system of *Olifants Estuary* follows.

8.1.2. Case Study Context: The Olifants Estuary
8.1.2.1. The Ecological System

The Olifants estuary is one of South Africa's largest, and one of only four permanently open estuaries on the west coast (Turpie et al., 2002). It represents a unique and productive ecosystem, and has been ranked the second most important South African estuary from a conservation perspective (Turpie et al., 2002; van Niekerk et al., 2017). The estuarine system stretches approximately 36 km upstream from its mouth to a low water causeway, and drains the Olifants-Doring Catchment (i.e. one of the largest catchments in the country). However, the Clanwilliam and Bulshoek dams regulate water flow in this catchment (Brown et al., 2010; SAR, 2017).

The ecological system provides a habitat for many fish, invertebrates and bird species (SAR, 2017; BirdLife International, 2019), and has long been identified as an estuary requiring PA status (Turpie et al., 2002). Due to low annual rainfall, the area mainly comprises dwarf succulent and species-rich scrubland vegetation (Veldkornet et al., 2015). However, it is also home to the country's largest supratidal and floodplain salt marshes (Turpie et al., 2002; SAR, 2017). The estuary experiences a summer marine-dominated state for about 6 months of the year (i.e. November to April), as saline water extends further upstream, followed by a freshwater-dominated state during the remaining winter months, though this is subject to precipitation levels and the frequency with which sluice gates are opened upstream (SAR, 2017). Increasing agricultural irrigation demands upstream have led to deterioration of water quality, and weed proliferation in the upper reaches, gradually altering flora and fauna (Brown et al., 2010; SAR, 2017).

8.1.2.2. The Social System
<u>An overview</u>

The *Olifants Estuary* is located approximately 350 km north of Cape Town, on the South African west coast (refer to *Chapter 2: Figure 2.5.*). The area remains relatively undeveloped with a rural poor population characterized by low education and high unemployment levels (EcoAfrica, 2013; Williams, 2013; Sowman, 2017). The local *Ebenhaeser* community comprises approximately 1200 households found within *Ebenhaeser* (inclusive of the settlements/ villages of *Nuwestasie*, *Nuwepos*, and *Olifantsdrif*) and *Papendorp*, though collectively the community is considered the Ebenhaeser community (Williams, 2013). An estimated 120 of these households derive a livelihood from fishing, the remainder are largely reliant upon State social grants and *ad-hoc* employment, for example within the agricultural or road works sectors (Williams, 2013; Matzikama, 2016).

Ebenhaeser was originally founded as a mission station by the Rhenish Missionary Society in 1834, and is one of the oldest settlements in the district (Matzikama, 2016). The community is descendant of indigenous *Khoi-San* groups – a unifying name for two historical ethnic groups of southern Africa (i.e. the hunter-gatherer

San and the pastoral *Khoi*) that settled in the area in the seventeenth century (Sowman, 2011; Williams, 2013). However, in 1926, in an act of racial discrimination the community was forcibly removed from their farmlands upstream, and relocated to lands adjacent to the estuary (Williams, 2013). Due to poor soils and lack of fresh water (for drinking and irrigation), many community members became and remain almost exclusively reliant on fishing for their livelihoods (Sowman, 2003, 2009). Whilst the current community consists mainly of farmers, fishing activities have been described as part of the "cultural landscape" (Matzikama, 2016: p6-7). This is most notable in *Olifantsdrif* and *Papendorp*. Consequently, considering high levels of unemployment and food insecurity within the area, the *Olifants Estuary* remains vital to local socio-economic livelihoods and well-being (SAR, 2017; Sowman, 2017).

As introduced previously (in *Chapter 4*), and above, the issue of land claims is central to the implementation of the proposed CCA. At 23,755 ha, this represents the provinces largest land reform project, and was identified as a rural "flagship" project in 2009 by the erstwhile minister of land reform (Schreiber, 2017). The Ebenhaeser community initiated a land claim in 1996 for previously dispossessed land under the apartheid regime (EcoAfrica, 2013). The community was assisted in the complex land claim application process, under the *Restitution of Land Rights Act of 1994,* by the *Legal Resource Centre* (LRC), a public interest and human rights legal NGO. An initial offer of an R 20 million cash settlement by the State in 1998 was rejected by the community to pursue land ownership (Schreiber, 2017). However, the land claim process stalled, initially over a lack of community consensus regarding who the rightful beneficiaries were, and subsequently over specific land ownership, and an approved land-use plan (Schreiber, 2017). Since apartheid, the area has become a valuable commercial wine-grape producing region, and therefore, the land claim required these commercial farmers to sell their land to the State under its "willing buyer, willing seller" policy (Schreiber, 2017). However, initially only half these farmers were 'willing sellers', with the remainder opting to challenge the ruling in court, thus further complicating and delaying the land settlement (Schreiber, 2017).

The community formed a CPA, and later a Land Trust to manage its finances, and was required to produce business/ land-use plans, which was assisted by *Phuhlisani Solutions*, and *EcoAfrica Environmental Consultants* (Schreiber, 2017). By the end of 2014, the community had signed a land settlement agreement with the erstwhile *Minister of Rural Development and Land Affairs*, which was subsequently concluded by the hand-over of title deeds in March 2019 (Cape Times, 2017a&b, 2019; RSA, 2019a). Consequently, there is a need for greater collaboration amongst diverse stakeholders in the land claim process, including various national and provincial State departments and local government, commercial farmers, and the Ebenhaeser community with support from other partners.

<u>The Fishery</u>
Much research on the local gillnet fishery, and the Olifants estuary more generally, has been conducted including the topics of fisher's local ecological knowledge (Hushlak, 2012; Rice, 2015), co-management and community-based monitoring (Sowman, 2003; Carvalho et al., 2009; Soutschka, 2014), the issue of bycatch (Rice et al., 2017), and land and natural resource *rights* (Williams, 2013). The latter including issues of 'coastal grabbing' of land and resources, specifically

concerning recent mining interests near the *Olifants Estuary* (Bavinck et al., 2017 - see also GroundUp, 2018).

Local fishers mainly fish at night using rowboats and gillnets and are most active during the summer period between November and April (Rice et al., 2017). The main target species is *Liza richardsonii* (i.e. *southern mullet*, known locally as *harder)* but there is also an incidental catch or "bycatch" comprising a few commercially valuable line-fish species, most notably elf (*Pomatomus saltatrix*), silver kob (*Argyrosomus inodorus*) and white stumpnose (*Rhabdosargus globiceps*). Fishers are prohibited to catch or sell any bycatch species by law, but acknowledge consuming them (Rice et al., 2017).

Fisheries Management

The gillnet fishery comprises 45 legal permit holders, with an additional crew member allowed on each boat, resulting in a total of 90 legal fishers (Rice et al., 2017). Furthermore, since 2005, 30 Interim Relief Marine Permits have been issued to local fishers to fish at sea, which may have reduced fishing effort, although marine fishers are known to fish in the estuary and river when conditions prevent them going to sea (Sowman, 2017). Mesh size, net length and prohibited harvesting or retention of any bycatch species are regulated (Rice et al., 2017). Furthermore, there is currently a no-take fishing area at the mouth of the estuary which is recognized under a provincial ordinance between local fishers, *Cape Nature* (i.e. the parastatal conservation agency), and other stakeholders, which is largely supported by fishers (Sowman, 2003, 2009).

Research shows sustained pressure on estuarine-dependent fish resources may be gradually reducing and changing the nature of fish stocks (SAR, 2017).

Furthermore, specific concerns exist over juvenile-capture (i.e. individuals not having reached sexual maturity), and the status of certain commercially valuable line-fish species caught as bycatch species, which has led to several conservation and fisheries scientists proposing the closure of the gillnet fishery in all west coast estuaries (including the Olifants estuary) (Rice et al., 2017). Attempted closures have strained relations between the fishing community, and fishery scientists and management authorities (Sowman, 2017). However, these closure efforts have been unsuccessful thus far due to opposition from fishers, and support from their *social partners*, namely: The *Environmental Evaluation Unit* (EEU) at the *University of Cape Town* (UCT), and two NGOs in the form of *Masifundise Development Trust* (MDT) and the *Legal Resource Centre* (LRC). These social partners have sought recognition of the community's cultural and socio-economic rights to estuarine resources (Jackson et al., 2013; Sowman, 2017). The aforementioned contention over closure attempts necessitates an alternative approach to managing these resources. Accordingly, the *National Small-Scale Fisheries Policy* (SSFP) promulgated in 2012 (see *Chapter 4: section 4.1.3.3.*), recognizes small-scale fisher's rights and promotes their participation in management (Sowman et al. 2014).

Since 1996, several initiatives have involved local fishers in conservation efforts. These include the establishment of a community-based monitoring programme, and developing a co-management arrangement (Sowman, 2003; 2009; Carvalho et al., 2009; Soutschka, 2014). Furthermore, fishers were involved in the seven-year EMP development process in collaboration with their social partners (Jackson et al., 2013; Sowman, 2017). Likewise, the proposed CCA, which arose out of the EMP development process, was driven by fishers with support from their social partners and *Cape Nature* (i.e. the provincial parastatal conservation agency) (Jackson et al., 2013). The Olifants Fisher Committee (OFC) represents local community interests in both the EMP and proposed CCA. Moreover, once the land claim was settled it became necessary to get the support of the Ebenhaeser CPA since the CCA will include riparian land that falls under the land claim. However, the land claim process has overshadowed all other community objectives and specifically slowed down the CCA declaration process. This is due to a lack of community-wide agreement required for the land-use plan. In addition, the *Olifants Estuary Management Forum* (OEMF) represents a multi-stakeholder advisory forum (though without legal decision-making power) associated with the planning and implementation of the Olifants EMP, proposed CCA, and other community development projects. In addition to *Cape Nature*, numerous departments represent the State across the various management activities associated with the EMP and CCA, including local government (i.e. the *Matzikama Local Municipality* and the *West Coast District Municipality*), multiple national and provincial State departments including *DEA, Department of Water and Sanitation, DAFF,* and the *Department of Rural Development and Land Reform*, as well as the *Western Cape Provincial Department of Environmental Affairs and Development Planning* (DEADP)(see Jackson et al., 2013).

8.2. Findings

Calls for the establishment of the CCA commenced in 2013-2014, and were supported by the fishers and most stakeholders, but lengthy delays have 'derailed' the process. Therefore, it is useful to explore *contextual change triggers*, *enablers*, *actions* taken thus far, as well as the *external influences*, and 'issues arising' to gain a better understanding of *how* to promote progress within this CBC change process, and prevent other CCA implementation attempts being stalled in the future.

8.2.1. Contextual Change Triggers

An overarching local-level contextual issue, as confirmed by all respondents and discussed in the two regional cases, is high levels of poverty. Furthermore, all fishers specifically confirmed a lack of alternative livelihoods due to low levels of education and employment. This has implications for the land claim and proposed CCA, for as one community respondent states, "a lack of education also affects the understanding of community for the CCA and land use, especially short term versus long term gains" (OD21).

As described above local fishers initiated the CCA implementation process, with support from their social partners, which originated from the EMP development process. Therefore, the EMP development process can be considered the preliminary CCA contextual change trigger. Respondents acknowledged this stemmed from another major contextual change trigger, namely fisher's recognition of decreased catches, and a desire to protect the ecological system for future generations. As a partner respondent noted, "[fishers] realised [the need to protect the environment] since the late 80's" (OPO9). In addition, numerous partner and community respondents highlighted concerns for the negative ecological impacts of recreational (e.g. camping, dune-vehicles, fishing and ski-boat use), and extractive commercial activities (e.g. mining), as a contextual change trigger. Consequently, these ecological concerns led to fisher's support for a no-take area at the mouth of the river, which represents another initial 'trigger' that laid the groundwork for the potential finalization of both the EMP, and subsequently the CCA. Lastly, the CCA implementation process is affected by the land claim settlement, which therefore, presented a potential enabling action; however, this process has lost momentum due to both a lack of community consensus, and the State dragging their feet. Consequently, the land claim process represents an initial contextual issue for the proposed CCA.

8.2.2. Review of Enablers

As with other case study chapters this section analyses the extent to which the 14 enablers identified in *Chapter 3* are present in the Olifants Estuary case study. Since the CCA is yet to be proclaimed this list is incomplete and constantly 'evolving'. A summary of the *absence, partial presence, partial absence,* or *presence* of these enablers can be found in *Figure 8.1.* Furthermore, *Table 8.1.* provides a brief overview of the findings that influenced these enabler presence ratings. These findings are discussed further below.

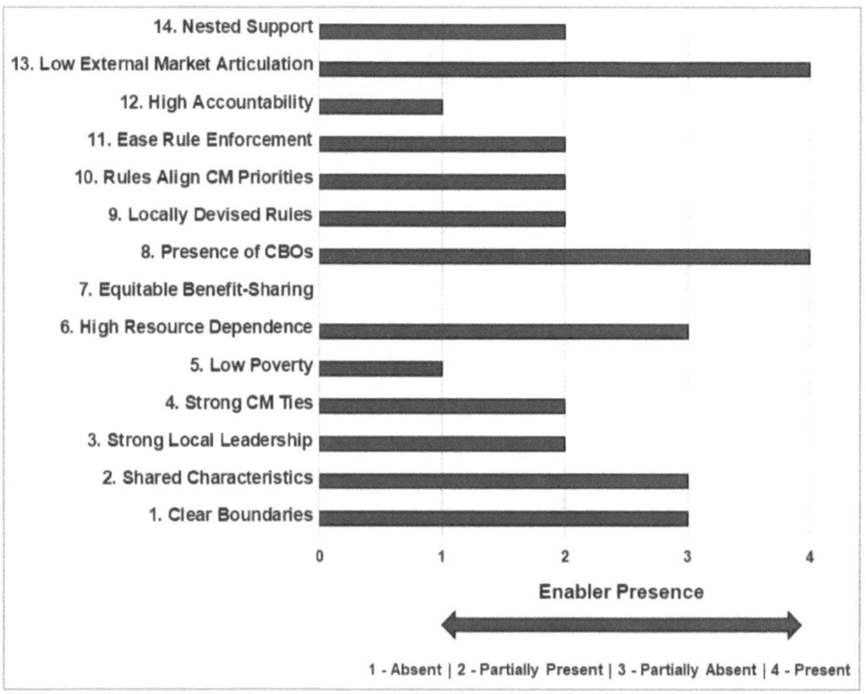

Figure 8.1.: A graphical summary of the presence of the 14 enablers in the Olifants Estuary. **Note:** CM – Local Community.

	Enabler	Presence	Explanation
Resource System & Users	1. Clearly-defined resource system & user boundaries	**Partially Absent**	Established no-take area near at the mouth marked by a beacon and well known to the fishers. CCA boundaries of the proposed have been broadly agreed to, but have not been finalised due to on-going procedural issues. The geographical resource-user boundaries are clearly defined. While fishery still important, is heterogenous community.
	2. Shared norms, values, interests & identities	**Partially Absent**	Long history of customary fishery with shared norms and values regarding the importance of the estuary and river for food, livelihood and culture. However, heterogeneous community with diverse interests and many now seek alternative livelihoods (outside fishing) or survive on social grants (elders). A decrease in the elder fisher population, has decreased the strength of social norms in the community.
	3. Strong local leadership	**Partially Present**	Community respondents identified issues with local leaders inclusive of the OFC and CPA associations. A change in leadership and inaction has weakened the initially strong OFC. Strong community perception of elite-capture of knowledge and highly vested interests in both the CPA and present OFC. Amongst fishers, there is still a culture of sharing and providing for each other, especially the elderly.
	4. Strong community ties	**Partially Present**	Strong family and in intra-village ties. However, weaker inter-village ties, notably between predominantly fisher (i.e. Olifantsdrif and Papendorp) and non-fisher villages (Nuwestasie & Nuwepos). Furthermore, questions over the strength of ties between the two fisher villages. Ties also affected by poor community-perceived CBO representation.
	5. Low levels of poverty	**Absent**	High levels of poverty emphasized by all community members.
	6. High levels of dependence on resources	**Partially Absent**	High levels of dependence amongst fishers, although highly heterogeneous community with diverse livelihood strategies.
	7. Equitable distribution of benefits from common property resources	**N/A**	No benefit structure in place yet, such as a local co-operative. High fisher-perceived potential benefits to fish catches due proposed protection of fish nursery area at the mouth,
Institutional Structure & Externalities	8. Presence of community institutions	**Present**	Established OFC and CPA committee, the latter legally recognized, yet numerous representational issues.
	9. Locally devised access and management rules	**Partially Present**	Initial community-involvement in decision-making meetings was relatively high, however, has declined with the protracted process and no visible outcomes or progress.
	10. Rules strongly align with local priorities/ needs	**Partially Present**	Poverty alleviation is the main local priority. Fishing remains key priority and desire exists to protect this source of food and livelihood, and cultural tradition. Fishing also key to the broader community who buy their fish. Some fishers wish for access to the estuary as better perceived catches.
	11. Ease in enforcement of rules, and conflict resolution	**Partially Present**	State driven rules and enforcement, although limited State enforcement due to capacity constraints. Community respondents emphasized difficulty controlling 'outsiders' since they lack legitimacy of a legally declared PA or State approved EMP.
	12. High levels of accountability	**Absent**	Community-based monitoring system collapsed in recent years and is no longer operational. Community perceived accountability of local representatives, and especially absent State officials, is very low. Both characterized by limited feedback.
	13. Low levels of articulation with external markets	**Present**	Only allowed to sell one target species. Lack of market-access due to isolated location. Limited access to buyers such as local farmers and other community members.
	14. Presence of 'nested' governance with high levels of initial external support	**Partially Present**	Numerous organizations have been involved with, and supported initial planning phases of the proposed CCA. However, since then support from the State has been sporadic and inefficient. Furthermore, long delays with institutional procedures have led to disillusionment. Continued active support is limited to a select few individuals from local government, researchers and the private sector.

<u>8.2.2.1. Clearly-defined resource system & user boundaries:</u>

The established no-take area located near the estuary mouth possesses clear boundaries, and these are well known to the fishers. However, the boundaries of the proposed CCA, which include surrounding salt marsh and shrub-land, have not been finalized due to on-going procedural issues (as introduced previously in *section 8.1.2.2.*). Furthermore, whilst the community can be considered clearly-defined geographically, and is largely racially homogenous, belief systems differ (e.g. Christianity and Rastafarianism are two dominant belief systems). Moreover, livelihood strategies have diversified within the community over the years. As a community respondent stated:

> "Ebenhaeser is a divided community. It is different now than in the past. In the past, men wanted to work on your boat. Now they look for other options. This has led to division in the community. We are not one community. If a bakkie[9] comes with a farmer who says there's work, are you going to fish and maybe catch, or are you going to take a job?" (OD22)

However, in reference to the on-going importance of the fishery, one community respondents emphasized, "Everyone wins with fish, even those involved in agriculture" (OD2). Fisher respondents clarified this was because other community members buy fish from them, which represents a key and affordable protein source. Accordingly, community respondents – inclusive of OFC and CPA members – noted the need to 'speak' to all community interests and livelihood options. Therefore, while social-cultural boundary challenges exist, and the proposed CCA boundaries are unfinalized, due to broad agreement on these CCA

---

[9] A South African colloquial term for a light truck or pickup truck.

boundaries this enabler is considered *partially absent*. Consequently, these findings mirror socio-cultural boundary concern that emerged in the two regional cases.

<u>8.2.2.2. Shared norms, values, interests & identities</u>

While many respondents acknowledged the partial erosion of customary practices, this fishery has a long history and most fishers still possess a shared desire to live off the river. Accordingly, as one fisher stated, "Life is from the river!" (OD1). However, all community members confirmed poverty has caused fishing to be viewed by many as more a survival strategy than a customary activity, as one stated, "after all a fisher needs to put some food on the table" (OD21). Furthermore, as established, the community is not homogeneous, and possess diverse interests, with many of the younger generation especially now seeking alternative livelihoods, and many elders now resigned to surviving on social grants. Moreover, many elders have now migrated out of the community to areas with better health care. In addition, numerous elder fishers are recently deceased, especially those from *Papendorp*. Lastly, notwithstanding community divisions the vast majority of those interviewed perceived the proposed CCA positively, this not surprisingly strongest amongst the predominantly fisher population of Olifantsdrif (*Figure 8.2.*). Therefore, most respondents emphasized these factors may have decreased the strength of social norms and values. Consequently, this enabler is considered *partially absent*.

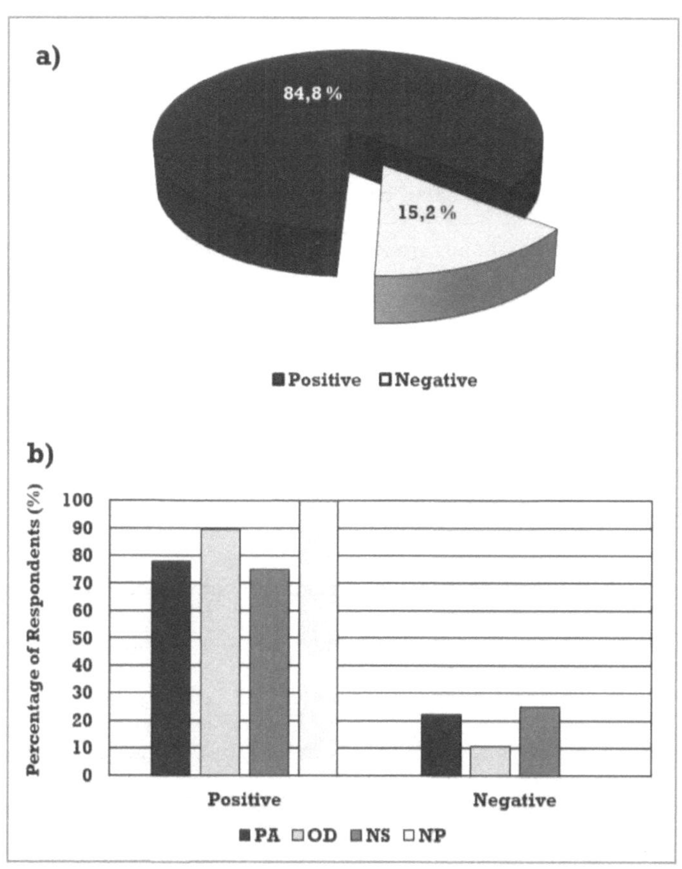

Figure 8.2.: Community perceptions regarding the proposed CCA by percentage of respondents: **a)** overall, and **b)** by village. Note: **PA** – Papendorp; **OD** – Olifantsdrif; **NS** – Nuwestasie; and **NP** – Nuwepos..

<u>8.2.2.3. Strong local leadership</u>

The topic of local leadership emerged prominently in all community interviews. Community respondents raised major concerns with local leaders, inclusive of both the OFC and CPA, notably perceptions of elite-capture. This particularly pertained to knowledge, but also perceived (potential) benefits. Accordingly, one community respondent stated that within both CBOs "there's no feedback, they make decisions, we don't know what's going on!" (OD11). All 'non-representative' community respondents described the issue of their institutional procedures being guided by the highly vested interests of local leaders (in addition to interests of partners). More specifically, some community responses specifically noted that, "poor management of money received, and land issues continue to cause problems in the community" (OD20), this since as respondent NS1 stated, "all take their chance to score [money]". This led some to specifically emphasize that, "we need to be able to trust one another with money" (NS1). Consequently, most community members described high levels of mistrust regarding representatives of both CBOs.

Almost all 'non-CPA' respondents noted the land claim process has not been well managed. Accordingly, as one community member stated the "community is unhappy; therefore, there is no attendance at CPA meetings" (OD20). Numerous partner and community respondents alike emphasized the need for a strong local leader or 'champion' to drive the land claim process. This finding in accordance with those of *Chapter 3*, and more specifically the South African enablers identified *Chapter 4*. As was the case in the Bay of Ranobe community respondents noted local leaders were often appointed to positions of power without the necessary skills. Accordingly, most community and all partner respondents noted the need for targeted local capacity building. This since as one partner respondent stated, "if [a project is] working it is because of strong leaders with knowledge" (PO6).

Therefore, as one community respondent stated, "we need to figure out who is holding back the [land claim] process within the CPA." (OD20). CPA respondents noted two reasons for the lack of progress at the time of interviews, firstly, that the title deeds to the land had not been handed over, and secondly, as mentioned throughout, conflicts between diverse community interests, which have created community division. Furthermore, one CPA respondent suggested in reference to South Africa's ruling political party (i.e. the African National Congress – ANC), and lead opposition and provincial ruling party (i.e. the Democratic Alliance – DA), "Politics plays a big role. DA people want money, but ANC people want land" (CPA3). Moreover, another CPA member noted, "the State are happy to sit back and let us fight" (CPA2). However, since these interviews were conducted title deeds to most of the land have been handed over (in March 2019), with only a few 'unwilling sellers' (i.e. commercial farmers still challenging the land ruling).

The OFC is also constantly criticised for a lack of feedback and perpetuating their own self-interests. As a fisher stated, "we had a lot of representation and it was good [in the past] and we worked well together, we need to get back to that" (OD18). Current OFC representatives responded by acknowledging that communication is *poor* due to problems assembling fishers for meetings. Furthermore, some fishers noted conflicts exist between the old and new fisher chairperson, between river and marine fishers, as well as between *Papendorp* and *Ebenhaeser* fishers. Therefore, as with the CPA, a lack of trust and progress has decreased community interest among many regarding the declaration of the proposed CCA.

Consequently, most community respondents emphasized an overwhelming desire for reduced corruption, and increased consultation with their local leaders, who they perceived to possess the only 'real' connection to partners. Unsurprisingly, community respondents largely perceived representation to be *average* to *poor*

(*Figure 8.3.*). The only moderate exception coming from Olifantsdrif, which contributes the majority of these CBO representatives, with approximately 47% of respondents describing representation as *good*. Therefore, due to overwhelming community concerns for local representation on both CBOs, this enabler is considered *partially present*. Consequently, as with the findings in the two regional cases, these findings reinforce the importance of the need for strong local leadership to enable CBC interventions, which also confirms those of previous studies (Pomeroy et al., 2001; Agrawal, 2002; Galvin et al., 2018; Biggs et al., 2019). Lastly, the related topic of decision-making power is discussed subsequently in *section 8.2.2.9* in relation to degree to which rules are devised locally.

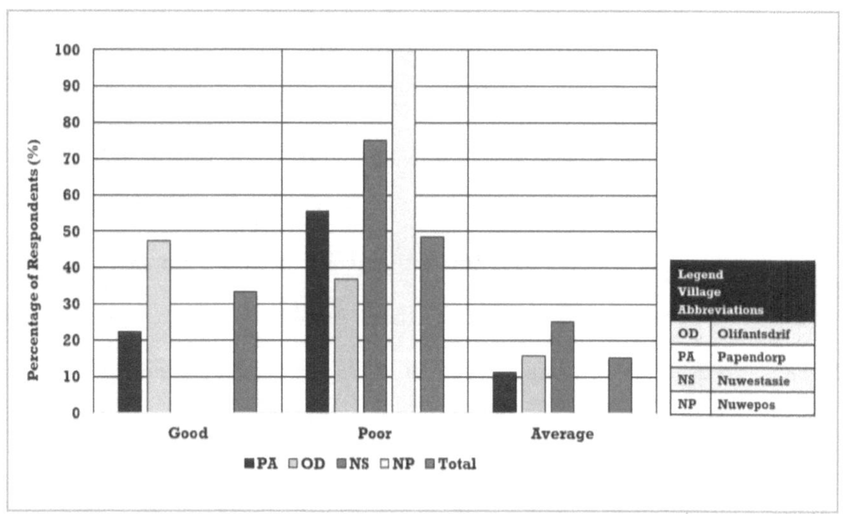

Figure 8.3.: Community perceived representation by local leaders within the *Ebenhaeser CPA* and *OFC* by percentage of respondents. See legend for village abbreviations.

<u>8.2.2.4. Strong community ties</u>

As with the previous two regional cases strong family-ties and intra-village ties were consistently identified by community respondents, which was most notable amongst the fishers in *Papendorp* and *Olifantsdrif*, and especially pertained to perceived levels of knowledge acquisition and diffusion. This finding is also consistent with recent local research conducted in the community (e.g. Hushlak, 2012; Rice et al., 2015). Moreover, it strongly resonates with the findings that emerged from the two regional cases. However, once again in accordance with the regional cases, most community respondents emphasized intra-community ties were *weak* due to the aforementioned community divisions. The main causal factors to emerge regarding this division among the fishers specifically is a perceived lack of representation of *Papendorp* fishers on the OFC. Furthermore, as one of these fishers stated, "Ebenhaeser take all the benefits" (OPO6). This was often rationalised by other respondents, most notably partners, that this was due to Ebenhaeser – which once again incorporates *Olifantsdrif, Nuwestatsie* and *Nuwepos* – being the larger settlement and therefore able to benefit a larger population. The topic of perceived equitable benefit distribution is discussed further in *section 8.2.2.7.*

Whilst few fishers come from *Nuwestatsie*, those interviewed did comment on the need for greater communication to improve relations with other fishers, in particular those in the dominant fisher settlement of *Olifantsdrif*. Likewise, this was proposed by fisher respondents regarding the need to improve relations between *Papendorp* and *Olifantsdrif* fishers. Moreover, increasing internal conflict amongst river and marine fishers was also commonly noted, especially within *Olifantsdrif* since they hold the majority of the river and marine permits, unlike *Papendorp* who are almost exclusively river fishers. This conflict largely stems from marine fishers fishing in the river when the sea does not allow it or for their own consumption. Consequently, many elder river fishers commented that, "you can't have it both ways", in terms of the right to fish in both the river and the sea. Therefore, this enabler is considered *partially present*, and requires actions to strengthen community relations.

8.2.2.5. Low levels of poverty

High levels of poverty were confirmed by all community respondents and acknowledged by partners. As one fisher stated, "If the river has no fish, then what?" (OD12). Furthermore, all respondents confirmed poverty and unemployment were a primary motivating factor for non-compliance within the established no-take zone, and potentially within the proposed CCA. Moreover, community members note that local CBO-representative elite-capture is largely due to poverty (*section 8.2.2.3.*). Therefore, as with the regional case studies, high levels of poverty were shown in this case to be (potentially) undermining efforts to pursue CBC, since putting food on the table is the priority. Accordingly, as one partner suggested there is a need to change community perceptions away from "thinking of today [and] not tomorrow" (PO6), especially as it relates to conservation. Consequently, not surprisingly within a developing nation context, and again in accordance with the two regional cases, this enabler is considered *absent*.

8.2.2.6. High levels of resource-dependence

As established community livelihood strategies are diverse, yet many fishers still rely heavily on the fishery and limited alternative livelihoods exist, or when they do these are seasonal or sporadic. Accordingly, community respondents noted "unemployment is high, so we need fishing" (NP1), and "there is only fish, it's our source of income, if we don't have it we starve" (OD10). However, as mentioned previously many do engage in *ad-hoc* employment. Interviews conducted in *Nuwestasie* specifically confirmed many of their members in particular engage in alternative livelihoods or relied solely of social grants (the latter also notably emphasized in *Papendorp*). However, even amongst fisher families a growing realization has emerged that fishing may not be a sustainable long-term livelihood. Lastly, numerous partners and community respondents alike consistently proposed ecotourism initiatives as potential long-term livelihood strategy. Therefore, based upon the findings discussed throughout high levels of resource-dependency are considered *partially absent*.

8.2.2.7. Equitable distribution of benefits from common property resources

As mentioned previously in *section 8.2.2.3.*, several respondents raised concerns about the equitable management of potential benefits by the two CBOs. Once again, community frustration at a lack of progress in deriving benefits strongly emerged. Accordingly, many community respondents commented for example that, "we need outcomes to motivate participation and action" (PA2), and that, "the challenge is keeping interest, we need to see progress to maintain interest" (NS4). Furthermore, some community respondents – notably from outside of *Olifantsdrif* – did note inequality within the OFC, notably associated with the allocation of highly sought after marine fisher permits. Nevertheless, since both the land claim and CCA processes are still incomplete it is difficult to further comment on the presence of this enabler.

<u>8.2.2.8. The presence of community institutions</u>

The CPA and OFC represent community interests concerning the EMP and land claim process, and subsequently, the proposed CCA declaration. However, as discussed throughout weak local governance exists. Consequently, whilst this enabler is considered *present*, as discussed throughout community representation and capacity related to these CBOs needs to be strengthened.

<u>8.2.2.9. Locally devised access and management rules</u>

A main constraint identified by most respondents relates to *who* has authority. Community-perceived involvement in management activities associated with both the land claim and CCA is *low* (*Figure 8.4.*). As with the regional case studies, many community members reasoned that while their local representatives were more involved, they were not, since as established above representation is perceived to be *poor* (refer to *Figure 8.3.* previously). Therefore, community respondents emphasized the need for increased participation beyond that of their representatives, noting that this is necessary to avoid only representatives' interests and opinions being represented in decision-making.

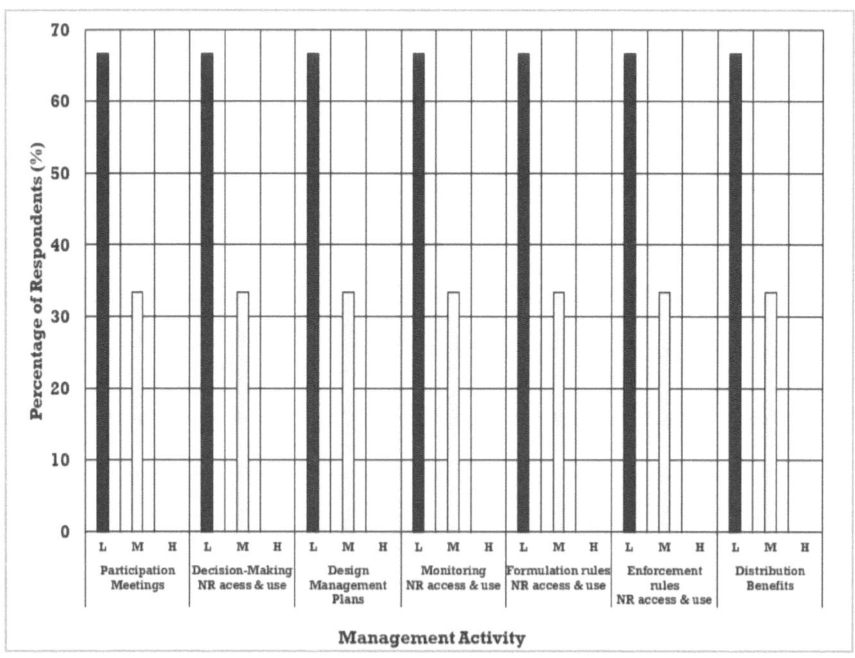

Figure 8.4.: Community perceived involvement in land claim and CCA implementation processes by percentage of respondents.

However, the topic of community participation in decision-making in this case study highlights further community divisions. Accordingly, while fishers perceived higher involvement in the EMP development process, and initial CCA talks (but not decisions), they perceived lower involvement in the land claim process. The inverse appears to be true regarding CPA meetings, with higher perceived involvement by non-fishers and the land claim process. Furthermore, based on respondents, the minutes of OEMF meetings, and personal attendance, community attendance has decline drastically at OEMF meetings. This is especially noteworthy since 2014 when the EMP and CCA processes had the most momentum and support. Moreover, even the attendance of community representatives at both

CPA and OFC meetings has decreased. All community respondents noted a lack of visible progress, and resultant community frustrations, as well as strained relations between community members and their various representatives, as causal to these attendance levels. Consequently, this lack of progress is a similar finding to that of the *Urok Islands* case. Lastly, this has led to a lack of trust and ultimately consensus within the community related to both the land-use zones associated with the land claim, as well as proposed regulations of the CCA.

Numerous community responses emphasized the desire to manage the CCA independently. For example, community responses noted, "we can, we already do, we won't not protect our resource" (OD22). Accordingly, community responses perceived the *Ultimate Decision-Making Power* to reside with the community and especially the fishers (*Figure 8.5.*). However, most of these respondents reasoned that this is how it *should be*, but is not the *de facto* situation. Accordingly, as established above regarding local leadership, a community respondent stated, "that would be best, but leadership must improve" (NS1). Therefore, even those that responded positively stated the community currently lacked the power, legitimacy and management capacity. This largely due to low levels of education. Consequently, most community respondents acknowledged a need for State support, specifically regarding monitoring and enforcement, since they suggested this would increase the legitimacy of proposed CCA. Accordingly, a community respondent stated, "we need laws and the State, they have the power to legitimise [the CCA]" (OD1). Finally, a partner respondent suggested that whilst "the community needs to get to the point of managing the [CCA] themselves", they specifically noted this must happen before support and funding lapse, otherwise the initiative will inevitably collapse (PO6). This in accordance with findings from *Chapter 4* related to South African CBC initiatives in general. Therefore, due to the need for greater community participation, this enabler is considered only *partially present*. Consequently, these findings echo those in the two regional cases, and confirm the importance of greater and more inclusive community participation in decision-making, as emphasized by extensive literature (as discussed in *Chapter 3*).

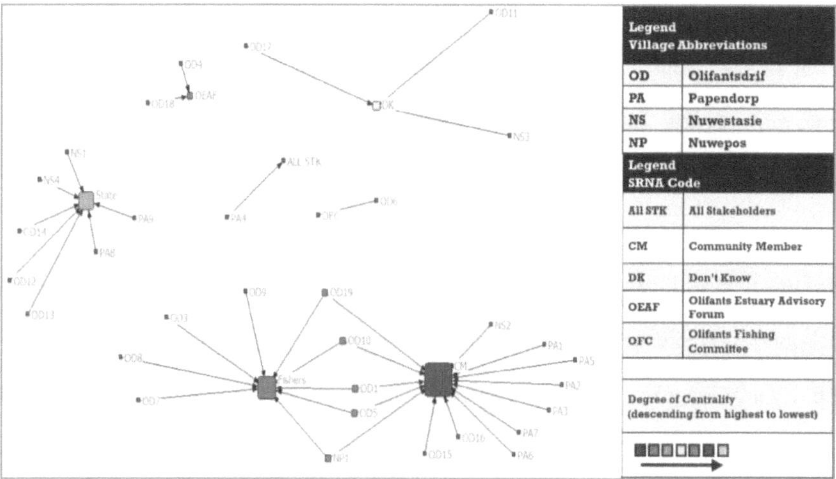

Figure 8.5.: A social network map depicting actors perceived by community respondents to have the 'Ultimate Decision-Making Power' related to the proposed Olifants Estuary CCA. **Note:** Degree of centrality is indicated by size (the bigger the square icon the higher the centrality, i.e. the more powerful perceived actor) and colour (see legend). See legends for village abbreviations and SRNA code.

8.2.2.10. Rules strongly align with local priorities/ needs

All community respondents emphasized the need for greater alignment of CBC initiatives, and local community development projects in general, with local priorities. However, as all respondents also emphasized these priorities are

328

diverse, especially as it pertains to the land use associated with claimed land (as discussed throughout). It can be inferred from fisher responses specifically that this required alignment is two-fold. Firstly, in relation to the cultural connection to fishing and the river, and secondly, the ability to alleviate poverty. As established previously, notwithstanding acknowledged partial customary erosion, fishers identified a strong cultural connection to the river and fishing, and supported the proposed CCA in a desire to conserve their local resources. Furthermore, whilst interviews conducted, and past research (e.g. Soutschka, 2014; Rice, 2015; Sowman, 2017), emphasize that community members possess a wealth of knowledge about conserving their natural resources and realise the need to change, poverty dictates conservation is not always locally prioritised (as discussed in *section 8.2.2.5.*). Once again, this reflects findings from the two regional cases. Since the land claim and CCA is unsettled, and rules are yet to be finalised, it is difficult to comment further on this enabler. However, the majority of respondents considered alignment of community projects with local priorities poorly implemented in practice, and emphasized this is something the State needs to address. Nevertheless, as widespread agreement exists for present regulations concerning the no-take zone, proposed CCA boundaries, this enabler is considered *partially present*, although further research will be required once the CCA is implemented and regulations are established.

8.2.2.11. Ease in enforcement of rules, and conflict resolution

As introduced above in *section 8.2.2.9*, and as a partner stated there is "no law enforcement for the local natural environment!" (PO2). Accordingly, all respondents noted a lack of enforcement of current fishery regulations. Once again, all community respondents noted the need to declare the CCA to have the legal right to establish, monitor and enforce rules against activities having negative environmental impacts, particularly at the mouth. These activities include gillnet and recreational fishing, dune driving infringements, and illegal camping at the mouth. Moreover, numerous community and partner responses specifically raised concerns related to associated with a lack of legitimacy required to tackle increasingly encroaching commercial extractive industries such as mining. Therefore, this enabler is considered only *partially present*, and requires attention of the State if it is to improve. Consequently, this finding strongly resonates with those in the *Urok Islands* case.

8.2.2.12. High levels of accountability

Monitoring of the fishery currently falls predominantly to the state fisheries department DAFF. Previously a community-monitoring system was in place, but due to a lack of sustained funding this has been 'on-and-off'. Whilst some monitoring of catches exists, DAFF's focus in the area is marine fishery landings at *Doringbaai*. Accordingly, fishers confirmed monitoring is sporadic on the river. Furthermore, respondents once again consistently emphasized a lack of legal recognition of the CCA to date prevents the ability of the community to hold officials accountable to the community and the fishery, as well as the ability of community members to enforce rules, particularly related to 'outsiders' related to the aforementioned negative environmental activities at the mouth. Therefore, this enabler can only be considered *absent*, and needs to be addressed. More specifically, as emerged strongly in the *Urok Islands* case, the issue of *downward accountability* requires urgent action.

<u>8.2.2.13. Low levels of articulation with external markets</u>

Low levels of external market-articulation exist due to the community's distance from urban centres. While fishers do at times sell their catches to local farmers on a limited scale (Rice, 2015), community respondents acknowledged most of their catch is consumed in-house or sold/ traded to fellow community members. The market for fish is further diminished by the fishery being limited legally to the one target species. Many fishers noted that if they could get more money for their catch, and be able to sell additional species, this could stem overfishing. However, this was not a unanimous response, and many acknowledged this might also attract greater fishing effort. Consequently, though this enabler is deemed *present*, the topic requires further research to determine the effect on the resource if markets were to be introduced, as is the desire of all fishers.

<u>6.2.2.14. The presence of nested governance with high levels of initial external support</u>

As mentioned previously all respondents acknowledged the need for external support, notably from the State, due to a lack of local management capacity, and a need for greater required legitimacy. Accordingly, both partner and community respondents noted a persistent lack of legitimacy owing to a lack of legal recognition of the CCA.

However, while all respondents acknowledge the need for partner support, the majority of community respondents emphasized a general lack of relations with partner organizations. More specifically, this was noted for State partners DEA, DAFF, DEADP, the Matzikama Municipality (i.e. local government), and the provincial parastatal conservation agency Cape Nature. In particular, the vast majority of community respondents expressed concern for the lack of support and involvement of the Matzikama Municipality, though it does play a significant administrative role in OEMF. It should be acknowledged that negative community responses describing their relations with the local municipality relate almost exclusively to the community-perceived lack of sufficient basic service delivery.

Communication between the aforementioned partners and the community was characterized by the majority of both respondent groups as *poor*. For example, in reference to *Cape Nature* one partner respondent specifically noted there is a "need [for] greater support from conservation agencies" (PO1). Furthermore, regarding the lack of attendance at meetings by State representative, another partner stated, "I won't go to meetings because nothing is happening, I am waiting for the land" (PO9). However, in contrast to the above *poor* community-perceived partner relations, relations with social partners (introduced in *section 8.1.2.2.*) emerged more positive amongst community respondents. This was confirmed by these respondents to be the result of relations of trust built through the various aforementioned institutional processes (as introduced in *section 8.1.2.2.*). Nevertheless, some respondents did identify 'champions' from not only the civil and private sector, but also the *West Coast District Municipality* who have been instrumental in attempting to drive the CCA implementation process forward.

While most community respondents stated strong community support remains for community projects like the land claim and CCA, and that the community is happy to work with partners, they did emphasize that they trust some partners more than others. More specifically, many community members questioned whether *Cape Nature* is the best role player concerning the proposed CCA, emphasizing a lack of support from this partner. Nevertheless, most community respondents stated they still trusted Cape Nature more than DAFF. Consequently, while the support of the social partners has kept the proposed CCA alive, since many challenges exist with specific partners (notably the State) regarding support, and which require urgent attention, this enabler is considered only *partially present*.

8.2.3. Actions

A brief discussion follows on key actions taken to facilitate change toward a CBC mode of governance within the Olifants Estuary case. As with the two preceding regional case studies (*Chapters 6 & 7*), legal recognition can be considered a key preliminary and foundational action for the implementation of the CCA through the creation of enabling legislation. However, unlike the prior cases this has not yet been utilized. An additional key action, as observed in the regional cases, was social partners building relations of trust with the community over many years and through many meetings and projects. A notable example being the EMP development process from which the CCA implementation process originates. More specifically, these many meetings helped to raise awareness of issues and proposed projects. The EMP process first required the formation of a CBO in the form of the *OFC*, itself a foundational and necessary action toward the CCA's implementation. Likewise, the formation of *OEMF* was an equally important action for the EMP development, and subsequently CCA implementation process. However, whilst *OEMF* attempts to provide a collaborative multi-stakeholder advisory forum, as a partner respondent noted, "[it] is just a forum and therefore has no decision-making power" (OPO4). Consequently, whilst many meetings have taken place to progress implementation of the CCAs, especially more recently involving conservation agencies, nothing has materialized.

8.2.4. External Influences

A key emerging external influence that is outside the purview of the conservation sector but influential to achieving the present desired impact, is a lack of State inter-departmental communication and collaboration related to the land claim and CCA implementation process. Accordingly, one local government partner stated with regards to national government departments and the CCA declaration process that, "[The State] is not capable of handling these issues and I don't think they even knew what was happening", and characterized State inter-departmental communication in such processes as "not much talk, even less doing, [and it] needs to be the other way around" (OPO4). Consequently, numerous partner respondents highlighted the need for greater State inter-departmental communication.

Connected to State inter-departmental collaboration issues, most non-State partner respondents, and some community respondents, noted the negative external influence of conflicting objectives of conservation and other more economically lucrative sectors, most notably mining. More specifically, these respondents identified mining as a major external threat due to concerns for environmental degradation, specifically its effects on water flow and quality in the estuary mouth, and subsequently impacts on local fish stocks. As a partner respondent stated there are "issues with mining and getting the Department of Mineral Resources to hold companies accountable" (PO4), further alluding to inequitable management by stating, "Department of Mineral Resources is very strict on some and not others" (PO4). Furthermore, as one community respondent stated, "mining is a priority of the State, not the people!" (OD19). However, this view was not unanimous amongst all community members, as some also said they would embrace mining if it provided jobs. Nevertheless, the major external influence that emerged from respondents is that the DMR does not communicate with the communities, or even other State departments such as DEA, DEADP, and local government, or Cape Nature, concerning issues affecting the estuary. Consequently, this represents a major external influence on objectives of the proposed CCA.

An additional and related external influence is the country's recent embracing of an ocean's economy (i.e. blue economy) approach to the management of its coastal national resources, through *Operation Phakisa* (RSA, 2019b). The State estimates that this oceans economy project will potentially contribute up to R177 billion to Gross Domestic Product (GDP) by 2033 (compared to R54 billion in 2010), and create approximately 1 million jobs (compared to 316 000 in 2010) (RSA, 2019c). However, such State initiatives have the ability to externally influence CBC objectives. As this initiative is still developing, time will tell what influence this 'economic agenda' will have on CBC initiatives in general, and the Olifants Estuary in particular. Lastly, as with other case studies international institutional commitments, in this case the CBD, FAO Voluntary SSF Guidelines, and regional commitments including those of the AU and SADC, will also influence the achievement of the desired result.

8.2.5. Issues Arising

Whilst many community respondents remain positive and supportive of the CCA (see *Figure 8.2.* previously), as mentioned throughout delays have led to a decline in the favourable conditions of these institutional processes, and has ultimately resulted in 'institutional inertia' and subsequently a frustrated community. As one community respondents emphasized, "Nothing came out of the CCA process, it should have happened a long time ago" (OD22). Accordingly, respondents noted conditions for finalising the EMP, the land claim, and subsequently the CCA processes were more favourable around 2014 when negotiations were most positive, decisions were widely supported by all actors, and momentum was greatest within these processes. However, the main 'issue arising' amongst most community responses is a lack of State action and participation, which has led to frustrations and a lack of community participation. Accordingly, numerous social partner respondents commented that whilst they had invested a large amount of money and time into the EMP and CCA processes, State inaction has derailed these processes, and strained actor relations. More specifically, non-State partner and community respondents described the lack of State support as "feet-dragging" (PO1). Accordingly, the most notable 'foot dragging' activity referred to by respondents concerned the land claim process via the National *Department of Rural Development and Land Reform*. Moreover, one partner respondent noted at an *OEMF* meeting that whilst the required EMP had been prepared it was yet to receive provincial departmental approval by *DEADP*. Accordingly, one local government partner suggested the "biggest obstacle is the [State] chain of command!" (PO1). In addition, one community respondent emphasized that, "community objectives and priorities change, and you need to keep up with that" (PA6). Consequently, whilst the involvement of fishers in the Olifants EMP development process, and subsequently the recent *Small-Scale Fisheries Policy* (refer to *section 8.1.2.1.*), led to their support for the proposed CCA, all respondents identified the slow and drawn out institutional processes as a key 'issue arising'.

Therefore, respondents emphasized that both the land claim and CCA declaration processes have been constrained by ineffective governance characterized notably by a lack of State support and capacity at all levels, and a lack of local-level capacity. At a State-level, as discussed previously, a particular 'issue arising' that constrains community management of the proposed CCA, as identified by both partner and community respondents, is a lack of legitimacy due to a lack of legal recognition of the CCA. On a related note, another key external 'issue arising' from both partner and community respondents is the onerous *multi-step* and *multi-actor* land claim and CCA institutional processes. An additional and related 'issue arising' is a lack of clarity over which legal mechanism to use to proclaim the CCA, as well as which state agency to 'drive' the process. Moreover, some partner responses emphasized a lack of State commitment is largely due to conservation being a low priority in comparison to other sectors. Consequently, a lack of political will emerged as the major State-level 'issues arising' from the majority of respondents.

However, since the land claim process requires agreement amongst the broader community for the land-use plan (i.e. over and above the fishers), this local-level discord also emerged as an issue affecting both the land claim and CAA institutional processes. Moreover, a lack of community-wide participation, collaboration and an accommodation of different views amongst community members in both processes is a key 'issue arising'. However, as established above issues with community attendance are also negatively affecting progress. An additional local-level issue emerged concerning a lack of local representative capacity and accountability, and community-perceived inequitable management of (potential) benefits emerged. More specifically, community respondents stated in reference to representatives that there are, "vested interests in declaring the CCA" (OD21), and "corruption is key, the government manipulates, and [community] representatives are vulnerable" (OD6). Consequently, the majority of partner respondents emphasized the need to develop *strong local leaders*. Notwithstanding the above 'issues arising', most respondents believed approval of the land-use plan and subsequent conclusion of the land claim process should (re)trigger the CCA implementation process.

As established in *section 8.2.2.* an additional local-level issue that emerged was community division. This pertains to diverse community objectives associated with the land claim, and secondly, intra-community divisions and the aforementioned local leadership issues (refer to *sections 8.2.2.3. & 8.2.2.4.*). Accordingly, both community and partner respondents consistently emphasized that all of the above divisions inhibit community ownership over and potential management of the proposed CCA. Consequently, as one *Papendorp* respondent stated, "We need one community, not Papendorp versus Ebenhaeser" (PA8). Notwithstanding these community 'divisions', as one partner stated, "whilst there is conflict between different groups, [they are] united in fighting issues" (PO7).

8.3. Conclusion

A critical investigation of the implementation progress regarding the proposed Olifants estuary CCA, has identified *contextual change triggers*, the presence or absence of *enablers*, the effect of *actions* taken to date, and *external influences* and key 'issues arising' that must be addressed to progress efforts to implement CBC.

Contextual change triggers included local fisher concern for their declining resource, and early involvement of fishers in the EMP, from which fishers initiated the idea of the proposed CCA. Findings emphasize the need to address and improve some key enablers, notably local leadership, community ties (i.e. current divisions), levels of locally devised access and management rules, ease of enforcement (requiring greater support from State departments and *Cape Nature*), and whilst high levels of initial external support were provided, the need for greater on-going support emerged, particularly from the State.

Enabling actions for the CCA's implementation include the development of relations of trust and support between social partners and the community. However, failure to promote greater involvement of the broader community in the EMP, land claim and CCA processes from the outset has constrained progress within all institutional processes.

Several key 'issues arising' were observed and include a divided community – characterised by diverse interests and a lack of consensus – possessing local leaders with weak governance capacity characterized by perceived elite-capture of knowledge and decision-making. Furthermore, due to high levels of poverty, and a lack of alternative livelihood for the fishers specifically, requires poverty alleviation and implementation of the CCA to be parallel processes. Therefore, the Olifants Estuary case appears to mirror similar constraints emphasized by established South African wildlife CCAs (refer to *Chapter 4*).

At a State-level, incapacity and political will (characterised by inaction) to recognize and declare the CCA, perhaps represents an even greater constraint to the CCA's implementation. Whilst the State has provided an enabling legal framework for the CCA's implementation (*Chapter 4*), a lack of participation and local capacity building by the State continues to constrain the CCA's implementation. Furthermore, a lack of clarity over and overlapping legal mechanisms represent further 'issues arising'. Consequently, since the CCA is a multi-actor initiative greater communication and collaboration between all governance actors is deemed key moving forward. Nevertheless, as one partner respondent stated, "The community are motivated, and CBC can work!" (OPO6). *Figure 8.6.* provides an infographic summary of this chapter's key findings.

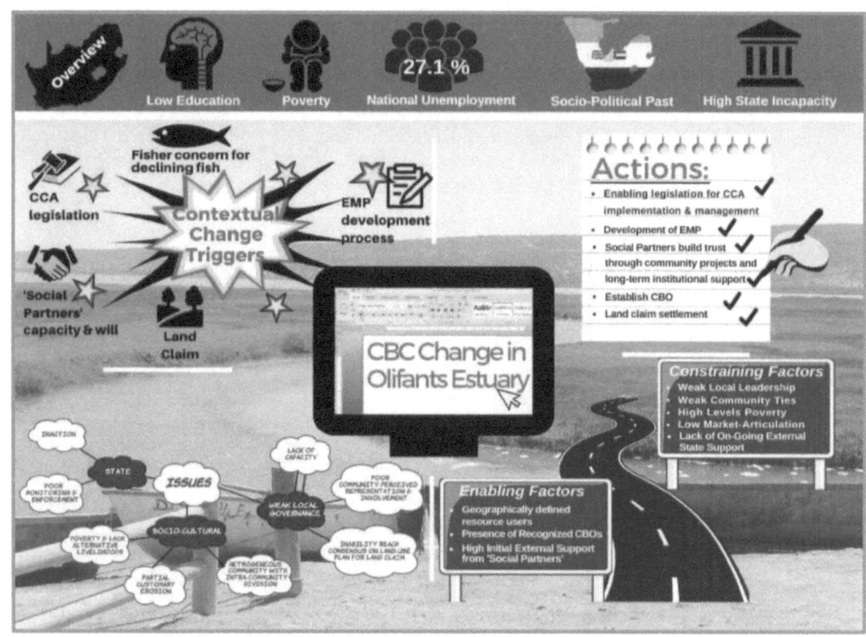

Figure 8.6.: An infographic summary of the key CBC change process findings in the Olifants Estuary proposed CCA, South Africa.

341

Chapter 9

Discussion and Conclusion

9.1. Introduction

In spite of a plethora of legislation in South Africa promoting participation of local communities in resource management and providing mechanisms to give effect to community-based conservation, there are no formally recognised CCAs in the coastal environment. *Chapter 1* introduced the research rationale and the background to, and context of CBC both globally, and in South Africa. It presented the key research question guiding the study, namely;

> *What factors, conditions and processes are required to facilitate a shift toward CBC initiation, implementation and governance in South Africa, when contextually appropriate, so as to realize desired social and ecological outcomes?*

Chapter 1 proposed the benefit of looking at CBC initiation, implementation and governance as a change process in order to address this question. Furthermore, *Chapters 2-8* contributed to exploring this change process. *Chapter 2* outlined the methodology employed to explore the change process in three case studies. *Chapters 3* and *5* provided the theoretical foundations informing this exploration. More specifically, *Chapter 3* explored key governance factors and conditions that enable CBC initiation, implementation and governance. Furthermore, it culminated by proposing a list of 14 enablers, which are considered to positively influence this change process. *Chapter 5* explored the theoretical underpinnings of the *Theory of Change (ToC) approach,* and building upon ideas on enabling factors for CBC presented in *Chapter 3,* developed a *Generic CBC ToC Pathway.* This *pathway* represents the dissertation's conceptual framework, and as such informed data analysis and interpretation in the three case studies, and ultimately informed the development of a *South African Empirical ToC Pathway* presented in this chapter. *Chapter 4* provided a context-specific review of CBC progress in South Africa, specifically emphasizing the enabling and constraining factors, conditions and processes experienced in trying to pursue CBC.

Therefore, *Chapters 3-5* took the first step in this change process exploration. The next step involved investigating the *change elements* in the three case studies (in *Chapters 6, 7,* and *8)*. To reiterate these change elements included the *contextual change triggers, actions, external influences,* and *issues arising* from an analysis of this change process within each case. Consequently, this analytical process yields lessons learned across the cases that shed light on the *policy-praxis disjuncture* observed in CBC initiation, implementation and governance in South Africa's coastal environment.

This chapter is structured as follows. *Section 9.2.* presents a consolidated discussion of the case study findings. Thereafter, *section 9.3.* builds upon this discussion and includes the insights gained from reviewing South Africa's progress with CBC in *Chapter 4.* This analysis leads to proposing a *South African Empirical CBC ToC Pathway* (i.e. in *section 9.3.6.*), which depicts how shifting to a community-based approach to conservation governance *might* take place as a change process. Consequently, this chapter addresses the final two objectives (i.e. **objectives 6** and **7** – *Box 9.1.*). Thereafter, *section 9.4.* discusses the contribution of this book to CBC theory and practice, and presents this dissertation's conclusion in *section 9.5.*

9.2. Case-Study Findings

Developing a *South African Empirical CBC ToC Pathway* requires consolidating

common findings and lessons learnt relevant to the established change elements

that emerged within the three case studies presented in *Chapters 6-8*. A discussion

of these change elements follows in *sections 9.2.1.-9.2.5.*

9.2.1. Common Contextual Change Triggers

To recap a *contextual change trigger* represents either a contextual issue or action

that stimulates initiation and/ or maintenance of the CBC change process. Triggers

can stimulate either sudden change or motivate more long-term actions. The

primary contextual change trigger for pursuing CBC that emerged in all cases, not

surprisingly was concern for the degradation of natural resources. This concern

emerged from all actors, ranging from livelihood concerns in local communities,

to economic, social and ecological concerns amongst partner organizations. More

specifically, this concern related to coastal and marine resource degradation of

important and valuable species, for example, octopus in the *Bay of Ranobe*, shellfish and fish in the *Urok Islands*, and fish in the *Olifants Estuary*. This finding is in accordance with the literature on *ecological indicators* (e.g. stock status of a marine resource) can 'trigger' a conservation management decision (Bie et al., 2018).

The second common contextual change trigger identified across the three cases was the creation of enabling legislation for community conservation. This was most noticeable in the two regional cases where legislation had been developed to support initiation and implementation of CCAs. In contrast, enabling legislation in South Africa has not yet resulted in the implementation of a coastal CCA. Nevertheless, this legislation has promoted implementation of a few terrestrial CCAs in the country. Furthermore, having this legislation in place has supported calls to declare a CCA at the *Olifants Estuary* (i.e. the South African case-in-progress). One possible reason for greater progress with declaring CCAs in the terrestrial environment is that the legislation governing terrestrial CCAs is less complex, as opposed to legislation relevant to coastal and marine areas. For example, in the Olifants Estuary case, legislation exists pertaining to managing fish, estuaries, and PAs, and at local, provincial and national government level. Lastly, all respondents considered the presence of enabling legislation necessary for the legitimacy of their CCAs and the ability to enforce regulations, principally exclusion of 'outsiders', which aligns with research elsewhere (e.g. Seixas & Davy, 2008).

In both *Madagascar* and *Guinea-Bissau* legal recognition of community conservation originated with a State-driven PA expansion strategy which led to the promotion of a community-based mode of governance in the form of legally recognized NGO-CBO partnered CCAs. Furthermore, the established CBOs are a legally recognized management authority for their CCAs, though these initiatives still rely heavily on partnerships. Notwithstanding the importance of legal recognition in all three cases, the majority of respondents considered their respective legislation to be complex and onerous, and emphasized that the slow and drawn-out processes are constraining progress in legal declaration of a CCA. This emerged most notably in the *Olifants Estuary* where support for declaring a CCA was confirmed by all actors in 2013 at an OEMF meeting, yet the legal process to declare the CCA only began to be explored by the State in earnest in 2019. Moreover, the title deed associated with the land claim connected to the Olifants Estuary area was only recently handed over (in March 2019), and according to respondents this process constrained progress of the proposed CCA. Therefore, in all three cases legislative complexity necessitates support from the State and other partners to navigate a way forward. Consequently, whilst enabling legislation is an important *contextual change trigger*, respondents confirmed in all cases the process requires streamlining. In the case of the Olifants estuary, political support and better coordination amongst government departments and stakeholders was required to give effect to the legislation. The topic of CBC enabling legislation is discussed further throughout.

On a related note, political will to enable a shift to CBC governance emerged as an additional key contextual change trigger. In *Madagascar* and *Guinea-Bissau,* the State acknowledged its fragility and lack of capacity to manage coastal resources, particularly in remote areas, and supported NGO-CBO partnerships in facilitating implementation and governance of CBC interventions. In contrast, as alluded to above, political will has been slow to emerge at the *Olifants Estuary*, although recent commitments by the State suggests all levels of government are finally supportive of CBC governance at this site, which is perhaps largely in response to many actors' objections to the threat of mining. Nevertheless, in all three cases non-State respondents confirmed State 'meddling' often constrained the implementation of CBC. Consequently, the presence, willingness and capacity (both technical and financial) of non-State partners to support initiation, implementation and governance of CBC, emerged as an important contextual change trigger across the three cases. Thus involvement of partners such as NGOs in *Madagascar* (e.g. Reef Doctor in the *Bay of Ranobe*) and *Guinea-Bissau* (e.g. Tiniguena in the *Urok Islands*), and the social partners (e.g. the EEU, Masifundise and LRC) at the *Olifants Estuary*, has been crucial to progress. Accordingly, initial external support from partners appeared 'catalytic' for CCA implementation, especially regarding navigating onerous legislation, a finding which is supported by past research (Berkes & Seixas, 2004; Seixas & Davy, 2008).

Lastly, poverty emerged as a dominant *contextual issue*, and the potential to alleviate it as a key *contextual change trigger,* in all three cases. Accordingly, poverty alleviation is often employed as a strategy to 'trigger' community participation in decision-making, and subsequently garner community support of, and compliance with, conservation regulations. The topic of poverty is discussed further in subsequent sections.

9.2.2. Common Enablers

This section consolidates the findings related to the presence of the 14 CBC *enablers* proposed in *Chapter 3,* in the three case studies. To reiterate, an *enabler* is a factor or condition assisting success of CBC institutions. A graphical summary of the presence of these 14 enablers within each case study can be found in *Figure 9.1*. Moreover, once again these 14 enablers are not 'set-in-stone', and based upon the consolidated findings discussed below, will be amended in section 9.3.3. with specific reference to the South African CBC context.

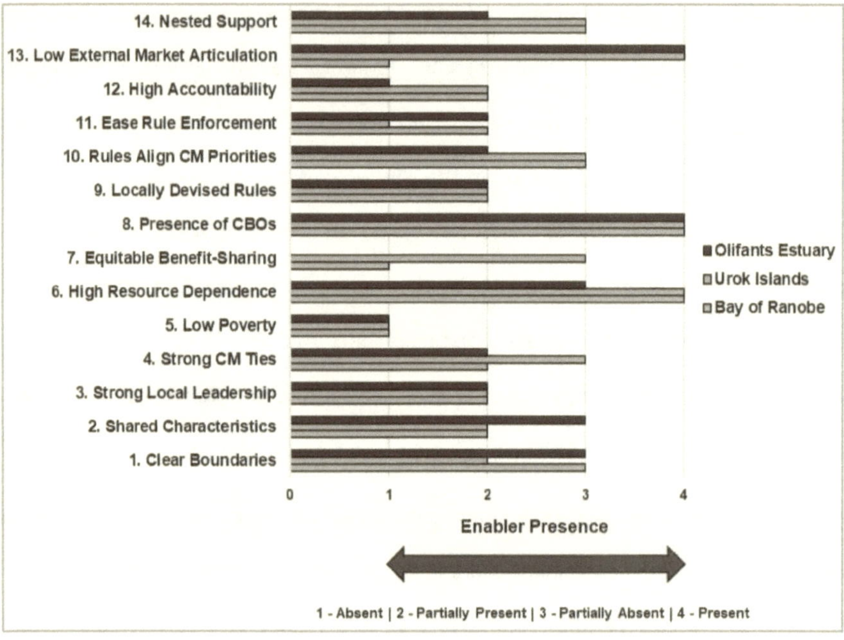

Figure 9.1.: A graphical summary of the presence of the 14 enablers across the three case study sites.
Note: CBOs – *Community-Based Organizations*; and CM – *Community*.

The first two enablers associated with resource-system and -user characteristics, emerged as highly interconnected, since while the geographical boundaries of both the resource system (i.e. the CCAs), and the resource users appeared clearly-defined, it emerged from responses across the three cases that greater socio-cultural diversity within the communities may be constraining effective CCA management. Accordingly, analysis of *enabler 2* (i.e. shared norms, values, interests and identities) indicated increasing socio-cultural diversity in the three cases. Respondents in the *Bay of Ranobe* and the *Urok Islands* noted this is predominantly caused by the increased 'in-migration' of 'outsiders', which is contributing to the partial erosion of customary systems. In particular *Bay of Ranobe* respondents emphasized, increasingly less productive agricultural activities are augmenting migration of groups such as the *Masikoro* towards the coast. Likewise, in the *Urok Islands* respondents noted an increase in the presence of small-scale fishing crews on the islands from other West African nations is negatively influencing this enabler. Moreover, in the *Olifants Estuary* the community is characterized by diverse livelihood strategies, however, those that consider themselves *fishers* do possess higher levels of shared norms, values, interests and identities. Therefore, these findings reinforce those of previous studies as introduced in *Chapter 3,* and therefore, the presence of these first two enablers is confirmed to be enabling for CBC, at least within the context of these three cases. Lastly, comparison of these findings to those of Biggs et al. (2019) provides interesting insights. Biggs et al. (2019: p3), in reviewing Zimbabwe's CBC program CAMPFIRE, emphasize several conditions that enable the emergence of new rules in groups, these include the need for a community to possess collective recognition and shared understanding of the problem, and a collective interest in adopting new rules to address the problem. Accordingly, these conditions are not completely present in the three cases investigated, which constrains their effectiveness.

Strong local leadership (i.e. *enabler 3*) emerged as a crucial enabler in all three cases. More specifically, respondents across the three cases considered strong local leadership key to legitimacy and community support required to positively influence compliance. This is a well-established finding in the conservation literature (e.g. Crona et al., 2017; Steenbergen & Warren, 2018; Biggs et al., 2019). However, analysis of *enabler 3* highlighted *poor* representation across all three cases, largely due to weak local capacity and high levels of poverty which are stimulating local elite-capture (i.e. inequitable benefit sharing – in contrast to *enabler 7*). Accordingly, in all three cases this enabler emerged only *partially present*, and community and partner respondents alike emphasized this as a key area requiring improvement.

Notwithstanding the above local leadership concerns, customary leaders were commonly considered more legitimate and effective in leading the CCA intervention than other local leaders. This was specifically emphasized in the *Bay of Ranobe* and *Urok Islands*, where respondents noted the negative effect of local representatives often being selected outside of local customary leadership structures. For example, *Bay of Ranobe* respondents identified their *village presidents*, most notably those from the villages of *Ifaty* and *Beravy,* as key to CCA management. However, these actors were not represented on *FIMIHARA* (i.e. the CCA-related CBO). Furthermore, *Urok Islands* respondents, particularly those within villages located further from the 'project hub' of *Abu* village, emphasized customary leadership retains strong influence. Accordingly, greater inclusion of customary leadership was noted by respondents as important to the development of both strong local leadership (i.e. *enabler 3*) and strong community ties (i.e.

enabler 4), and subsequently the ability to more equitably distribute benefits (i.e. *enabler 7*). This corresponds with extensive CBC literature highlighting the important role of local leadership to enabling governance processes (e.g. Lyons & Cavaye, 2016; Crona et al., 2017; Steenbergen & Warren, 2018). Moreover, this specifically correlates with commons scholars (e.g. Pomeroy et al., 2001; Agrawal, 2002), and in particular findings from African CBC studies (Galvin et al., 2018; Biggs et al., 2019), reviewed in *Chapter 3.* Consequently, the presence of *strong local leadership* is confirmed as a CBC enabler within the context of these three cases.

Respondents in all cases confirmed *strong community ties* (i.e. *enabler 4*) are crucial to fostering collaborative intra-community relations characterized by trust, which they deemed necessary for CBC management. However, this enabler was not *present* in all three cases, and largely attributed to weak local representation. While strong *family* and *intra-village* ties emerged, weaker *inter-village* ties were common and acknowledged to be affecting CCA management. Accordingly, *Bay of Ranobe* respondents confirmed that weaker inter-village ties were resultant of a lack of involvement of many villages in the initial stages of establishing their CBC-related CBO (i.e. *FIMIHARA*), and also subsequently the implementation of the two LMMAs. Likewise, *Urok Islands* respondents located in villages further from *Abu*, perceived higher levels of exclusion and poor representation negatively affected inter-village ties. Similarly, respondents in *Olifants Estuary* noted a lack of involvement of fishers in the land claims process, and a lack of non-fishers in the previous EMP development process, and subsequent CCA planning process, has negatively influenced intra-community ties. In addition, inter-fisher group issues and diverse community livelihood strategies, are also negatively affecting the strength of these ties in this case. Consequently, it can be concluded from respondents that the presence of *strong community ties* would (potentially) be a CBC enabler if present in these three cases.

The socio-economic *enablers 5, 6* and *7* emerged as highly interconnected in all three cases. Accordingly, all respondents across the three cases unanimously confirmed that *high levels of poverty (i.e. contrary to enabler 5 – low levels of poverty)* are negatively affecting management of their interventions. As a result of high levels of poverty, and a lack of alternative livelihoods, *high levels of resource-dependence (i.e. enabler 6)*, were common to all three cases and considered to have contributed to user-perceived declines in resources. This was particularly noted in the two regional cases. The relationship between poverty, resource-dependence and the state of natural resources is well documented in the literature (e.g. Barbier, 2010; Thondhlana & Muchapondwa, 2014; Wilson et al., 2016). Furthermore, research specifically suggests that high levels of poverty may cause resource users to 'discredit' future incomes which may potentially influence overharvesting of resources (e.g. Ostrom, 1990; Raycraft, 2019). Moreover, coastal and marine CBC literature in particular indicates that the main impetus for creating CCAs is often a desire to maintain or improve livelihoods and food insecurity and vice-versa for stimulating local conservation interventions (e.g. Govan et al., 2009; Charles et al., 2016).

High levels of poverty in all three cases were also commonly cited by respondents to be stimulating elite capture by local elites (i.e. negatively affecting *enabler 7 - equitable distribution of benefits from common resources*). This in accordance with past CBC studies (e.g. Lund et al., 2013; Persha & Andersson, 2014). The issue of elite capture links back to concerns regarding *poor* representation (i.e. strength of local leadership – *enabler 3*), which most respondents confirmed to be undermining not only the equitable distribution of benefits, but in particular knowledge dissemination. Nevertheless, respondents in the two regional cases suggested customary leadership could potentially facilitate increased equitable sharing of resources. Accordingly, greater alignment of local CBO representation and customary leadership structures once again emerged as a key enabling factor to address concerns, most notably in the two regional cases. Consequently, it is concluded that the equitable distribution of benefits should still be considered a CBC enabler in these three cases.

Therefore, based upon the common findings emerging from the three cases high poverty levels are highly influential to the effectiveness of CBC interventions, More specifically, from an 'enabler' perspective high levels of poverty affect the presence of other commonly cited CBC enablers. However, within the African (i.e. developing world) context CBC interventions commonly take place in poor population groups. Therefore, within this context, and specifically the three cases investigated, the reality of achieving Agrawal's (2002) enabler of the presence of low poverty levels comes into question. This notion is discussed further in *section 9.3.3.* as it relates specifically to the South African context.

The topic of resource-dependence also interacted strongly with *enabler 13* (i.e. low levels of articulation with external markets). As introduced initially in *Chapter 3: section 3.3.2.2.*, and discussed subsequently in the case study chapters, the notion of directionality and nonlinearity arose strongly within the topic of resource-dependence, as well as levels of external market articulation. For example, low levels of external market articulation, and once again high resource-dependence, were found in the *Urok Islands*, which appeared to be limiting harvesting activity. That said, the UMC's proposed plans to increase market articulation may affect harvesting activity. In contrast, the *Bay of Ranobe* possessed high levels of external market articulation (and resource-dependence), and respondents raised concerns (as did secondary fishery landings data) that increased access to these markets may have increased harvesting activity, and may have led to further resource declines. Furthermore, in the *Olifants Estuary*, which possessed low external market articulation, it emerged among most respondents that this was leading to greater resource-dependence and potentially to declines in the resource. Consequently, these findings highlighted the context-specific nature, and bidirectionality, of these two enablers in particular.

It has been extensively shown that neoliberal conservation agendas (i.e. the promotion of 'free-markets' and thus local economic development - see Igoe & Brockington, 2007), may not be well received by communities, may lack local-level legitimacy, may be negatively affected by institutional and political contexts, and may exacerbate established community inequalities, and ultimately affect the management of CBC interventions. Of relevance to the present context, these effects have specifically been shown in various CBC studies in Madagascar (e.g. Desbureaux & Brimont, 2015; Brimont & Karsenty, 2015). Nevertheless, the locality and sophistication of institutional arrangements and processes will influence this enabler.

Therefore, within these three cases I would question Agrawal's (2002) critical enabler of "low levels of articulation with external markets", though further research would be required to substantiate this. Nevertheless, in accordance with past conservation research (e.g. Wilson et al., 2016), I would imagine that greater research into these three cases would find a 'tipping-point' associated with the presence of these two enablers. Accordingly, I propose that *moderate levels of resource-dependence* and *sustainable levels of articulation with external markets* may better enable successful CBC, whilst more extreme conditions of either may constrain, within the context of these three cases. That said, these two enablers represent 'ripe' and evolving topics for future research into these (and other) CBC cases, which I strongly encourage. These two enablers are discussed further in *section 9.3.3.* as it relates specifically to the South African context.

Whilst in all three cases the presence of CBOs related to CBC management was noted (i.e. *enabler 8*), issues associated with local CCA leadership – notably poor representation characterized by *poor* feedback and benefit-distribution – constrained their management effectiveness. Accordingly, respondents noted greater legitimacy of CBOs was contingent on State recognition, decreased internal corruption, and increased rotation of representatives. Nevertheless, as some *Bay of Ranobe* respondents suggested, effective CBC management also appeared to be dependent on retaining knowledge and capacity within *FIMIHARA*. Furthermore, in the *Olifants Estuary*, many community respondents acknowledged that a rotation of representatives could merely manifest in everyone taking their chance to benefit. Therefore, rotation of local representatives may in certain contexts actually increase elite capture, and this would be a context-specific action. Moreover, the enabling effect of rotating local representatives would probably only be feasibly realized within contexts where local leaders are less constrained by high levels of poverty. Nevertheless, the presence of community institutions commonly emerged as a CBC enabler across the three cases. However, this is contingent on the presence of democratic procedures resulting from increased rotation and accountability of representatives, and the appropriate transfer of local capacity (i.e. strong institutional memory). Lastly, as mentioned above, a commonly suggested enabler was the greater alignment of local CBO and customary leadership structures.

In all three cases, *enablers 9* (i.e. locally devised access and management rules) and *10* (i.e. rules strongly align with local priorities) were confirmed as key enablers, but emerged highly contingent once again upon the strength of local leadership (i.e. *enabler 3*), and more specifically local representation and institutional capacity. Accordingly, Biggs et al. (2019) note that the perceived legitimacy of decision-making structures are key to enabling communities to adopt new rules, but this did not emerge strongly from the three cases. Not surprisingly, many respondents confirmed low levels of *active* involvement in designing CCA-associated rules in all three cases. However, community participation appeared to be influenced not only by local leadership, but also the ability to derive tangible benefits from the intervention.

Community participation in governance activities such as the design of, and subsequently compliance with CCA rules, emerged highly dependent on CCA rule-alignment with local socio-economic and cultural priorities (i.e. *enabler 10*). This finding resonates with discussions throughout on the need for *social institutional fit* within CBC interventions, as reviewed in *Chapter 3*. In both regional cases coastal and marine resources had strong cultural importance, for example amongst the traditional *Vezo* fishers of *Bay of Ranobe*, and the importance of shellfish to cultural ceremonies amongst the *Bijagó* in the *Urok Islands*. Rules were found to strongly align with these cultural aspects in these two cases, for example the use of *Dina* (i.e. a socio-cultural norm) in the *Bay of Ranobe*, and culturally-aligned zonation allowing access to shellfish in the *Urok Islands*. Moreover, fishers of the *Olifants Estuary* also possess a long tradition of fishing, and although no longer a community-wide livelihood strategy, fishing remains a source of cultural pride, and is economically crucial to those still heavily reliant on the fishery.

However, while cultural-alignment of rules emerged relatively strongly in these cases, due to high levels of poverty, the alignment of CCA rules with local socio-economic strategies was less so. An example of a lack of alignment with local livelihood priorities emerged in the *Bay of Ranobe* with many community respondents explicitly calling for temporarily opening the permanently no-take CCAs. This they emphasized would increase benefits, and subsequently promote a sense of community ownership, and be potentially sustainable if focused on specific species and well monitored, as has been shown in coastal and marine CBC interventions elsewhere in south-west Madagascar (e.g. Harris, 2007; Oliver et al., 2015). Consequently, it can be concluded that a strong alignment of CCA rules with local socio-economic and cultural priorities, with necessary improvements, will be a key CBC enabler in these cases.

Both *enabler 11* (i.e. ease in enforcement of rules, and conflict resolution) and *12* (i.e. high levels of accountability) are interdependent enablers, and both emerged, as has been a theme throughout, highly contingent on the legitimacy of local leadership, and the levels of community participation in CBC governance. Furthermore, effective monitoring and enforcement, as well as conflict resolution, appeared to be dependent on not only the capacity, but also the accountability of both local leaders and partners (*enabler 12*). For example, respondents in Guinea-Bissau, including many State representatives, acknowledged a lack of State capacity and accountability with regard to monitoring and enforcement of fisheries, and noted that it is not surprising that the country is plagued by globally recognized high levels of *Illegal, Unreported and Unregulated fishing* (IUU) (see also Denton & Harris, 2019; Okafor-Yarwood, 2019).

Related to increasing issues with IUU, the lack of access to external markets (i.e. *enabler 13*), as introduced above, emerged key in the *Urok Islands* and *Olifants Estuary* cases. However, once again, whether access to external markets is an *enabling* or *constraining* factor appeared to be highly context-specific and representative of both the potential bidirectionality and nonlinearity of a CBC enabler (as discussed above).

In accordance with *enabler 14* (i.e. the presence of 'nested' governance with high levels of initial external support) all three cases confirmed a lack of local capacity necessitates not only initial, but also ongoing external support. This enabler requires a two-way engagement characterized by collaborative systems of communication and knowledge dissemination at all levels, amongst all actors, from community to non-State partners and the State. The importance of such communication is emphasized by recent coastal and marine CBC literature (e.g. Mascia & Mills, 2018). Moreover, this also links back strongly to the need for *high levels of accountability* (i.e. *enabler 12*), and in particular, the presence of both *downward* and *upward* accountability (e.g. Bluwstein et al., 2016; Wright, 2017). In reference to this, respondents in all three cases specifically noted a lack of communication across actor-groups as a key constraint to the effectiveness of external support.

The issue of power emerges key to 'nested' governance in all three cases. Firstly, while the *enabling* effect of the presence and power of non-State partners in all three cases is acknowledged – be it NGOs in the two regional cases or the 'social partners' in the *Olifants Estuary* case – the notion of how much power should be afforded to these partners becomes a key consideration. In Madagascar, the power and responsibilities of NGOs in conservation is highly 'enabled' by the State through political will, and specifically, legally recognition. This since the State acknowledges its own lack of capacity, and perhaps more importantly the capacity of NGOs and their ability to source international funding required for conservation interventions. In contrast, in Guinea-Bissau, it appears the empowerment of NGOs is largely due to severe political and environmental *fragility* (i.e. a lack of capacity and stability within the State), which essentially leads to a *de facto*, though not *de jure*, relinquishing of power and responsibility for conservation interventions to partner organizations. However, in South Africa – as emerged from both the findings of national interviews *in Chapter 4*, and the specific case of the *Olifants Estuary* – the ability of the State to relinquish power is not nearly as evident. Consequently, while the role and power afforded to non-State partners requires further insight, the case studies illustrate the importance of these actors to enabling CBC interventions.

All respondents confirmed that external support is especially key to enforcement (i.e. *enabler 11* – ease of enforcement), and in particular this requires ongoing State support to promote greater legitimacy. In the *Bay of Ranobe* many respondents emphasized – especially within the context of *Vezo* culture being inherently polite and non-confrontational – the necessity of partner assistance with enforcement. Furthermore, in the *Urok Islands* respondents noted that a lack of local capacity, as well as a lack of livelihood opportunities and market access, has led to an enduring high level of 'partner-dependency'. Moreover, while many in the two regional cases raised some concerns related to their interactions and relations with NGO partners, in contrast, in the *Olifants Estuary* non-State 'social partners' and their support was favourably perceived by all fishers.

Whilst external support emerged as a key enabler, most respondents, notably non-State partners, confirmed long-term successful CBC initiatives ultimately require that external support eventually be withdrawn, allowing communities to be both the *de facto* and *de jure* CCA management authority. This has also been shown in other CBC initiatives (e.g. Olsen & Christie, 2000). However, some have recommended this initial external support needs to be long-term before its withdrawal (e.g. Gurney et al., 2014). Additionally, in the *Olifants Estuary* a partner respondent noted that community 'self-managing' efficiency needs to be reached before funding cycles end, otherwise the collapse of the intervention is all but inevitable. The effect of short-term funding cycles on the effectiveness of CBC interventions is a well-established in the literature (e.g. Ostrom, 2000; Wells et al., 2010; Biggs et al., 2019).

Consequently, a key finding emerging from the above discussion is not only how context-specific, but interdependent these enablers are. Accordingly, there are many instances discussed above where the presence or absence of one enabler increased or decreased the chances of others being present. Therefore, in the case of *absent enablers*, urgent and targeted actions are required to improve their *presence* in order to strengthen other factors, conditions and processes required so as to increase the management effectiveness of a CBC intervention. Consequently, *common actions* taken thus far in the three cases to promote CBC governance, are now discussed.

9.2.3. Common Actions

As introduced in *Chapter 5*, the term *actions* is used here to be inclusive of an event, a project or programme, a policy or strategy, or even formation of an organization. Essentially, it is an activity implemented to achieve the desired result. It should be noted that these actions can, and in all three cases did, influence each other, as well as the ability to achieve the desired result. Furthermore, it is acknowledged that not all actions may have been documented in each case study, and other community development projects by 'non-conservation' partners could have contributed positively or negatively toward the CBC intervention's *actions* (and therefore presence of enablers) and their ability to achieve the desired result. Moreover, those actions identified were not evaluated for their success at achieving a desired result, but rather the focus lay with understanding common actions employed, and the extent to which they may potentially contribute to a successful shift to a CBC mode of governance.

9.2.3.1. Action Category 1: Strengthen Actor Relations

One of the common actions identified in the three CCAs investigated is the need to build strong actor relations both within communities, and between communities and partners. This action links strongly to discussions above regarding the need to strengthen *enablers 3* (i.e. *strong local leadership*), *4* (i.e. *strong community ties*), and *14* (i.e. *presence of external 'nested' support*).

Strengthen participation and representation
The first action taken to develop networks of support at a local-level in each case, was the establishment of their respective CBOs, which was facilitated by external partners. This action is critical to ensure that community interests are represented in CCA implementation and management. Examples of actions in this regard include holding local community meetings with their representatives, inclusive of customary representatives. These meetings were initially facilitated by an external partner, however, in all cases the CBOs have since taken this responsibility over. Furthermore, respondents noted these meetings were important for building trust between the two groups of actors. These included village assemblies in the *Urok Islands*, local fisher association meetings at the *Olifants Estuary*, and *Dina* committee meetings in the *Bay of Ranobe*. Nevertheless, respondents confirmed that despite meetings, problems with these relationships persist. Accordingly, many respondents confirmed rotation of local representatives was necessary for greater democratic decision-making. Whilst community meetings currently take place within all three cases, most respondents emphasized it should be more regular. Moreover, concerns were raised over the 'capture' of the local CBO-representative election process in all three cases.

In all three cases the exclusion of marginalized groups was emphasized to be negatively affecting the building of strong networks of intra-community relations (i.e. *enabler 4*). For example, *Bay of Ranobe* respondents noted how a lack of initial inclusion of additional villages (i.e. beyond *Ifaty* and *Mangily*), negatively affected the success of their CCAs. However, many partner respondents considered the increased inclusion of representatives from each region of the bay on *FIMIHARA* since the start of the process, to have rectified this concern. Furthermore, gender-based exclusion was specifically noted in the regional cases. For example, in the *Bay of Ranobe,* women, and especially *Masikoro* women, commented on their perceived exclusion from CCA management activities. Whilst this also emerged as a concern in the *Urok Islands,* these respondents emphasized this issue has to some extent been improved by UMC annual general assemblies, and more specifically CCA-related monitoring projects that have included and empowered women. A particular example is a recent shellfish community-monitoring project, initiated by Tiniguena, which is culturally significant since it is a resource used in various ceremonies and is principally harvested by women. Greater inclusion of women in a community-monitoring project in the gillnet-fishery at the *Olifants Estuary,* facilitated by the EEU, was also deemed an important action in strengthening actor relations.

Build Relations of Trust

In addition to intra-community relations, community-partner relations also emerged key to enabling CBC in all three cases. Accordingly, actions undertaken in all three cases centred upon external partners working to build relations of trust and respect with the community. Attempts to accomplish this most notably focused on, firstly, initiating other community development projects, and secondly, arranging various multi-actor CCA-related meetings. Regarding the former, most *Bay of Ranobe* respondents confirmed Reef Doctor has contributed to improving relations of trust through for example facilitating the development of local aquaculture activities (i.e. seaweed cultivation), as well as their long-term presence. Likewise, in the *Urok Islands* respondents confirmed Tiniguena has facilitated the development of several community development projects – in addition to community-monitoring projects. Furthermore, respondents in both regional cases confirmed that partners have also earned trust through supporting projects specifically aimed at improving basic services, notable examples being provision of water through the installation of wells, and the building of a primary school and clinic in the *Urok Islands*, restoration of a school and the building of a library in the *Bay of Ranobe*. In the *Olifants Estuary* social partners such as the University of Cape Town have built relations of trust through assisting the community in claiming their legal rights to fisheries resources, including facilitating community-input into the development of an *Estuarine Management Plan* (EMP), and the recent *Small-Scale Fishing Policy* (SSFP), as well as establishing a community-monitoring project. Additionally, preliminary investigations involving fishers and scientists were undertaken to assess the feasibility of growing *harders* (i.e. the target fish species of the gillnet-fishery) in the shallows of the wetlands in the *Olifants Estuary*. Although this project was deemed unfeasible, many perceived this as a helpful learning experience, and to have contributed to improved relations across a broader network of actors. Consequently, in all three cases most respondents confirmed that these actions have gone a long way towards building relations of trust and respect between community members and their external partners. These actions are also highly relevant to improving the alignment of CBC initiatives in terms of addressing local priorities (i.e. *actor category 2* – refer to *section 9.3.4.2.*).

An additional action employed to build relations of trust and collaboration between various actors is centred on improving multi-actor communication. In all three cases, this has taken place at multiple levels. *Firstly*, the aforementioned community-level gatherings, and *secondly*, multi-actor meetings, which in all cases involved both community representatives and partners enhanced communication amongst actors. Examples include *UMC annual general assemblies*, and *UMC representative meetings* in the *Urok Islands*, *FIMIHARA* meetings in the *Bay of Ranobe*, and several multi-actor meetings including the *OEMF* meetings at the *Olifants Estuary*. These gatherings are largely between community representatives and non-State partners, with most respondents and meeting minutes confirming a general lack of attendance by certain State-representatives, especially noted in the *Olifants Estuary*. Nevertheless, a decline in attendance of community members at these meetings in all three cases is an 'issues arising' that requires urgent attention. Consequently, two common and key interdependent actions related to *strengthening actor relations* that emerged from the three cases, emphasize firstly, improving local representation on management structures and subsequently increasing multi-actor participation, and secondly, the building of trust amongst actors to strengthen social networks and improve multi-actor collaboration. These actions align with recent conservation governance studies that emphasize the importance of leadership, trust and collaboration for positive conservation outcomes (e.g. Young et al., 2016; Crona et al., 2017; Baird et al., 2019a&b; Dressel et al., 2020). More specifically, Garcia et al. (2014) describe the multilevel manifestation of these attributes specifically within diverse marine conservation contexts.

9.2.3.2. Action Category 2: Increase Socio-Cultural Alignment

Improve Cultural Alignment

Notable actions targeting cultural alignment of the CBC initiative to the cultural context in the three CCA cases included the strengthening, or revitalizing, of customary institutions and practices. This was most notable in the two regional cases. For example, in the *Bay of Ranobe* a *Dina* (i.e. a socio-cultural norm) prohibiting fishing in the two CCAs was established. This approach had been shown to be effective elsewhere in the country (e.g. Harris, 2007, 2011). The official establishment of the CCAs in the Bay of Ranobe, and more recently the commencement of a coral restoration project, incorporated customary elders performing cultural ceremonies. Reef Doctor respondents confirmed that this was an attempt to generate greater legitimacy through cultural alignment.

Likewise, partner respondents in the *Urok Islands* confirmed that the zonation of the CMPA purposefully acknowledges resident *Bijagó* customary practices, notably providing access to shellfish for ceremonial consumption, and exclusive access to fishing using traditional methods within the inner most zone. Furthermore, the State and other partners have attempted to acknowledge the rights of the traditional gillnet fishers in the proposed *Olifants Estuary* CCA. For example, by consenting to artisanal salt mining and sheep grazing in times of drought.

In all three cases, most community respondents confirmed that actions promoting cultural alignment had positively affected their perceptions of the initiatives and their management, although acknowledged that issues do exist. Lastly, most respondents confirmed a lack of alignment of CBO representatives with customary institutional structures, which they emphasized requires future actions.

Improve Socio-Economic Alignment
A common socio-economic action, which emerged particularly strongly in the two regional cases, involved parallel actions focused on alleviating poverty. This entailed identifying and promoting suitable supplementary or alternative livelihoods that substitute or have a lower impact on natural resources and therefore, positively influence social and ecological outcomes of the CCAs. Whilst it can be inferred from most respondents that 'pro-conservation' behaviour has been promoted by other actions, they confirmed that community perceptions of and participation in CBC management decision-making emerged highly contingent on deriving benefits. Furthermore, most respondents suggested the 'success' of this action category is highly linked with the actions of *improved knowledge dissemination* and *strengthened institutional capacity* (discussed subsequently). A notable example of this action includes the development of local aquaculture activities in the *Bay of Ranobe,* and vegetable gardens in the *Urok Islands,* which appear to have had some positive local socio-economic impact. Furthermore, entrance fees from tourists to the CCAs in the *Bay of Ranobe* do contribute to the running costs of *FIMIHARA*, as well as fund community development projects implemented thus far (e.g. building of a clinic).

However, many community respondents in these two regional cases acknowledged these activities involve a limited number of community members, which minimizes their community-wide impact. Whilst not a direct action implemented within the *Urok Island* CMPA initiative, community respondents confirmed alternative livelihood sources such as cashew and palm fruit, which are sold or traded within the community, has indirectly aided the CMPA by reducing pressure on coastal and marine resources. However, some respondents did emphasize the irony of how increased focus on cashew production has become a common method of trade for rice (i.e. a former staple traditional crop), as shown elsewhere on in the archipelago (see Madeira, 2016). Lastly, in the *Olifants Estuary*, and more specifically the village of *Papendorp,* has received funding, and work has begun on refurbishing the Papendorp Guesthouse, which if accomplished can increase tourism and income to members of the community. However, in general partner respondents acknowledged in this case, there has been a lack of direct involvement and support from the State at a national level, though respondents confirmed local government has been involved.

9.2.3.3. Action Category 3: Improve Knowledge Dissemination

Increase recognition & inclusion of LEK
Conservation requires coupling an adequate knowledge base, inclusive of Local Ecological Knowledge (LEK), to appropriate institutional structures and behaviour (e.g. Mascia et al., 2003; Rands et al., 2010; Aswani et al., 2018). Accordingly, *knowledge dissemination* within communities, and between communities and partners, has been shown fundamental to enhancing CBC (e.g. Bodin & Crona, 2009; Ruiz-Mallen et al., 2014; Aswani et al., 2018). Knowledge dissemination actions identified in the three cases have focused on identifying, and if necessary, revitalizing, and improving alignment of LEK with conventional scientific knowledge. This action therefore links to cultural alignment actions. Furthermore, the importance of local-level representatives to the ability of community members to access and disseminate knowledge, and specifically disseminate LEK to partners is noted here, in accordance with past CBC research (e.g. Berdej &

Armitage, 2016; Steenbergen & Warren, 2018). Moreover, this action, like *strengthening actor relations*, requires multi-actor collaboration characterized by improved communication. Accordingly, the two common knowledge dissemination actions that emerged from the three cases focused on improving community awareness of the CCAs and their governance arrangements and rules through multi-actor meetings. Notable examples in the three cases include the aforementioned *FIMIHARA* and *UMC* meetings in the two regional cases, and OEMF meetings in the *Olifants Estuary*. This action also relates to the ability to build local representative capacity (discussed below).

Improve Partner-Community Knowledge Dissemination
Dissemination of knowledge from partners to the community was also an important action that helped to raise awareness of broader issues, and enabled community participation from a more informed position. The establishment and subsequent meetings of the CBOs were a key action in this regard, as in all three cases both the former and later actions hinged on local representative feedback. Most respondents confirmed that in all three cases, these actors possess the most, and often only, contact with partners and their technical knowledge. Partner actions focused on knowledge dissemination to representatives in all three cases involving multi-actor meetings (examples of which have been described above).

9.2.3.4. Action Category 4: Strengthen Institutional Capacity
Institutional capacity building has been shown to better predict 'win–win' CBC outcomes (Brooks, 2016). Once again, *capacity* refers here to an ability to combine various forms of capital within CBC governance to produce the desired result. Capacity building activities in the three cases are strongly linked to the ability to disseminate knowledge (as discussed above). Furthermore, institutional capacity

building promotes *actor interactions* that can *strengthen actor-relations*, which are considered important for multiple CBC outcomes (Brooks, 2016; Baird et al., 2019a; O'Connell et al., 2019).

Strengthen Local Capacity and Skill Development
The first action regarding local capacity building relates to external partner support of the establishment of CBOs and their institutional structures and responsibilities. This involved Reef Doctor in the *Bay of Ranobe*, Tiniguena in the *Urok Islands*, and the social partners, notably the EEU and some civil society members, at the *Olifants Estuary*. Moreover, in all three cases local capacity building actions occurred through external partner support of the multi-actor meetings.

The second action related to local capacity is local skill development through community development projects. This emerged as a particularly important action across various service delivery and alternative livelihood projects in all three cases. For example, skill development facilitated by Reef Doctor has taken place within local aquaculture activities in the *Bay of Ranobe*. Furthermore, community-monitoring initiatives have occurred in both the *Urok Islands*, and the *Olifants Estuary*, facilitated by Tiniguena, and the University of Cape Town respectively. Tiniguena's local capacity building actions also extend to specific skill development through the employment of local staff.

Strengthen Partner Capacity
Notwithstanding local CBOs characterized by incapacity, respondents confirmed capacity building actions need to identify and develop not only local but partner institutional capacity through targeted capacity building initiatives. This was predominately related to the State departments. Many respondents also noted this

action could increase State representative participation in their respective CBC institutions.

<u>Increase Capacity Retention</u>
A lack of capacity and high turnover of State representatives was deemed by respondents in all cases to be constraining in this regard. Accordingly, partner respondents specifically noted a need to retain not only community, but State capacity by developing strategies for the transfer of capacity to newly elected individuals, as otherwise a 'capacity-vacuum' persists. This was noted as especially important considering the rotation of local representatives, and community respondent's calls for greater rotation of these representatives. This has been attempted through the various multi-actor collaborative meetings in the respective cases (as discussed previously).

9.2.4. Common External Influences
As introduced in *Chapter 5, external influences* refer to consideration of the external factors and conditions that can either enable or constrain achievement of the desired result. Common external influences that emerged from the three cases included State instability characterized by high levels of *fragility,* notably in the two regional cases. Political will was highlighted as an additional key 'State-centric' external influence, which can be considered largely enabling in the two regional cases. In both regional cases this is evidenced by recent national PA expansion and an NGO enabling agenda in both, which has positively influenced the promotion and recognition of CCAs. In contrast, political will was considered to being constraining the declaration of the proposed CCA at the *Olifants Estuary*.

An additional State-centric' external influence is inter-State departmental collaboration, which can either constrain or enable CCA implementation and governance. For example, in the *Olifants Estuary* partner respondents raised concerns about conflicting interests amongst conservation and other more economically lucrative industries like mining, and specifically noted that the DMR (i.e. the national mining department) is not being held accountable to the same standards as other departments regarding environmental impact. Therefore, recently awarded mining prospecting rights in vicinity of the *Olifants Estuary* could negatively affect both social and ecological outcomes of the proposed CCA.

External market strategies can also influence CCA management. For example, Guinea-Bissau's Bijagós Archipelago is a rich fishing area, and IUU by foreign artisanal and commercial fishing vessels represents an increasingly serious problem for the State, and more specifically a major constraint on management of the CMPA in the *Urok Islands*. Similarly, in Madagascar some respondents raised concerns over commercial fishing, notably Chinese vessels encroaching on the LMMAs of the *Bay of Ranobe*, which they considered to be constraining effective management of fish stocks. Furthermore, the introduction of CCA enabling legislation – often a result of external pressure from international commitments such as the CBD's *Aichi Targets* and *Post-2020 Global Biodiversity Framework* and the *FAO's* 2014 *Voluntary SSF Guidelines* – can potentially influence CCA implementation and governance, and has done so in the two regional cases. In contrast, while these international commitments are acknowledged to have promoted creation of CBC enabling legislation in South Africa, this has not managed to catalyse implementation of this enabling legislation, especially in the coastal and marine realm.

Lastly, international donor agendas, often closely linked to State stability and international commitments, may fluctuate, and therefore, if withdrawn can affect CBC implementation and governance. Non-State partner respondents commonly noted this in the two regional cases. Madagascar perhaps provides a most notable example of the potential vulnerability of a nation's conservation sector to international donor agendas (as introduced in *Chapter 6: section 6.1.1.*).

9.2.5. Common Issues Arising

As established previously (*Chapter 5: section 5.3.1.*) a key function of a ToC pathway is identifying a "perceived course whereby wrongs might be put to rights, deficiencies of behaviour corrected, [and] inequalities of condition alleviated" (Pawson & Tilley, 2004: p2). Accordingly, 'issues arising' from the three cases represent those requiring attention through future actions so as to adapt and ultimately develop a 'robust' ToC pathway. To reiterate, the term 'issues arising' relate to those identified after analysing the implementation of a CBC intervention's initial actions. Furthermore, this can include both newly arising issues and persistent *initial contextual issues,* which continue to constrain achievement of the intervention's *desired result.* A full assessment of the effectiveness of the initial actions implemented is beyond the present scope or focus. Nevertheless, in the case study analyses several common 'issues arising' were identified. Since 'issues arising' have been discussed throughout, particularly as they relate to absent *enablers* (*section 9.2.2.*), a brief summative discussion follows.

A key 'issue arising' that affects governance in the three cases relates to the capacity of both local and State representatives. Local incapacity in the three cases was characterized by ineffective community representation, plagued by issues with feedback, and at times inequitable distribution of benefits. Most respondents in all three cases specifically emphasized *poor* representation, and subsequently their involvement in decision-making, as a key 'issues arising'. Accordingly, across the three cases most respondents highlighted that low levels of local education as contributing factor toward ineffective local governance capacity. Furthermore, conflict between customary and CCA-related CBO institutional structures, caused by 'culturally-misaligned' election of representatives, which was emphasized by respondents in the two regional cases. However, these respondents also acknowledged that customary institutions have partially eroded, largely as a result of the increased in-migration of 'outsiders', and that this negatively influences the effectiveness of CBC governance. Moreover, in the case of the *Olifants Estuary,* increasingly diversified livelihood strategies emerged as key issue effecting CBC governance. In addition, respondents across the three cases also commonly noted exclusion of marginalized groups in decision-making as a 'issue arising', this most notably related to both village- and gender-based exclusion, which requires urgent attention.

'Issues arising' associated with the institutional capacity of State representatives concerned most notably weak governance participation in CBC institutions, and *poor* inter-departmental collaboration. This in accordance with recent research emphasizing ineffective management persistently plaguing all types of PAs (e.g. Coad et al., 2019), including specifically MPAs (e.g. Gill et al., 2017). More specifically, this has been shown to effect the effectiveness of managing CCAs (Garcia et al., 2014; Charles et al., 2016). Furthermore, issues also pertain to communities-partner collaboration. Notwithstanding a lack of intra-community *trust* – notably caused by weak local representation and inter-village community ties in all three cases – issues of community-partner trust emerged. Consequently, as established above, a lack of *trust* is a key 'issue arising' in all three cases that continues to constrain the effectiveness of collaboration associated with both intra-community and community-partner relations. The latter perceived as largely negatively regarding the State, and largely positively regarding other partners, notably the two NGOs in the regional cases, and social partners in the *Olifants Estuary*.

Whilst all cases possessed enabling CBC legislation, unlike CCA implementation in the regional cases, this has not materialized in South Africa. More specifically, in the *Olifants Estuary* case, while national CCA enabling legislation (potentially) devolves *de jure* management authority to the community-level, its complexity and the onerous implementation processes emerged as a key 'issue arising'. The negative effect of onerous legislation was also emphasized by respondents from the two regional cases. Nevertheless, legislative support has the potential to promote increased active and meaningful local community involvement in CBC, and is thus deemed a crucial enabler. Consequently, respondents called for more 'streamlined' legislative procedures concerning CCA implementation and legal PA recognition, and increased collaboration of State representatives with other partners and the local community. Moreover, a lack of local and State capacity means community 'aid dependency' for other partners remains an 'issue arising' in all cases. This emerged particularly strongly regarding a lack of State financial and technical support. On a related note, the short-term project cycle of donor funding was acknowledged by non-State partner respondents as a key 'issue arising', as implementation of interventions like CCAs is a long-term endeavour.

Lastly, an overarching 'issue arising' that was emphasized explicitly within all three cases was the *high levels of poverty* and *natural resource-dependence*. This in accordance with Ostrom (1990) and Agrawal (2002). In all three cases the alleviation of poverty was deemed highly influential for 'pro-conservation' *mindsets*, *institutions*, and ultimately *behaviour*. Likewise, a lack of alternative livelihoods in all cases, and a lack of access to external markets in the case of the *Olifants Estuary* and the *Urok Islands* specifically, emerged especially constraining to effective CBC governance.

9.2.6. Concluding Remarks

Exploring the three cases as a change process has provided some key insights for developing a CBC ToC pathway to improve initiation, implementation and governance of CBC interventions in South Africa. A central observation emerging from the three cases is how the various *enablers* are supportive of each other, and if in place can improve the likelihood of others being present. For example, high levels of poverty dictates the need for poverty alleviation. If poverty levels are decreased, this may improve the presence of other enablers, and therefore increase the probability that other proposed actions will lead to the desired result.

Common actions were undertaken across the three cases to facilitate a shift to CBC. The need to build multi-actor relations and networks based on trust, inclusion and open communication was found to be key to enable greater collaboration amongst actors involved in CBC interventions. Strengthening these actor relations in turn appears linked to actions such as increasing socio-cultural alignment of CBC initiatives. This includes identifying and developing alternative livelihoods to deliver tangible benefits, increasing alignment of CBC institutions with customary institutions, knowledge and practices, as well as strengthening both local and partner capacity. Strengthening local capacity is also linked to the ability to increase recognition and inclusion of LEK, and improve knowledge dissemination amongst all CBC institutional actors.

Whilst developing a ToC pathway emphasizes the strategic and systematic implementation of actions to achieve the desired result, it can be inferred from all three cases that the shift to a CBC mode of governance was not necessarily viewed as a systematic and iterative change process. More specifically, whilst initial actions, such as the establishment of CBOs and community-partner partnerships – and in the two regional cases actions related to socio-cultural alignment (e.g. use of *Dina* in the *Bay of Ranobe*) – did attempt this, it can be inferred from respondents in all three cases that actions were largely implemented in an *ad hoc* manner and subject to arising issues. Therefore, it can be concluded these cases may have lacked what Mayne (2017: p159) refers to, in relation to developing a robust ToC pathway, as a complete "solid and plausible intervention design". Consequently, future efforts might be better served by considering CBC initiation, implementation and governance as a holistic, strategic, systematic and iterative change process. Moreover, due to the interaction of enablers, which overlap with other development sectors, an improved strategic approach to CBC interventions may be better served by integrating other relevant development sectors, perhaps most notably health. This has been noted by some in other scholars (e.g. Harris, 2012; Robson & Rakotozafy, 2015; Reed et al., 2020).

9.3. Proposing a South African Empirical CBC ToC Pathway

As introduced in *Chapter 5*, ToC enables practically and backwardly 'mapping' the logical pathways and sequences of actions toward desired result within an intervention's change process (Connell & Klem, 2000; Stein & Valters, 2012; Valters, 2015). Furthermore, developing a ToC pathway involves identifying, "those components most essential to the goals of the targeted change effort" (Foster-Fishman et al., 2007: p93). Therefore, this section consolidates findings related to the *change elements* emerging from the three cases studies (as discussed above), together with the findings from the review of South Africa's progress with CBC in *Chapter 4*. Accordingly, this led to the modification of the *change elements* proposed in *Chapter 5*, most notably the *enablers*, and *actions* proposed. Consequently, this leads to a proposed 'field-based' *South African Empirical CBC ToC Pathway* which provides a 'template' emphasizing those components deemed most essential to enable a shift to CBC in the South African context. Therefore, it should be acknowledged that this pathway represents merely a generic pathway deemed specific to the South African context, but that would still require context- and intervention-specific modifications.

9.3.1. Desired Result

A core step in developing a ToC pathway involves articulating the *desired result*, which collectively represents the intermediary *desired outputs and outcomes*, and ultimately the *desired impact* (i.e. the final desired outcome(s)). Whilst monitoring and evaluation of the desired result is crucial to adapting and reformulating actions, and ultimately improving ToC pathways, this was beyond the focus of this study.

The desired impact used within this pathway remains the same as that used in the *Generic CBC ToC Pathway* in *Chapter 5*, which is to facilitate successful initiation, implementation and governance of CBC, in this case in South Africa. It is worth reiterating that the South African conservation policy and legislative framework enables local community participation in decision-making and the devolution of power to local actors and communities, and therefore promotes the desired impact (i.e. community-based conservation management), *when* contextually appropriate. Furthermore, once again based on *Chapter 5,* the three broad desired results for South Africa are kept as strengthened/improved 'pro-conservation' *mindsets, institutions* and *behaviour.* These desired results would obviously need to be modified to reflect a specific interventions 'goals'.

9.3.2. Issues Arising and Proposed Actions

To reiterate, the *enablers* and *actions* can be considered to possess a two-way feedback, in that the actions are considered to positively influence the presence of the enablers, which in turn are assumed when present to positively influence the effectiveness of the actions to achieve the desired result. Whilst CBC enabling legislation exists in the country, implementation has been extremely slow due to numerous socio-cultural and institutional challenges which continue to constrain the ability of the country to deliver positive social and ecological conservation outcomes, perhaps most notably a lack of political will. Accordingly, actions for facilitating CBC in South Africa are proposed in response to common 'issues arising' as identified in the above discussion, as well as those that specifically emerged in *Chapter 4.*

Consideration of the socio-cultural (past and present) and ecological characteristics of South Africa's approach to CBC is key to enabling CBC implementation and governance. The country's exclusionary colonial and apartheid past continues to negatively affect collaborative, multi-actor conservation governance. At a local level the country's 'celebrated' multi-cultural diversity manifests itself in typically heterogeneous community groups, with diverse socio-political, cultural and economic priorities. Many years of colonial, and more recently apartheid rule, have at the very least partially eroded many customary systems of governance. This was confirmed by both the South African literature, and respondents in *Chapter 4*. Notwithstanding this 'erosion' there are still shared values and functioning customary institutions and practices in local communities that play an important role in conservation in South Africa (Sunde et al., 2013, Sunde, 2014; Mbatha, 2018). Consequently, in some contexts these customary institutions may be appropriate for CBC, if they are still functioning, otherwise it is key that actions in CBC interventions be implemented to accommodate and better align these customary institutions with local CBC-related CBOs.

At the national and provincial-level, State departments and parastatal conservation agencies lack alignment with customary institutions and objectives. This requires specific and urgent attention. Consequently, findings from the case studies and *Chapter 4* suggest that a shift to a CBC mode of governance in the country requires building collaborative multi-actor CBC relations, characterized by trust and respect, with a shared and formalized CBC vision and objective (Hauck & Sowman, 2001; Napier et al., 2005; Matose & Watts, 2010). An additional key action related to strengthening collaborative multi-actor relations in the country is increased presence, and involvement of partners. Unlike in the two regional cases, NGO involvement 'championing' and supporting CBC interventions in South Africa appears to still be limited, and should be further promoted. Furthermore, the support of conservation agencies (i.e. national, provincial and parastatal) to foster CBC is very limited and guidance on how this may be facilitated and sustained is key. In the case of the *Olifants Estuary* despite a plethora of legislation, and local institutions willing to embrace governance responsibilities, the lack of guidance and support from a conservation agency has constrained progress. Nevertheless, the intervention's 'social partners' are substantially involved in supporting this case.

Thus, an additional key action for improved CBC multi-actor relations in South Africa relates to strengthening State capacity to give effect to legislation, largely through improved inter-departmental communication and collaboration. The State still pursues a largely silo-based approach to environmental management, though this has improved over time (e.g. Meissner et al., 2013; Colenbrander & Bavinck, 2017). In the case of the *Olifants Estuary*, the lack of integration and communication between actors involved in the land claim process and conservation efforts to implement the proposed CCA, delayed action and undermined this CBC intervention. The challenge of settling land claims and conservation goals was also noted nationally, by both respondents and in the literature in *Chapter 4* (e.g. Paterson, 2011, Cundill et al., 2013, Kepe, 2018).

South Africa is characterized by persistent challenges associated with local elite and State-capture (Fabricius & Collins, 2007; Standing, 2008; Sutton & Rudd, 2014

Sundström, 2016). However, findings and the literature suggest that local leaders can become strong CBC change agents for the country, if high levels of 'nested' external support is available to promote targeted capacity-building and sustained long-term technical and financial support (Hauck & Sowman, 2001; Fabricius & Cundill, 2010; Pool-Stanvliet et al., 2018). Yet, this support also requires strengthened community-partner social relations and networks, characterized by multi-actor trust and respect, as noted in both South African (Napier et al., 2005; Matose & Watts, 2010), and global literature (Pretty, 2003; Young et al., 2016). Findings suggest that local leaders will need to become agents of change regarding both *knowledge dissemination* and *equitable benefit-sharing*. This is necessary since the provision of long-term, sustainable and tangible community benefits consistently emerged as necessary for CBC interventions to thrive in the country (see *Chapter 4*), as was echoed by case study findings.

Greater community participation in CBC governance and especially in the design of rules that more closely align with local socio-economic and cultural priorities is not possible without improved local representation. This notably pertains to how CBC interventions can align with customary institutions and endeavours to alleviate poverty, which are considered prerequisites to promote successful implementation and governance of CBC interventions (Ostrom, 1990; Fabricius & Collins, 2007). Strengthening local leadership also requires greater political will characterized by not only enabling legislation, but political and institutional support of, and active participation of the State in CBC institutions. This is necessary to legitimize 'true' devolution of management authority and natural resource rights to the local level, and thus increase the effectiveness of CCA implementation and governance (Boonzaaier, 2012; Cundill et al., 2013).

9.3.3. Enablers

As has been stated throughout a set of enablers should never be considered to be 'set-in-stone', and like the ToC pathway development process, are never perfect, and should be open to constant adaptation. Accordingly, an expanded set of 20 enablers is now proposed from a South African perspective (*Table 9.1.*). These are based on the initial 14 enablers identified in *Chapter 3*, and all the findings discussed above. A notable change relates to addressing the enabler *low levels of poverty*. Whilst both Ostrom (1990) and Agrawal (2002), and many scholars since, have stipulated the importance of this enabler in achieving CBC outcomes, poverty is prevalent in Africa and South Africa, an unlikely to be addressed in the short term. This is especially the case in South African fishing communities (see Isaacs, 2011; Isaacs, & Witbooi, 2019). Therefore, *Strategies in place to address high levels of poverty* (i.e. *enabler 5* below) is introduced as a key enabler in the South African context. Moreover, Agrawal's (2002) prescribed need for the presence of *high levels of resource dependence* has been questioned throughout the case study findings, and specifically in discussions. These discussions notably related to potential *bidirectionality* and *nonlinearity* of this enabler. Therefore, based on the present context, and in accordance with tipping-point literature reviewed (e.g. Wilson et al., 2016), *Moderate levels of dependence on resources with the presence of alternative livelihood strategies* (i.e. *enabler 5* below), is introduced as a key enabler for South African CBC interventions.

Additional key revised and / or additional enablers include *enablers 8, 10, 11, 12, 14, 18* and *19* (see *Table 9.1.*). *Enabler 8* emphasizes the need for *CBC enabling legislation,* which emerged necessary to provide the required legitimacy for a CBC intervention to enforce its rules. *Enabler 10* reflects the case study findings which highlighted the need for *strong democratic processes and accountable local representatives.* In connection to local representation *enabler 11*, based upon a common finding within the case studies, emphasizes the need for *high levels of institutional alignment with customary institutions, knowledge and practices.*

An additional 'leadership' related enabler added here is the need for *The presence of a strong and motivated change agent from the community or external partner organizations* (i.e. *enabler 12*). In accordance with *Chapters 4 and 5*, the presence of *strong change agents* emerged as a particularly key enabler. These change agents represent either a community leader or other influential actors that can 'drive' change toward the desired result. However, these change agents can emerge from outside local leadership structures. That said, many partner respondents across the three cases, emphasized the importance of a strong State change agent for the potential success of implementing the many actions required to achieve a CBC intervention's desired result. This is specifically necessary within the South African CBC context, has been stated in past South African CBC literature (e.g. Shackleton, 2009).

An additional institutional enabler is represented by *enabler 14,* i.e. *High levels of gender mainstreaming and inclusion in all institutions, and decision-making processes.* The lack of women in CBC decision-making emerged in all three cases, and since these actors are key to many cultural processes and involved almost exclusively in certain harvesting activities, this was deemed a necessary addition. The topic of greater gender mainstreaming has also recently been strongly emphasized in the *Post-2020 Global Biodiversity Framework* (CBD, 2020). While the need for *high levels of accountability* is well-established in CBC literature, *enabler 18* seeks to accentuate its role, together with the importance of relations amongst actors built on *trust*, in promoting improved *collaboration* in CBC interventions. Finally, *Sustainable levels of articulation with external markets* (i.e. *enabler 19*) has been amended from Agrawal's (2002) critical factor of " Low levels of articulation with external markets" based upon the discussions throughout. In particular, this enabler, like that of *enabler 6* above related to resource-dependence, depicted the potentially *bidirectionality* and *nonlinearity* of an enabler. More specifically, based upon the case study findings, and given the lack of alternative livelihoods in many South African (and African) contexts, it is proposed that when external market articulation is more moderate this may enable greater community support for, and more effective CBC interventions. Nevertheless, this enabler will be extremely context-specific, and requires extensive research, which I strongly encourage.

		Enablers
Resource System & Users	1.	Clearly-defined resource system & user boundaries
	2.	Shared norms, values, interests & identities
	3.	Strong local leadership
	4.	Strong community ties
	5.	Strategies in place to address high levels of poverty
	6.	Moderate levels of dependence on resources with the presence of alternative livelihood strategies
	7.	Equitable distribution of tangible benefits from common property resources
Institutional Structure & Externalities	8.	CBC enabling legislation
	9.	The presence of a community-based institution with devolved decision-making power
	10.	Strong democratic processes, and accountable local representatives
	11.	High levels of institutional alignment with customary institutions, knowledge and practices
	12.	The presence of a strong and motivated change agent from the community or external partner organizations
	13.	Locally devised access and management rules
	14.	High levels of gender mainstreaming and inclusion in all institutions, and decision-making processes
	15.	Rules strongly align with local socio-economic and -cultural priorities
	16.	Ease in enforcement of rules, and conflict resolution
	17.	Ability to continuously monitor, learn and adapt institutions
	18.	High levels of accountability and collaboration characterized by relations built on trust amongst actors
	19.	Sustainable levels of articulation with external markets
	20.	The presence of 'nested' governance with high levels of initial and ongoing external financial and technical support

9.3.4. Assumptions

As introduced in *Chapter 5*, assumptions ultimately consider, "the needs, interests and behaviour of stakeholders and other key actors" (van Es et al., 2015: p16). The underlying assumptions made relate to the potential of actions to lead to the desired result, which is a crucial change element within any ToC pathway. As in the *Generic CBC ToC Pathway* presented in *Chapter 5,* the *rationale assumption* is that implementation and governance of CBC initiatives may offer a much needed and viable alternative approach to centralized conservation within appropriate contexts. *Causal assumptions* are once again broken down into *causal pathway* and *causal link assumptions*. Accordingly, the *causal pathway assumptions* assume the presence of the expanded set of 20 key enablers (see *Table 9.1.*) to positively influence achieving the desired result, and the intervention actions will strengthen the status of these enablers. The *causal link assumptions*, however, relate to the specific actions proposed, and align with those introduced in *Chapter 5*. These assumptions discussed briefly below.

The ability to *strengthen actor relations* assumes that all actors are willing to engage in the process, and that actors possess a shared vision and objectives or the potential to achieve this. Furthermore, increasing the *socio-cultural-alignment* of CBC interventions primarily assumes the presence of customary institutions and practices linked to conservation, which can positively influence achieving the desired result, or if in an eroded state have the potential to be revitalized. Furthermore, it assumes the presence of customary leaders willing to support the CBC intervention, and therefore, better align it with customary governance systems. Moreover, (potentially) suitable and socially acceptable alternative livelihoods that do not constrain achieving ecological conservation outcomes are assumed to decrease poverty levels, and therefore, promote community support for the CBC intervention. Additionally, a key assumption is a willingness for local resource users to supplement, modify, or find alternatives to their present livelihoods, and participate in and build capacity related to engaging in the newly introduced livelihood.

The central assumption regarding the ability to *improve knowledge dissemination* amongst CBC actors is the acceptance of, and willingness to incorporate LEK. Furthermore, it assumes the ability to reconcile both LEK and scientific knowledge. Moreover, it assumes the willingness of those that possess the knowledge to share it, where these 'knowledge holders' are inclusive of local community members and representatives, and external partners. These assumptions are also relevant to *strengthening institutional capacity*. Practical assumptions associated with the ability to *strengthen institutional capacity* are again the presence of willing and motivated actors. In addition, the availability of technical and financial resources of partners to facilitate the capacity building process is assumed. Lastly, political will and participation of the State is assumed for the CBC capacity building process, and equally relevant to the ability to *strengthen actor relations*.

9.3.5. External Influences

Once again it should also be acknowledged that external influences can either positively or negatively affect progression through *ToC pathways,* and facilitators of these change processes need to be aware of these influences. Therefore, the proposed *South African Empirical CBC ToC Pathway* should be considered as a 'model of contribution' and not a 'model of causation' to the desired result (*Cf.* Mayne, 2015).

A key external influence, in the case of South Africa's transition to CBC, as discussed above, is a lack of inter-departmental communication and collaboration amongst State actors related to the land claims process, and the CCA implementation process. Furthermore, issues with inter-departmental communication extend to other horizontal and vertical departmental relations. For example, horizontal issues of communication and collaboration at a national level between those responsible for mining and conservation (e.g. DEA, DAFF and the DMR), and vertical issues between relevant local and national departments. Therefore, on a related note negative additional external influences in the country relate to the State's support for economically lucrative sectors such as commercial fishing and mining, which is particularly pronounced in view of South Africa's economic crises (Rice et al., 2017; Bond, 2019). Accordingly, the implementation of *Operation Phakisa*, South Africa's blue economy initiative – which is largely focused on commercial industries such as oil and gas – is a major potential negative external influence in the context of coastal CBC initiatives (Bond, 2019; RSA, 2019b). Lastly, the State's support of various international agreements, for example, the CBD's *Aichi Targets* and proposed *Post-2020 Global Biodiversity Framework*, the FAO's *Voluntary SSF Guidelines*, as well as regional commitments including those of the AU and SADC, will also potentially positively influence the achievement of the desired result. Consequently, all actors involved in CBC interventions in the country need to be mindful of these external influences, as well as others that may arise, that can constrain or enable achievement of a CBC intervention's desired result.

9.3.6. A South African Empirical CBC ToC Pathway

As van Es et al. (2015: p14) state, ToC pathways can't promise delivery of the desired result, but should rather be considered a means to "better understand the system they are part of without oversimplifying it, in order to support change in a strategic and responsive way...[and] learn from how the process evolves in reality, so that strategies can be reviewed and adjusted along the way." Accordingly, *Figure 9.2.* proposes a *South African Empirical CBC ToC Pathway* that provides a generic approach to transitioning to a community-based mode of conservation governance in South Africa. Furthermore, this pathway attempts to promote strategic action based on the findings of the case studies, as well as the review of South Africa's progress with CBC in *Chapter 4*. While the two regional cases differ in certain instances to the South African context, many overlapping aspects exist. This notably includes the high levels of poverty, and low levels of State capacity, and onerous legal and institutional processes. Moreover, in particular the South African context can benefit from learnings emerging from the two regional cases in terms of the enabling effect of the presence of empowered non-State partners, which due to a reluctance of the State to relinquish power is not the case in South Africa.

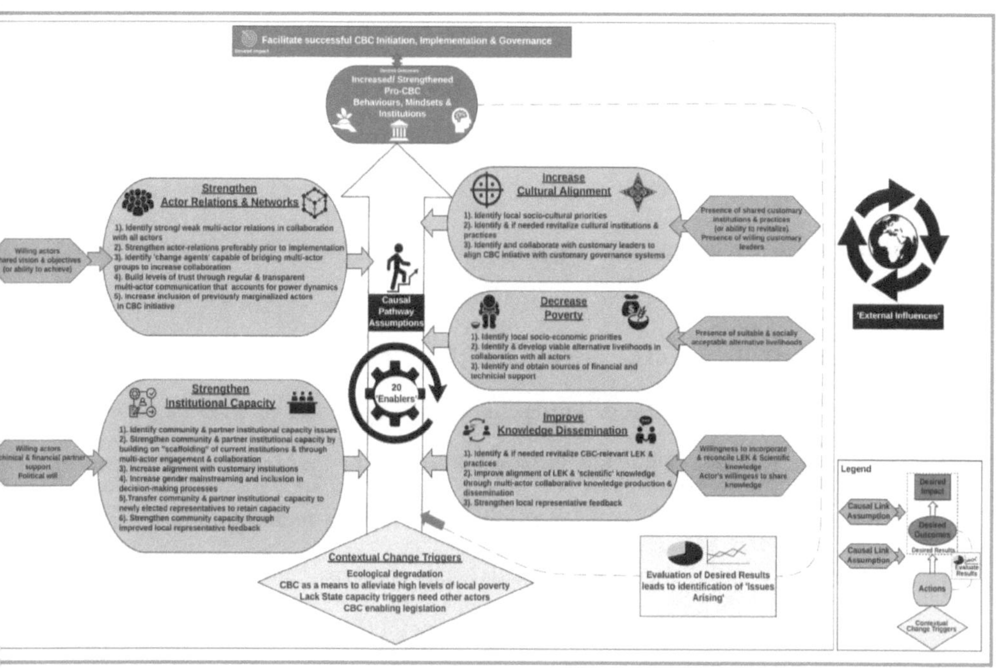

Facilitate successful CBC Initiation, Implementation & Governance

Increased/ Strengthened Pro-CBC Behaviours, Mindsets & Institutions

Strengthen Actor Relations & Networks
1). Identify strong/ weak multi-actor relations in collaboration with all actors
2). Strengthen actor-relations preferably prior to implementation
3). Identify 'change agents' capable of bridging multi-actor groups to increase collaboration
4). Build levels of trust through regular & transparent multi-actor communication that accounts for power dynamics
5). Increase inclusion of previously marginalized actors in CBC initiative

Willing actors
shared vision & objectives
(or ability to achieve)

Increase Cultural Alignment
1). Identify local socio-cultural priorities
2). Identify & if needed revitalize cultural institutions & practices
3). Identify and collaborate with customary leaders to align CBC initiative with customary governance systems

Presence of shared customary institutions & practices (or ability to revitalize). Presence of willing customary leaders

Causal Pathway Assumptions

Decrease Poverty
1). Identify local socio-economic priorities
2). Identify & develop viable alternative livelihoods in collaboration with all actors
3). Identify and obtain sources of financial and technical support

Presence of suitable & socially acceptable alternative livelihoods

'External Influences'

Strengthen Institutional Capacity
1). Identify community & partner institutional capacity issues
2). Strengthen community & partner institutional capacity by building on "scaffolding" of current institutions & through multi-actor engagement & collaboration
3). Increase alignment with customary institutions
4). Increase gender mainstreaming and inclusion in decision-making processes
5). Transfer community & partner institutional capacity to newly elected representatives to retain capacity
6). Strengthen community capacity through improved local representative feedback

20 'Enablers'

Willing actors
technical & financial partner support
Political will

Improve Knowledge Dissemination
1). Identify & if needed revitalize CBC-relevant LEK & practices
2). Improve alignment of LEK & 'scientific' knowledge through multi-actor collaborative knowledge production & dissemination
3). Strengthen local representative feedback

Willingness to incorporate & reconcile LEK & Scientific knowledge
Actor's willingness to share knowledge

Contextual Change Triggers
Ecological degradation
CBC as a means to alleviate high levels of local poverty
Lack State capacity triggers need other actors
CBC enabling legislation

Evaluation of Desired Results leads to identification of 'Issues Arising'

Legend
Desired Impact
Causal Link Assumption
Desired Outcomes
Causal Link Assumption
Desired Results
Evaluate Results
Actions
Contextual Change Triggers

Therefore, this pathway differs from, and builds upon the *Generic CBC ToC pathway* presented in *Chapter 5* by offering more detailed and specific information on the change elements. The changes include the expanded set of 20 enablers (*section 9.3.3.*), as well as a revised and expanded list of South Africa-specific proposed actions, which are all based upon all the empirical chapters of the book. These actions can be considered to focus on addressing two key and overarching 'issues arising' within the South African context, namely the need to improve land tenure processes and security, as well as levels of poverty and multi-actor collaboration. Furthermore, this finding has recently been shown elsewhere in Africa (e.g. Davis & Goldman, 2019). In doing so this ToC pathway attempts to offer a useful tool for guiding *how* to facilitate successful initiation, implementation and governance of CBC in South Africa. Lastly, the proposed actions work interdependently, and can therefore strengthen one another and subsequently the presence of the 20 enablers. Furthermore, it is important to note that this ToC pathway is not sequential but iterative, allowing for 'relapses' in the form of learning and adaptation, and progression can involve multiple different routes towards the desired result. In particular, this involves constant modification of actions so as to address 'issues arising', most notably through attempts to improve the presence of the enablers.

9.4. Contribution to theory and practice of CBC governance and proposed future research

In view of the increasing calls for greater involvement of communities in conservation management and governance (e.g. Brechin et al., 2003, Brockington et al., 2010), further empirical evidence on how to facilitate successful CBC implementation and governance, *when* contextually-appropriate, to realize desired social and ecological outcomes, is required. Agrawal (2001) and Ostroms' (1990) enablers for successful CPRM are widely recognized as necessary conditions for CBC initiatives. However, this book proposed from the outset that these conditions may not be sufficient to affect change toward a CBC mode of governance. Accordingly, this study has sought to improve understanding of how to enable this shift to CBC governance by exploring *how* initiation, implementation and governance of CBC takes place as a change process.

A critical exploration of selected coastal CBC case studies in *Madagascar (Chapter 6)*, *Guinea-Bissau (Chapter 7)*, and *South Africa (Chapters 4 & 8)* – guided by the theoretical foundations established in *Chapters 3* and *5* – identified common and potentially useful insights into *contextual change triggers, enablers, actions* and *assumptions,* and *issues arising.* This *change perspective* offers conservation scholars and practitioners a practical way of thinking about the process of transitioning to a community-based mode of governance, and a framework comprising various change elements to guide initiating and implementing CBC interventions. Firstly, a literature inspired *ToC Pathway Design Framework* and *Generic CBC ToC Pathway* was presented in *Chapter 5*, and secondly, a *South African Empirical CBC ToC Pathway* is presented in this final chapter. Consequently, this book makes a specific contribution towards understanding and potentially addressing the *policy-praxis disjuncture* observed in South Africa's coastal CBC efforts, by providing a context-specific guide on *how to* navigate this impasse, given the presence of enabling legislation, and other enabling and constraining factors, conditions and processes. In addition, the process, and the proposed *ToC pathways*, also offer a practical framework to study, promote and appraise CBC implementation and ongoing governance practices, and could be modified to apply to other regional and global contexts. More specifically, this *ToC Pathway Design Framework* and the proposed *ToC pathways* are useful to both the planning of future CBC interventions, as well as post-implementation adaptation of established CBC interventions. Consequently, whilst acknowledging the CBC change process will always be context-specific and influenced by external factors, conditions and processes, the *ToC pathways* presented offer a 'template' for CBC change which can be adapted and applied within a specific context for a specific CBC intervention, i.e. which is the express goal of the book.

A key finding within all three case studies has been that CBC implementation and governance has not been considered, or taken place as a strategic, systematic and iterative change process. Rather change has been dealt with largely in an *ad hoc* manner so as to address arising issues. This may be as a result of the importance of other overlapping goals (e.g. community health or education projects). Nevertheless, this book strives to highlight the advantages of viewing CBC as a collaborative, strategic and systematic change process (as introduced in *section 9.2.6.*). More specifically, it emphasizes the need to develop a context-appropriate, systematic and iterative set of actions to support the change to CBC governance; a process, which is more likely to achieve CBC governance and associated outcomes, especially when integrated with the goals of other development sectors. Thus, this change perspective aligns with calls for a more holistic, nuanced and systems view of CBC implementation and governance, which is able to learn and adapt, and therefore, is deemed better able to tackle wicked complex conservation problems by facilitating positive social and ecological conservation outcomes (e.g. Game et al., 2014).

Lastly, this book has reconfirmed and contributed to the thinking on some of the fundamental ideas underpinning *governance* and *commons theory*. Concerning *governance theory* this book has reinforced that robust collaboration within CBC institutions is crucial to the CBC change process, and requires improved relations of trust and channels of communication, and acknowledges the significance of power and its diverse representations and applications, and holds those possessing various forms of power accountable. More specifically, whilst governance literature refers to different modes of governance (*Chapter 3: section 3.2.4.*), it does not fully consider the processes involved in making the shift from one mode of governance to another. Moreover, African CBC literature has shown very mixed results with making the shift towards a CBC mode of governance. Consequently, this book emphasizes that greater attention to *how* this shift takes place, and how it can contribute to guiding future CBC endeavours, is required. Finally, this research also advances knowledge in *commons theory* by contributing new insights relating to common enabling and constraining factors, conditions and processes for successful CPRM by proposing a set of 14 key socio-institutional enablers based on an extensive review of CBC literature, and subsequently, offering a set of 20 specific enablers that would support CBC efforts in South Africa, although these may be more widely applicable. Consequently, whilst these theoretical contributions come specifically from an African perspective, findings potentially have global theoretical and practical relevance to diverse actors and across diverse CBC contexts.

9.5. Conclusion

The increasing recognition of the shortcomings of 'protectionist' conservation approaches in tackling contemporary wicked conservation problems, emphasize that more nuanced, participatory and holistic governance approaches, better able to deliver social and ecological outcomes, are required. CBC provides a viable conservation approach to address the dual conservation objectives of protecting biodiversity and improving social well-being, within certain contexts. This is also true within many contexts in South Africa. However, global and South African CBC efforts thus far have produced mixed results. Accordingly, greater theoretical and empirical understanding of CBC initiation, implementation and governance, informed by practical evidence, is required. To this end this book has attempted to explore and better understand the factors, conditions and processes that enable CBC initiation, implementation and governance. This has been informed by findings from practical work in two African countries and one case-in-progress in South Africa.

In doing so this book has provided specific insights into addressing the *policy-praxis* disjuncture currently experienced in South African CBC initiatives, most notably the implementation and governance of coastal CCAs. More specifically, it has highlighted common *contextual change triggers*, key *enablers* and *'issues arising'*, which emphasize important avenues for proposing relevant and practical future *actions*, though these actions will be highly context-specific. Moreover, it has produced a potentially useful regional and global *ToC Pathway Design Framework,* and *Generic CBC ToC Pathway,* and has subsequently through empirical research proposed a *South African Empirical CBC ToC Pathway* that outlines how change towards CBC can happen in the country. However, these *change pathways* should have wider application. Consequently, these research outputs can contribute to improving conservation planning from a more people-centred perspective in South Africa, but also in the region and beyond. In this regard, the book makes a valuable contribution to CBC theory and practice. It also provides ideas for further research on factors and processes for improving the 'success' of CBC interventions moving forward, which are specifically inspired by viewing these interventions as collaborative, strategic and systematic change processes.

References:

1. Abreu, A.J.G.D. (2012). Migration and development in contemporary Guinea-Bissau: a political economy approach. Doctoral book, SOAS, University of London. Available from: https://eprints.soas.ac.uk/14243/1/Abreu_3401.pdf

2. Acheson, J. (2000). Clearcutting Maine: Implications for the theory of common property resources. *Human Ecology*, 28(2): 145-169.

3. Acheson, J. (2011). Ostrom for anthropologists. *International Journal of the Commons*, 5(2): 319–339.

4. Acheson, J.M., & McCloskey, J. (2008). Causes of deforestation: The Maine case. *Human ecology*, *36*(6): 909-922.

5. Ackrill, R., Kay, A., & Zahariadis, N. (2013). Ambiguity, multiple streams, and EU policy. *Journal of European Public Policy*, 20(6): 871-887.

6. Adams, W.M. (2003). Nature and the Colonial Mind. In: *Decolonizing Nature: Strategies for Conservation in a Postcolonial Era.* Adams, W.M. & Mulligan, M. (Eds). London: Earthscan. 16–50.

7. Adams, W.M. (2004a). *The global conservation regime.* In: *Against Extinction: The story of conservation.* Adams, W.M. (Ed.). London: Earthscan publication. 43-66.

8. Adams, W.M. (2004b). *Against Extinction: The Story of Conservation.* London: Earthscan.

9. Adams, W.M., & Hulme, D. (2001a). *Conservation and communities: Changing narratives, policies and practices in African conservation.* In: *African Wildlife and Livelihoods: The Promise and Performance of Community Conservation.* Hulme, D. & Murphree, M. (Eds.). London: James Currey. 9-23.

10. Adams, W.M., & Hulme, D. (2001b). If community conservation is the answer, what is the question? *Oryx*, 35: 193-200.

11. Adato, M., Carter, M.R., & May, J. (2006). Exploring poverty traps and social exclusion in South Africa using qualitative and quantitative data. *The Journal of Development Studies*, 42(2): 226-247.

12. Adhikari, B. (2005). Poverty, property rights and collective action: understanding the distributive aspects of common property resource management. *Environment and development economics*, 10(1): 7-31.

13. Adhikari, B., & Lovett, J.C. (2006). Transaction costs and community-based natural resource management in Nepal. *Journal of environmental management*, 78(1): 5-15.

14. Agrawal, A. (1999). Accountability in decentralization: A framework with South Asian and West African cases. *The Journal of Developing Areas*, 33(4): 473-502.

15. Agrawal, A. (2001). Common property institutions and sustainable governance of resources. *World development*, 29(10): 1649 -1672.

16. Agrawal A. (2002). *Common pool resources and institutional sustainability.* In: *The drama of the commons.* Ostrom, E., Dietz, T., Dolsak, N., Stern, P., Stonich, S., & Weber, E.U. (Eds.). Washington, DC: National Academy Press. 41–85.

17. Agrawal, A. (2005). *Environmentality: Technologies of Government and the Making of Subjectivities.* Durham, NC, USA and London, UK: Duke University Press.

18. Agrawal, A. (2014). Studying the commons, governing common-pool resource outcomes: Some concluding thoughts. *Environmental Science & Policy*, 36: 86-91.

19. Agrawal, A., & Gibson, C.C. (1999). Enchantment and disenchantment: The role of community in natural resource conservation. *World Development*, 27(4): 629–649.

20. Agrawal, A., & Gibson, C.C. (2001). *The role of community in natural resource conservation. Communities and the environment: Ethnicity, gender, and the state in community-based conservation.* New Brunswick: Rutgers University Press. 1-31.

21. Agarwal, B. (2009). Rule making in community forestry institutions: the difference women make. *Ecological Economics*, 68(8–9): 2296– 2308.

22. Agrawal, A., & Redford, K. (2009). Conservation and displacement: An overview. *Conservation & Society*, 7(1): 1-10.

23. Aheto, D.W., Kankam, S., Okyere, I., Mensah, E., Osman, A., Jonah, F.E., & Mensah, J.C. (2016). Community-based mangrove forest management: Implications for local livelihoods and coastal resource conservation along the Volta estuary catchment area of Ghana. *Ocean & Coastal Management*, 127: 43-54.

24. AISSR (Amsterdam Institute of Social Science Research). (2017). *AISSR Ethical Procedure and Questions, 17 February 2017.* Available from: http://aissr.uva.nl/binaries/content/assets/subsites/amsterdam-institute-for-social-science-research/map-1/aissr-ethical-review-procedure-and-questions.pdf?1487234676823

25. Ajzen, I. (1991). The theory of planned behavior. *Organizational Behavior and Human Decision Processes*, 50: 179-211.

26. Akamani, K., Wilson, P.I., & Hall, T.E. (2015). Barriers to collaborative forest management and implications for building the resilience of forest-dependent communities in the Ashanti region of Ghana. *Journal of Environmental Management*, 151: 11-21.

27. Alcántara-Salinas, G., Hunn, E. S., & Rivera-Hernández, J.E. (2015). Avian Biodiversity in Two Zapotec Communities in Oaxaca: The Role of Community-Based Conservation in San Miguel Tiltepec and San Juan Mixtepec, Mexico. *Human Ecology,* 43(5): 735-748.

28. Alexander, S.M., Andrachuk, M., & Armitage, D. (2016). Navigating governance networks for community-based conservation. *Frontiers in Ecology and the Environment*, 14(3): 155-164.

29. Alexander, S.M., Epstein, G., Bodin, Ö., Armitage, D., & Campbell, D. (2018a). Participation in planning and social networks increase social monitoring in community-based conservation. *Conservation Letters*, 11(5): e12562. Available from: https://doi.org/10.1111/conl.12562

30. Alexander, S.M., Bodin, Ö., & Barnes, M. (2018b). Untangling the drivers of community cohesion in small-scale fisheries. *International Journal of the Commons*, 12(1): 519–547.

31. Alexander, S.M., Jones, K., Bennett, N.J., Budden, A., Cox, M., Crosas, M., Game, E.T., et al. (2019a). Qualitative data sharing and synthesis for sustainability science. *Nature Sustainability*. Available from: https://doi.org/10.1038/s41893-019-0434-8

32. Alexander, S.M., Provencher, J.F., Henri, D.A., Taylor, J.J., Lloren, J.I., Nanayakkara, L., et al. (2019b). Bridging Indigenous and science-based knowledge in coastal and marine research, monitoring, and management in Canada. *Environmental Evidence*, 8(1), 1-24.

33. Algotsson, E. (2006). Wildlife conservation through people-centred approaches to natural resource management programmes and the control of wildlife exploitation. *Local Environment*, 11(1): 79-93.

34. Allison, E.H., Perry, A.L., Badjeck, M-C., Adger, W.N., Brown, K., Conway, D., Halls, A.S. et al. (2009). Vulnerability of national economies to the impacts of climate change on fisheries. Fish and fisheries, 10(2): 173-196.

35. Almeida, O.T., Lorenzen, K., & McGrath, D.G. (2009). Fishing agreements in the lower Amazon: for gain and restraint. *Fisheries Management and Ecology*, 16(1): 61-67.

36. Andersson, K. (2013). Local governance of forests and the role of external organizations: Some ties matter more than others. *World Development*, 43: 226-237.

37. Anderson, D., & Grove, R. (1987). *The scramble for Eden: Past, present and future in African conservation.* In: *Conservation in Africa: People, Policies and Practice.* Anderson, D., & Grove, R. (Eds.). Cambridge: Cambridge University Press. 1-12

38. Andersson, K., & Agrawal, A. (2011). Inequalities, institutions, and forest commons. *Global environmental change*, 21(3): 866-875.

39. Andersson, K., Benavides, J.P., & León, R. (2014). Institutional diversity and local forest governance. *Environmental Science & Policy*, 36: 61-72.

40. Andrew, N.L., Béné, C., Hall, S.J., Allison, E.H., Heck, S., & Ratner, B.D. (2007). Diagnosis and management of small-scale fisheries in developing countries. *Fish and Fisheries*, 8: 227–240.

41. Angelsen, A., Jagger, P., Babigumira, R., Belcher, B., Hogarth, N.J., Bauch, S., Börner J, Smith-Hall, C., & Wunder, S. (2014). Environmental income and rural livelihoods: a global-comparative analysis. *World Development,* 64(1): S12–S28. https://doi.org/10.1016/j.worlddev.2014.03.006

42. Annecke, W., Masubele, M. (2016). A review of the impact of militarisation: the case of rhino poaching in Kruger National Park, South Africa. Conservation & Society, 14(3): 195–204.

43. Antona, M., Bienabe, E.M., Salles, J.M., P'Echard, G., Aubert, S., & Ratsimarison, R. (2004). Rights transfers in Madagascar biodiversity policies: achievements and significance. *Environment and Development Economics*, 9: 825–847.

44. Amador, R., Casanova, C., & Lee, P. (2015). Ethnicity and perceptions of bushmeat hunting inside Lagoas de Cufada Natural Park (LCNP), Guinea-Bissau. *Journal of Primatology*, 3(121). Available from: http://dx.doi.org/10.4172/2167-6801.1000121

45. Araral, E. (2014). Ostrom, Hardin and the commons: A critical appreciation and a revisionist view. *Environmental Science & Policy*, 36: 11-23.

46.   Araral, E. (2016). Reply to: Design principles in commons science: A response to "Ostrom, Hardin and the commons"(Araral). *Environmental Science & Policy*, 100(61): 243-244.

47.   Armitage, D., Berkes, F., & Doubleday, N. (2010). *Adaptive co-management: collaboration, learning, and multi-level governance*. Canada: UBC Press.

48.   Armitage, D., Berkes, F., Dale, A., Kocho-Schellenberg, E., & Patton, E. (2011). Co-management and the co-production of knowledge: Learning to adapt in Canada's Arctic. *Global Environmental Change*, 21(3): 995-1004.

49.   Armitage, D., de Loe, R., & Plummer, R. (2012). Environmental governance and its implications for conservation practice. *Conservation Letters*, 5(4): 245-255.

50.   Armitage, D., Charles, A., & Berkes, F. (2017). *Governing the coastal commons: Communities, Resilience and Transformation*. London, New York: Earthscan/Routledge.

51.   Armitage, D.R., Okamoto, D.K., Silver, J.J., Francis, T.B., Levin, P.S., Punt, A.E., Davies, I.P., et al. (2019). Integrating Governance and Quantitative Evaluation of Resource Management Strategies to Improve Social and Ecological Outcomes. *BioScience*, 69(7): 523-532. Available from: https://academic.oup.com/bioscience

52.   Armitage, D., Mbatha, P., Muhl, E.K., Rice, W.S., & Sowman, M. (2020). Governance principles for community-centered conservation in the post-2020 global biodiversity framework. *Conservation Science and Practice*, 2(2): e160. Available from: https://doi.org/10.1111/csp2.160

53.   Astuti, R. (1995). *People of the sea: Identity and descent among the Vezo of Madagascar*. Cambridge, UK: Cambridge University Press.

54.   Aswani, S. (2005). Customary sea tenure in Oceania as a case of rights based fishery management: does it work? *Reviews in Fish Biology and Fisheries,* 15: 285–307.

55.   Aswani, S., Albert, S., Sabetian, A. & Furusawa, T. (2007). Customary management as precautionary and adaptive principles for protecting coral reefs in Oceania. *Coral Reefs*, 26: 1009–1021.

56.   Aswani, S., Lemahieu, A., & Sauer, W. H. (2018). Global trends of local ecological knowledge and future implications. *PloS one*, 13(4): e0195440. Available from: https://doi.org/10.1371/journal.pone.0195440

57.   Baggio, J.A., Barnett, A.J., Perez-Ibara, I., Brady, U., Ratajczyk, E., Rollins, N., Rubiños, C., et al. (2016). Explaining success and failure in the commons: the configural nature of Ostrom's institutional design principles. *International Journal of the Commons*, 10(2): 417-439.

58.   Baird, J., Plummer, R., Schultz, L., Armitage, D., & Bodin, Ö. (2019a). How Does Socio-institutional Diversity Affect Collaborative Governance of Social–Ecological Systems in Practice?. *Environmental management*, 63(2): 200-214.

59.   Baird, J., Schultz, L., Plummer, R., Armitage, D., & Bodin, Ö. (2019b). Emergence of Collaborative Environmental Governance: What are the Causal Mechanisms?. *Environmental management*, 63(1): 16-31.

60.   Baker, L.R., Tanimola, A.A., & Olubode, O.S. (2018). Complexities of local cultural protection in conservation: the case of an Endangered African primate and forest groves protected by social taboos. *Oryx*, 52(2): 262-270.

61.   Baldursdóttir, S., Gunnlaugsson, G., & Einarsdóttir, J. (2018). Donor dilemmas in a fragile state: NGO-ization of community healthcare in Guinea-Bissau. *Development Studies Research*, 5(1): S27-S39. Available from: https://doi.org/10.1080/21665095.2018.1500143

62.   Balfour, D., Barichievy, C., Gordon, C., & Brett, R. (2019). A Theory of Change to grow numbers of African rhino at a conservation site. *Conservation Science and Practice*, 1(6): e40. Available from: https://doi.org/10.1111/csp2.40

63.   Balmford, A., & Cowling, R. (2006). Fusion or failure? The future of conservation biology. *Conservation Biology,* 20: 692–695.

64.   Baland, J.-M., & Platteau, J.-P. (1996). *Halting Degradation of Natural Resources: Is There a Role for Rural Communities?* Oxford: Clarendon Press.

65.   Balint, P.J., & Mashinya, J. (2006). The decline of a model community-based conservation project: Governance, capacity, and devolution in Mahenye, Zimbabwe. *Geoforum*, 37(5): 805-815.

66.   Bamberg, S., Rees, J., & Seebauer, S. (2015). Collective climate action: Determinants of participation intention in community-based pro-environmental initiatives. *Journal of Environmental Psychology*, 43: 155-165.

67.   Baral, N., & Heinen, J. T. (2005). The Maoist people's war and conservation in Nepal. *Politics and the Life Sciences*, 24(1): 2-11.

68. Barbier, E.B. (2010). Poverty, development, and environment. *Environment and Development Economics*, 15(6): 635-660.

69. Bare, M., Kauffman, C., & Miller, D.C. (2015). Assessing the impact of international conservation aid on deforestation in sub-Saharan Africa. *Environmental Research Letters*, *10*(12): 125010. http://dx.doi.org/10.1088/1748-9326/10/12/125010

70. Barendse, J., Roux, D., Currie, B., Wilson, N., & Fabricius, C. (2016). A broader view of stewardship to achieve conservation and sustainability goals in South Africa. *South African Journal of Science*, 112(5/6). Available from: http://dx.doi.org/10.17159/sajs.2016/20150359.

71. Barnes, M.L., Bodin, Ö., Guerrero, A.M., McAllister, R.J., Alexander, S.M., & Robins, G. (2017). The social structural foundations of adaptation and transformation in social–ecological systems. *Ecology and Society*, 22(4): 16. Available from: https://doi.org/10.5751/ES-09769-220416

72. Barnes, M.L., Mbaru, E., & Muthiga, N. (2019). Information access and knowledge exchange in co-managed coral reef fisheries. *Biological Conservation*, 238: 108198. Available from: https://doi.org/10.1016/j.biocon.2019.108198

73. Barnes-Mauthe, M., Oleson, K.L.L., & Zafindrasilivonona, B. (2013). The total economic value of small-scale fisheries with a characterization of post-landing trends: An application in Madagascar with global relevance. *Fisheries Research*, 147: 175-185.

74. Barnes-Mauthe, M., Oleson, K.L.L., Brander, L.M., Zafindrasilivonona, B., Oliver, T.A., & van Beukering, P. (2015). Social capital as an ecosystem service: Evidence from a locally managed marine area. *Ecosystem Services*, 16: 283-293.

75. Barr, A., Fafchamps, M., & Owens, T. (2005). The governance of non-governmental organizations in Uganda. *World Development*, 33(4): 657-679.

76. Barrett, C.B., Travis, A.J., & Dasgupta, P. (2011). On biodiversity conservation and poverty traps. *Proceedings of the National Academy of Sciences*, 108: 13907-13912.

77. Bates, D.G., & Plog, F. (1990). *Human adaptive strategies.* New York: Mcgraw-Hill.

78. Batterbury, S.P.J., & Fernando, J.L. (2006). Rescaling governance and the impacts of political and environmental decentralization: an introduction. *World Development*, 34: 1851– 1863.

79. Baum, H.S. (1999). Community organizations recruiting community participation: Predicaments in planning. *Journal of Planning Education and Research*, 18(3): 187-199.

80. Bavinck, M., Chuenpagdee, R., Jentoft, S., &, Kooiman., J. 2013. *Governability of Fisheries and Aquaculture: Theory and Applications.* MARE Publication Series 7. Available from: https://link.springer.com/book/10.1007%2F978-94-007-6107-0

81. Bavinck, M., Berkes, F., Charles, A., Dias, A.C.E., Doubleday, N., Nayak, P., & Sowman, M. (2017). The impact of coastal grabbing on community conservation–a global reconnaissance. *Maritime studies*, 16(1): 8. Available from: https://doi.org/10.1186/s40152-017-0062-8

82. Baynham-Herd, Z., Redpath, S., Bunnefeld, N., Molony, T., & Keane, A. (2018). Conservation conflicts: Behavioural threats, frames, and intervention recommendations. *Biological Conservation*, 222: 180-188.

83. Beckley, T.M., Martz, D., Nadeau, S., Wall, E., & Reimer, B. (2009). Multiple capacities, multiple outcomes: Delving deeper into the meaning of community capacity. *Journal of rural and community development*, 3(3): 56–75.

84. Beeton, R.J.S., & Lynch, A.J.J. (2012). Most of nature: A framework to resolve the twin dilemmas of the decline of nature and rural communities. *Environmental Science and Policy*, 23: 45-56.

85. Belhabib, D., Sumaila, U. R., and Pauly, D. (2015a). Feeding the poor: contribution of West African fisheries to employment and food security. *Ocean and Coastal Management*, 111: 72–81.

86. Belhabib, D., Nahada, V.A., Blade, D., & Pauly, D. (2015b). *Fisheries in troubled waters: a catch reconstruction for Guinea-Bissau. 1950–2010.* The University of British Columbia, Fisheries Centre Working Paper Series, 2015, no. 72. Available from: http://www.seaaroundus.org/doc/publications/wp/2015/Belhabib-et-al-Guinea-Bissau.pdf

87. Belhabib, D., Lam, V. W., & Cheung, W.W. (2016). Overview of West African fisheries under climate change: Impacts, vulnerabilities and adaptive responses of the artisanal and industrial sectors. *Marine Policy*, 71: 15-28.

88. Belle, E.M., Stewart, G.W., De Ridder, B., Komeno, R.J., Ramahatratra, F., Remy-Zephir, B., & Stein-Rostaing, R.D. (2009). Establishment of a community managed marine reserve in the Bay of Ranobe, southwest Madagascar. *Madagascar Conservation & Development*, 4(1): 31-37.

89. Benbow, S., Humber, F., Oliver, T.A., Oleson, K.L.L., Raberinary, D., Nadon, M., Ratsimbazafy, H., & Harris, A. (2014). Lessons learnt from experimental temporary octopus fishing closures in south-west Madagascar: benefits of concurrent closures. *African Journal of Marine Science*, 36(1): 31-37.

90. Béné, C., Riba, A., & Wilson, D. (2020). Impacts of resilience interventions–Evidence from a quasi-experimental assessment in Niger. *International Journal of Disaster Risk Reduction*, 43: 101390. Available from: https://doi.org/10.1016/j.ijdrr.2019.101390

91. Bennett, N.J., & Dearden, P. (2014). Why local people do not support conservation: community perceptions of marine protected area livelihood impacts, governance and management in Thailand. *Marine Policy*, 44: p107-116.

92. Bennett, N.J., & Satterfield, T. (2018). Environmental governance: A practical framework to guide design, evaluation, and analysis. *Conservation Letters*, 11: e12600. Available from: https://doi.org/10.1111/conl.12600

93. Bennett, N.J., Roth, R., Klain, S.C., Chan, K., Clark, D.A., Cullman, G., Epstein, G., et al. (2016). Mainstreaming the social sciences in conservation. *Conservation Biology*, 31(1): 56-66.

94. Bennett, N.J., Roth, R., Klain, S.C., Chan, K., Christie, P., Clark, D.A., Cullman, G., et al. (2017). Conservation social science: Understanding and integrating human dimensions to improve conservation. *Biological Conservation*, 205: 93–108.

95. Bennett, N.J., Whitty, T.S., Finkbeiner, E., Pittman, J., Bassett, H., Gelcich, S., & Allison, E.H. (2018). Environmental stewardship: A conceptual review and analytical framework. *Environmental management*, 61(4): 597-614.

96. Bennett, A., Acton, L., Epstein, G., Gruby, R.L., & Nenadovic, M. (2018). Embracing conceptual diversity to integrate power and institutional analysis: Introducing a relational typology. *International Journal of the Commons*, 12(2): 330-357.

97. Bennett, N.J., Di Franco, A., Calò, A., Nethery, E., Niccolini, F., Milazzo, M., & Guidetti, P. (2019). Local support for conservation is associated with perceptions of good governance, social impacts, and ecological effectiveness. *Conservation Letters*, 12(4): e12640. Available from: https://doi.org/10.1111/conl.12640

98. Bentler, P.M., Jackson, D.N., & Messick, S. (1971). Identification of content and style: a two-dimensional interpretation of acquiescence. *Psychological bulletin*, 76(3): 186-204.

99. Benzinho, J., & Rosa, M. (2015). *Tourist Guide: Discovering Guinea-Bissau. NGO Afectos com Letras.* Available from: https://eeas.europa.eu/sites/eeas/files/tourist_guide_guinea-bissau_eu_acl2018_en_web.pdf

100. Berdej, S.M., & Armitage, D.R. (2016). Bridging organizations drive effective governance outcomes for conservation of Indonesia's marine systems. *PloS one*, *11*(1), e0147142. Available from: https://doi.org/10.1371/journal.pone.0147142

101. Berge, E., & Van Laerhoven, F. (2011). Governing the commons for two decades: a complex story. *International Journal of the Commons*, 5(2): 160-187.

102. Berkes, F. (1989). *Common Property Resources: Ecology and Community Based Sustainable Development.* London: Belhaven Press.

103. Berkes, F. (2004). Rethinking community-based conservation. *Conservation Biology*, vol. 18, no.3, p621-630.

104. Berkes, F. (2007). *Community-based conservation in a globalized world. Proceedings of the National Academy of Sciences*, 104(39): 15188-15193.

105. Berkes, F. (2009a). Evolution of co-management: role of knowledge generation, bridging organizations and social learning. *Journal of Environmental Management*, 90(5): 1692–1702.

106. Berkes, F. (2009b). Community Conserved Areas: Policy Issues in Historic and Contemporary Context. *Conservation Letters*, 2(1): 20–25.

107. Berkes, F. (2011). *Restoring unity: the concept of social-ecological systems.* In: *World Fisheries: A Social-Ecological Analysis.* Ommer, R.E., Perry, R.I., Cochrane, K., & Cury, P. (Eds.). Oxford: Wiley-Blackwell. 9-28.

108. Berkes, F., & Folke, C. (1998). *Linking social and ecological systems: management practices and social mechanisms for building resilience.* Cambridge, UK: Cambridge University Press.

109. Berkes, F., & Seixas, C. (2004). *Lessons from community self-organization and cross-scale linkages in four Equator Initiative projects.* Equator Initiative Synthesis Report. Winnipeg: Natural Resources Institute, University of Manitoba. Available from: http://www.umanitoba.ca/institutes/natural_resources/nri_cbrm_projects_eiprojects.html

110. Berkes, F., Feeny, D., McCay, B.J., & Acheson, J.M. (1989). The benefits of the commons. *Nature*, 340(6229): 91-93.

111. Berkes, F., Colding, J., & Folke, C. (2003). *Navigating social–ecological systems: building resilience for complexity and change.* Cambridge, UK: Cambridge University Press.
112. Berkes, F., Seixas, C.S., Fernandes, D., Medeiros, D., Maurice, S., & Shukla, S. (2004). *Lessons from Community Selforganization and Cross-scale Linkages in Four Equator Initiative Projects.* Centre for Community-Based Resource Management, University of Manitoba, Winnipeg.
113. Berkes, F., Arce-Ibarra, M., Armitage, D., Charles, A., Loucks, L., Makino, M., Satria, A., et al. (2016). *Analysis of Social-Ecological Systems for Community Conservation.* Community Conservation Research Network, Halifax Canada. Available online: https://www.communityconservation.net/wp-content/uploads/2016/01/Analysis-of-Social-Ecological-Systems-for-Community-Conservation-CCRN-2.pdf
114. Bersaglio, B., & Cleaver, F. (2018). Green grab by bricolage-The institutional workings of community conservancies in Kenya. *Conservation and Society,* 16(4): 467-480.
115. Bertzky, B., Corrigan, C., Kemsey, J., Kenney, S., Ravilous, C., Besançon, C., & Burgess, N.D. (2012). *Protected Planet 2012: Tracking Progress Towards Global Targets for Protected Areas.* IUCN, Gland, Switzerland, and UN Environment Programme–World Conservation Monitoring Centre, Cambridge, UK. Available from: https://cmsdata.iucn.org/downloads/protected_planet_report.pdf
116. Beyers, C., & Fay, D. (2015). After restitution: Community, litigation and governance in South African land reform. *African Affairs,* 114(456): 432-454.
117. Biai, J., Campredon, P., Ducrocq, M., Henriques, A., & Ocante da Silva, A. (2003). *Plan de Gestion de la Zone Côtière des Iles Urok (Formosa, Nago and Chediã), 2004 – 2008, Biosphere Reserve on the Bolama/ Bijagós Archipelago, Guinea Bissau.* Bissau: IBAP.
118. Bie, K., Addison, P.F., & Cook, C.N. (2018). Integrating decision triggers into conservation management practice. *Journal of applied ecology,* 55(2): 494-502.
119. Biermann, F., & Gupta, A. (2011). Accountability and legitimacy: an analytical challenge for earth system governance. *Ecological Economics,* 70(11): 1854-1855.
120. Biggs, R., Westley, F.R., & Carpenter, S.R. (2010). Navigating the back loop: fostering social innovation and transformation in ecosystem management. *Ecology and society,* 15(2): 9. Available from: http://www.ecologyandsociety.org/vol15/iss2/art9/
121. Biggs, D., Abel, N., Knight, A.T., Leitch, A., Langston, A., & Ban, N.C. (2011). The implementation crisis in conservation planning: could "mental models" help?. *Conservation Letters,* 4(3): 169-183.
122. Biggs, D., Cooney, R., Roe, D., Dublin, H.T., Allan, J.R., Challender, D.W., & Skinner, D. (2017). Developing a theory of change for a community-based response to illegal wildlife trade. *Conservation Biology,* 31(1): 5-12.
123. Biggs, D., Ban, N.C., Castilla, J.C., Gelcich, S., Mills, M., Gandiwa, E., Etienne, M., Knight, A.T., Marquet, P.A., & Possingham, H.P. (2019). Insights on fostering the emergence of robust conservation actions from Zimbabwe's CAMPFIRE program. *Global Ecology and Conservation,* 17: e00538. Available from: https://doi.org/10.1016/j.gecco.2019.e00538
124. BirdLife International. (2019). *Important Bird Areas factsheet: Olifants river estuary.* Available from: http://datazone.birdlife.org/site/factsheet/olifants-river-estuary-iba-south-africa
125. Blaikie, P.M. (2006) Is small really beautiful? Community-based natural resource management in Malawi and Bostwana. *World Development,* 34: 1942–1957.
126. Blamey, A., & Mackenzie, M. (2007). Theories of change and realistic evaluation: peas in a pod or apples and oranges?. *Evaluation,* 13(4): 439-455.
127. Blue Ventures. (2018). *New research: The expansion of Madagascar's protected area network.* Available from: https://blueventures.org/new-research-review-of-the-expansion-of-madagascars-protected-area-network/
128. Blue Ventures. (2019a). *A theory of change: communities think critically about pathways to sustainable management.* Available from: https://blog.blueventures.org/en/theory-change-communities-think-critically-pathways-sustainable-management/
129. Blue Ventures. (2019b). *MIHARI Network wins prestigious conservation prize.* Available from: https://blueventures.org/mihari-network-wins-prestigious-conservation-prize/
130. Bluwstein, J., & Lund, J.F. (2018). Territoriality by conservation in the Selous–Niassa Corridor in Tanzania. *World Development,* 101: 453-465.
131. Bluwstein, J., Moyo, F., & Kicheleri, R. P. (2016). Austere conservation: understanding conflicts over resource governance in Tanzanian wildlife management areas. *Conservation and Society,* 14(3): 218-231.

132. Blythe, J., Cohen, P., Abernethy, K., & Evans, L. (2017). *Navigating the transformation to community-based resource management.* In: *Governing the Coastal Commons.* Armitage, D., Charles, A., & Berkes, F. (Eds.). London: Routledge. 141-156

133. Bockstael, E., Bahia, N.C., Seixas, C.S., & Berkes, F. (2016). Participation in protected area management planning in coastal Brazil. *Environmental Science & Policy*, 60: 1-10.

134. Bodin, Ö. (2017). Collaborative environmental governance: Achieving collective action in social-ecological systems. *Science*, 357(6352): eaan1114. https://science.sciencemag.org/content/sci/357/6352/eaan1114.full.pdf

135. Bodin, Ö., & Crona, B. (2009). The role of social networks in natural resource governance: what relational patterns make a difference? *Global Environmental Change*, 19: 366-374.

136. Bodin, Ö., Ramirez-Sanchez, S., Ernstson, H. & Prell, C. (2011). *A social relational approach to natural resource governance.* In: *Social networks and natural resource management: uncovering the social fabric of environmental governance.* Bodin, Ö, & C. Prell C. (Eds.). Cambridge: Cambridge University Press. 3–28.

137. Bond, P. (2019). Blue Economy threats, contradictions and resistances seen from South Africa. *Journal of Political Ecology*, 26(1): 341-362.

138. Boonzaaier, C.C. (2010). Rural people's perceptions of wildlife conservation—the case of the Masebe Nature Reserve in Limpopo Province of South Africa. *Anthropology Southern Africa*, 1-2(33): 55–64.

139. Boonzaaier, C.C. (2012). Towards a Community-Based Integrated Institutional Framework for Ecotourism Management: The Case of the Masebe Nature Reserve, Limpopo Province of South Africa. *Journal of Anthropology*, Article ID 530643. Available from: https://doi.org/10.1155/2012/530643

140. Boonzaaier, C.C., & Wels, H. (2016). Juxtaposing a cultural reading of landscape with institutional boundaries: the case of the Masebe Nature Reserve, South Africa. *Landscape Research*, 41(8): 922-933.

141. Boon, E.K., & Ahenkan, A. (2018). *Africa and environmental health trends.* In: *Oxford Textbook of Nature and Public Health: The role of nature in improving the health of a population.* Frumkin, H. (Ed.). Oxford: Oxford University Press.

142. Boonstra, W.J. (2016). Conceptualizing power to study social-ecological interactions. *Ecology and Society*, 21(1): 21. Available from: http://dx.doi.org/10.5751/ES-07966-210121

143. Borrini-Feyerabend, G., & Hill, R. (2015). *Governance for the conservation of nature.* In: *Protected Area Governance and Management.* Worboys, G.L., Lockwood, M., Kothari, A., Feary, S., & Pulsford I. (Eds.). Canberra: ANU Press. 169–206.

144. Borrini-Feyerabend, G., Farvar, M.T., Nguinguini, J.C., & Ndangang, V.A. (2000). *Co-management of Natural Resources: Organising, Negotiating and Learning-by-doing.* Heidelberg: IUCN and GTZ. Available from: https://portals.iucn.org/library/node/7839

145. Borrini-Feyerabend, G., Dudley, N., Jaeger, T., Lassen, B., Pathak Broome, N., Phillips, A., & Sandwith, T. (2013). *Governance of Protected Areas: From understanding to action.* Best Practice Protected Area Guidelines Series No. 20, Gland, Switzerland: IUCN. Available from: https://www.iucn.org/content/governance-protected-areas-understanding-action

146. Bordonaro, L.I. (2009). Culture Stops Development!: Bijagó Youth and the Appropriation of Developmentalist Discourse in Guinea-Bissau. *African Studies Review*, 52(2): 69-92.

147. Borgatti, S.P., Everett, M.G., & Freeman, L.C. (2002). Ucinet 6 for Windows: Software for Social Network Analysis. Harvard, MA: Analytic Technologies.

148. Borgatti, S., Mehra, A., Brass, D., & Labianca, G. (2009). Network analysis in the social sciences. *Science*, 323: 892-895.

149. Borgerson, C., Johnson, S.E., Louis, E.E., Holmes, S.M., Anjaranirina, E.J.G., Randriamady, H.J., & Golden, C.D. (2018). The use of natural resources to improve household income, health, and nutrition within the forests of Kianjavato, Madagascar. *Madagascar Conservation & Development, 13*(1): 45-52.

150. Bottrill, M.C., & Pressey, R.L. (2012). The effectiveness and evaluation of conservation planning. *Conservation Letters*, 5(6): 407-420.

151. Boudreaux, K., & Nelson, F. (2011). Community conservation in Namibia: Empowering the poor with property rights. *Economic Affairs*, 31(2): 17-24.

152. Brandon, K., Redford, K.H., & Sanderson, S.E. (1998). *Parks in Peril: People, Politics and Protected Areas.* Washington DC: Island Press.

153. Brass, J.N. (2012). Why do NGOs go where they go? Evidence from Kenya. *World Development*, 40(2): 387-401.

154. Bray, D.B., Duran, E., & Molina, O. (2012). Beyond harvests in the commons: multi-scale governance and turbulence in indigenous/community conserved areas in Oaxaca, Mexico. *International Journal of the Commons,* 6(2): 151–178.

155. Brechin, S.R., Wilshusen, P.R., Fortwangler, C.L., & West, P.C. (2003). *Contested nature: Promoting international biodiversity with social justice in the twenty-first century.* Albany: State University of New York Press.

156. Brenier, A., & Vogel, A. (2017). *Integrating conservation and development in Madagascar's marine protected areas.* FAO Fish Aquaculture Technical Paper 603: 85–98. Available from: https://www.gret.org/2017/04/integrating-conservation-and-development-in-marine-protected-areas-in-madagascar/?lang=en

157. Brenier, A., Emanuel Ramos, E., & Henriques, A. (2009). Live from Urok! Urok Islands Community Marine Protected Area: lessons learned and impacts. Available from: http://www.prcmarine.org/sites/prcmarine.org/files/8B Live from Urok islands lessons l earned and impacts.pdf

158. Brenier, A., Ferraris, J., & Mahafina, J. (2011). Participatory assessment of the Toliara Bay reef fishery, southwest Madagascar. *Madagascar Conservation & Development,* 6(2), 60-67.

159. Brewer, T.D., Cinner, J.E., Green, A., & Pressey, R.L. (2012) Effects of human population density and proximity to markets on coral reef fishes vulnerable to extinction by fishing. *Conservation Biology,* 27: 443-452.

160. Brimont, L., & Karsenty, A. (2015). Between incentives and coercion: the thwarted implementation of PES schemes in Madagascar's dense forests. *Ecosystem Services,* 14: 113-121.

161. Brockington, D., and Igoe, J. (2006). Eviction for conservation: A global overview. *Conservation & Society,* 4: 424-470.

162. Brockington, D., Duffy, R., & Igoe, J. (2010). *Nature Unbound.* London: Earthscan.

163. Brockman, C.F. (1962). *Supplement to the Report to the Committee on Nomenclature.* In: *First World Conference on National Parks.* Adams, A.B. (Ed.). Washington, DC: National Park Service.

164. Bromley, D.W. (1992). *Making the Commons Work: Theory, Practice and Policy.* San Francisco: ICS Press.

165. Brooks, J.S. (2016). Design Features and Project Age Contribute to Joint Success in Social, Ecological, and Economic Outcomes of Community-Based Conservation Projects. *Conservation Letters,* 10(1): 23-32.

166. Brooks, J.S., & Tshering, D. (2010). A respected central government and other obstacles to community-based management of the matsutake mushroom in Bhutan. *Environmental Conservation,* 37(3): 336-346.

167. Brooks, T.M., Bakarr, M.I., Boucher, T., Da Fonseca, G.A.B., Hilton-Taylor, C., Hoekstra, J.M., Moritz, T., et al. (2004). Coverage provided by the global protected-area system: is it enough? *Bioscience,* 54: 1081–1091.

168. Brooks, T.M., Mittermeier, R.A., da Fonseca, G.A.B., Gerlach, J., Hoffmann, M., Lamoreux, J.F., Mittermeier, C.G., et al. (2006). Global biodiversity conservation priorities. *Science,* 313: 58–61.

169. Brooks, J.S., Waylen, K.A., Borgerhoff-Mulder, M., & Brosius, P. (2010). The effect of non-local socio-political context on community-based conservation interventions: evaluating ecological, economic, attitudinal and behavioural outcomes. Systematic Review Protocol. *Collaboration for Environmental Evidence.* Available from: http://www.environmentalevidence.org/wp-content/uploads/2014/07/Protocol82.pdf

170. Brooks, J.S., Waylen, K.A., & Mulder, M.B. (2012). How national context, project design, and local community characteristics influence success in community-based conservation projects. *Proceedings of the National Academy of Sciences,* 109(52): 21265-21270.

171. Brooks, J.S., Waylen, K.A., & Mulder, M.B. (2013). Assessing community-based conservation projects: a systematic review and multilevel analysis of attitudinal, behavioral, ecological, and economic outcomes. *Environmental Evidence,* 2(1): 2. Available from: https://environmentalevidencejournal.biomedcentral.com/articles/10.1186/2047-2382-2-2

172. Brosius, J.P., Tsing, A.L., & Zerner, C. (1998). *Representing communities: Histories and politics of community-based natural resource management. Society & Natural Resources,* 11: 157-168.

173. Brousselle, A., & Champagne, F. (2011). Program theory evaluation: Logic analysis. *Evaluation and Program Planning,* 34: 69-78.

174. Brown, K. (2002). Innovations for conservation and development. *The Geographical Journal*, 168(1): 6-17.

175. Brown, F.P. (2009). *Participatory forest management (PFM) discourse in South Africa: ecological modernisation in the developing world*. PhD book, University of KwaZulu-Natal, South Africa. Available from: https://researchspace.ukzn.ac.za/handle/10413/938

176. Brown, H.C.P., & Lassoie, J.P. (2010). Institutional choice and local legitimacy in community-based forest management: lessons from Cameroon. *Environmental Conservation*, 37(3): 261-269.

177. Brown, C.A., Van der Berg, E., Sparks, A., & Magoba, R.N. (2010). Options for meeting the ecological Reserve for a raised Clanwilliam Dam. *Water SA*, 36(4). Available from: http://www.scielo.org.za/scielo.php?script=sci_arttext&pid=S1816-79502010000400002

178. Bruneau, T.C. (2017). *The Guinea-Bissau case*. In: *Security forces in African states: cases and assessment. Rapid Communications in Conflict and Security (RCCS) Series, March 2017*. Shemella, P., & Tomb N. (Eds.). Cambria Press. Available from: https://calhoun.nps.edu/bitstream/handle/10945/52412/Bruneau_Guinea-Bissau.pdf?sequence=2

179. Bruggemann, J., Rodier, M., Guillaume, M., Andréfouët, S., Arfi, R., Cinner, J., Pichon M., et al. (2012). Wicked social-ecological problems forcing unprecedented change on the latitudinal margins of coral reefs: the case of southwest Madagascar. *Ecology and Society*, 17(4): 47. Available from: http://dx.doi.org/10.5751/ES-05300-170447

180. Burger, J., Ostrom, E., Norgaard, R.B., Policansky, D., & Goldstein, B.D. (2001). *Protecting the Commons: A Framework for Resource Management in the Americas*. Washington, D.C.: Island Press.

181. Büscher, B. (2013). *Transforming the frontier: peace parks and the politics of neoliberal conservation in Southern Africa*. Duke University Press.

182. Büscher, B., & Dietz, T. (2005). Conjunctions of Governance: The State and the Conservation-development Nexus in Southern Africa. *The Journal of Transdisciplinary Environmental Studies*, 4(2): 1-15.

183. Büscher, B., & Dressler, W. (2012). Commodity conservation. *Geoforum*, 43: 367–376.

184. Büscher, B., & Ramutsindela, M. (2015). Green violence: Rhino poaching and the war to save Southern Africa's peace parks. *African Affairs*, 115(458): 1-22.

185. Büscher, B., Sullivan, S., Neves, K., Igoe, J., & Brockington, D. (2012). Towards a Synthesized Critique of Neoliberal Biodiversity Conservation. *Capitalism Nature Socialism*, 23(2): 4–30.

186. Butchart, S.H., Clarke, M., Smith, R.J., Sykes, R.E., Scharlemann, J.P., Harfoot, M., Buchanan, G.M., et al. (2015). Shortfalls and solutions for meeting national and global conservation area targets. *Conservation Letters*, 8(5): 329-337.

187. Butler, A. (2017). *Contemporary South Africa*. 3rd Edition. London: MacMillan.

188. Cabral, A.I., & Costa, F.L. (2017). Land cover changes and landscape pattern dynamics in Senegal and Guinea Bissau borderland. *Applied geography*, 82: 115-128.

189. Calfucura, E. (2018). Governance, Land and Distribution: A Discussion on the Political Economy of Community-Based Conservation. *Ecological Economics*, 145: 18-26.

190. Campbell, L.M., Haalboom, B.J., & Trow, J. (2007). *Sustainability of community-based conservation: sea turtle egg harvesting in Ostional (Costa Rica) ten years later. Environmental Conservation*, 34(2): 122–131.

191. Campbell, B.M., Sayer, J.A., & Walker, B. (2010). Navigating trade-offs: working for conservation and development outcomes. *Ecology and Society*, 15(2): 16. Available from: http://www.ecologyandsociety.org/vol15/iss2/art16/

192. Campese, J., Sunderland, T., Greiber, T., & Oviedo, G. (2009). *Rights-based Approaches: Exploring Issues and Opportunities for Conservation*. Center for International Forestry Research and IUCN, Bogor, Indonesia.

193. Campredon, P., & Catry, P. (2018). *Bijagós Archipelago (Guinea-Bissau)*. In: *The Wetland Book. II. Distribution, Description and Conservation*. Finlayson, C.M., Milton, G.R., Prentice, R., & Davidson, N.C. (Eds.). Dordrecht: Springer. 276–284.

194. Cape Times. (2017a). *Ebenhaeser land claimants 'frustrated, poor' after R350m settlement*. Available from: https://www.iol.co.za/capetimes/news/ebenhaeser-land-claimants-frustrated-poor-after-r350m-settlement-10016324

195. Cape Times. (2017b). *Ebenhaeser beneficiaries fear land claim won't be paid*. Available from: https://www.iol.co.za/capetimes/news/ebenhaeser-land-claimants-frustrated-poor-after-r350m-settlement-10016324

196. Cape Times. (2019). *Historic Khoi, Griqua land claim hailed. 25ᵗʰ March 2019.* Available from: https://www.iol.co.za/capetimes/news/historic-khoi-griqua-land-claim-hailed-20063039

197. Capistrano, D., Samper K.C., Lee J.M., & Raudsepp-Hearne C. (2005) *Ecosystems and human well-being: multiscale assessments.* Volume 4. Millennium Ecosystem Assessment. Washington, D.C.: Island Press.

198. Carbonetti, B., Pomeroy, R., & Richards, D.L. (2014). Overcoming the lack of political will in small scale fisheries. *Marine Policy*, 44: 295–301.

199. Carlsson, L., & Berkes, F. (2005). *Co-management: Concepts and methodological implications. Journal of Environmental Management*, 75: 65-76.

200. Carruthers, E. (1989). Creating a national park, 1910 to 1926. *Journal of Southern African Studies,* 15(2): 188-216.

201. Carruthers, J. (2008). Conservation and Wildlife Management in South African National Parks 1930s-1960s. *Journal of the History of Biology*, 41(2): 203–236.

202. Carvalho, A.R., Williams, S., January, M., & Sowman, M. (2009). Reliability of community-based data monitoring in the Olifants River estuary (South Africa). *Fisheries Research*, 96(2-3): 119-128.

203. Carvalho, J.S., Marques, T.A., & Vicente, L. (2013). Population status of Pan troglodytes verus in Lagoas de Cufada Natural Park, Guinea-Bissau. *PLoS One*, 8(8): e71527. Available from: https://doi.org/10.1371/journal.pone.0071527

204. Casanova, C., Sousa, C., & Costa, S. (2014). Are animals and forests forever? perceptions of wildlife at Cantanhez Forest National Park, Guinea-Bissau Republic. Memória-Special issue in anthropology and environment. *Sociedade de Geografia de Lisboa, Lisboa*, 16: 69-104.

205. Castro, P., & Mouro, C. (2011). Psycho-Social Processes in Dealing with Legal Innovation in the Community: Insights from Biodiversity Conservation. *American Journal of Community Psychology*, 47(3-4): 362-373.

206. Catalano, A.S., Lyons-White, J., Mills, M.M., & Knight, A.T. (2019). Learning from published project failures in conservation. *Biological Conservation*, 238: 108223. Available from: https://doi.org/10.1016/j.biocon.2019.108223

207. Catry, P., Barbosa, C., Paris, B., Indjai, B., Almeida, A., Limoges, B., Silva, C., & Pereira, H. (2009). Status, ecology, and conservation of sea turtles in Guinea-Bissau. *Chelonian Conservation and Biology*, 8(2): 150-160.

208. Catry, T., Figueira, P., Carvalho, L., Monteiro, R., Coelho, P., Lourenço, P. M., Catry, P., et al. (2017). Evidence for contrasting accumulation pattern of cadmium in relation to other elements in Senilia senilis and Tagelus adansoni from the Bijagós archipelago, Guinea-Bissau. *Environmental Science and Pollution Research*, 24(32): 24896-24906.

209. CBD. (Convention on Biological Diversity). (1996). *Programme of Work.* Available from: https://www.cbd.int/agro/pow.shtml

210. CBD. (Convention on Biological Diversity). (2011). *Strategic Plan for Biodiversity 2011–2020 and the Aichi Targets.* Available from: https://www.cbd.int/doc/strategic-plan/2011-2020/Aichi-Targets-EN.pdf

211. CBD. (Convention of Biological Diversity). (2020). *Zero Draft of the Post-2020 Global Biodiversity Framework.* Available from: https://www.cbd.int/doc/c/efb0/1f84/a892b98d2982a829962b6371/wg2020-02-03-en.pdf

212. CER (Centre for Environmental Rights). (2019). *Legal Toolbox.* Available from: https://water.cer.org.za/legal-toolbox

213. Chaffin, B.C., Garmestani, A.S., Gunderson, L.H., Benson, M.H., Angeler, D.G., Arnold, C.A., Cosens, B., et al. (2016). Transformative environmental governance. *Annual Review of Environment and Resources*, 41: 399-423.

214. Chan, K.M., Balvanera, P., Benessaiah, K., Chapman, M., Díaz, S., Gómez-Baggethun, E., Gould, R., et al. (2016). Opinion: Why protect nature? Rethinking values and the environment. *Proceedings of the National Academy of Sciences*, 113(6): 1462-1465.

215. Chandra, A., & Idrisova, A. (2011). Convention on Biological Diversity: a review of national challenges and opportunities for implementation. *Biodiversity and Conservation*, 20(14): 3295-3316.

216. Charles, A. (2001). *Sustainable Fishery Systems.* Oxford: Wiley-Blackwell.

217. Charles, A. (2011). *Good practices for governance of small-scale fisheries,* In: *World Small Scale-Fisheries: Contemporary Visions.* Chuenpagdee, R. (Ed.). Delft, The Netherlands: Eburon Publishing. 285-298.

218. Charles, A. (2013). Governance of tenure in small- scale fisheries: Key considerations. *Land Tenure*, 1: 9-37.

219. Charles, A., Westlund, L., Bartley, D. M., Fletcher, W. J., Garcia, S., Govan, H., & Sanders, J. (2016). Fishing livelihoods as key to marine protected areas: insights from the World Parks Congress. *Aquatic Conservation: Marine and Freshwater Ecosystems*, 26: 165-184.

220. Charles, A., Kalikoski, D. & Macnaughton, A. (2019). *Addressing the climate change and poverty nexus: a coordinated approach in the context of the 2030 agenda and the Paris agreement*. Rome. FAO. Available from: http://www.fao.org/3/ca6968en/CA6968EN.pdf

221. Charles, A., Loucks, L., Berkes, F., & Armitage, D. (2020). Community science: A typology and its implications for governance of social-ecological systems. *Environmental Science & Policy*, 106: 77-86.

222. Charmaz, K. (2008). *Constructionism and the grounded theory method.* In: *Handbook of constructionist research.* Holstein, J.A., & Gubrium, J.F. (Eds.). New York: Guilford. 397-412.

223. Cheng, A.S., & Randall-Parker, T. (2017). Examining the influence of positionality in evaluating collaborative progress in natural resource management: Reflections of an academic and a practitioner. *Society & Natural Resources*, 30(9): 1168-1178.

224. Chiesura, A., & De Groot, R. (2003). Critical natural capital: a socio-cultural perspective. *Ecological Economics*, 44(2-3): 219-231.

225. Child, B. (2004). *Parks in Transition.* 1st edition. London: Earthscan.

226. Child, B. (2009). *Conservation in transition.* In: *Evolution and Innovation in Wildlife Conservation. Parks and Game Ranches to Transfrontier Conservation Areas.* Suich, H., & Child, B. (Eds.). London: Earthscan. 3–18

227. Child, M.F. (2009). The Thoreau ideal as a unifying thread in the conservation movement. *Conservation Biology*, 23: 241–243.

228. Child, B. (2019). *Sustainable Governance of Wildlife and Community-based Natural Resource Management: From Economic Principles to Practical Governance.* New York: Routledge.

229. Child, B., & Barnes, G. (2010). The conceptual evolution and practice of community-based natural resource management in southern Africa: past, present and future. *Environmental Conservation*, 37(3): 283 – 295.

230. Child, G., & Child, B. (2015). The conservation movement in Zimbabwe: an early experiment in devolved community based regulation. *South African Journal of Wildlife Research*, 45(1): 1-16.

231. Chuenpagdee, R., Degnbol, P., Bavinck, M., Jentoft, S., Johnson, D., Pullin, R., & Williams, S. (2005). *Challenges and Concerns in Capture Fisheries and Aquaculture.* In: *Fish for life: interactive governance for fisheries.* Kooiman, J., Bavinck, M., Jentoft, S., & Pullin, R. (Eds.). Amsterdam: Amsterdam University Press.

232. Chuenpagdee, R., & Jentoft, S. (2019). *Transdisciplinarity for small-scale fisheries governance: Analysis and Practice.* Cham: Springer Nature.

233. Cinner, J.E. (2007). The role of taboos in conserving coastal resources in Madagascar. *Traditional Marine Resource Management and Knowledge Information Bulletin*, 22: 15-23.

234. Cinner, J.E. (2009). Poverty and the use of destructive fishing gear near east African marine protected areas. *Environmental Conservation*, 36(4): 321-326.

235. Cinner, J.E. (2018). How behavioral science can help conservation. *Science*, 362(6417): 889-890.

236. Cinner, J.E., & Aswani, S. (2007). Integrating customary management into marine conservation. *Biological Conservation*, 140: 201-216.

237. Cinner, J.E., Wamukota, A., Randriamahazo, H., & Rabearisoa, A. (2009a). Toward institutions for community-based management of inshore marine resources in the Western Indian Ocean. *Marine Policy*, 33(3): 489-496.

238. Cinner, J., Fuentes, M.M.P.B., & Randriamahazo, H. (2009b). Exploring social resilience in Madagascar's marine protected areas. *Ecology and Society* 14(1): 41. Available from: http://www.ecologyandsociety.org/vol14/iss1/art41/

239. Cinner, J.E., Basurto, X., Fidelman, P., Kuange, J., Lahari, R., & Mukminin, A. (2012a). Institutional designs of customary fisheries management arrangements in Indonesia, Papua New Guinea, and Mexico. *Marine Policy*, 36(1): 278-285.

240. Cinner, J.E., McClanahan, T.R., Graham, N.A.J., Daw, T.M., Maina, J., Stead, S.M., Wamukota, A., et al. (2012b). Vulnerability of coastal communities to key impacts of climate change on coral reef fisheries. *Global Environmental Change*, 22(1): 12-20.

241. Cinner, J.E., McClanahan, T.R., MacNeil, M.A., Graham, N.A., Daw, T.M., Mukminin, A., Feary, D.A., et al. (2012c). Comanagement of coral reef social-ecological systems. *Proceedings of the National Academy of Sciences*, 109(14): 5219-5222.

242. Cinner, J.E., Dawb, T.M., McClanahan, T.R., Muthiga, N., Abunge, C., Hamed, S., Mwakag, B., et al. (2012d). Transitions toward co-management: The process of marine resource management devolution in three east African countries. *Global Environmental Change*, 22(3): 651-658.

243. Cinner, J.E., Daw, T., Huchery, C., Thoya, P., Wamukota, A., Cedras, M., & Abunge, C. (2014). Winners and losers in marine conservation: fishers' displacement and livelihood benefits from marine reserves. *Society & Natural Resources*, 27(9): 994-1005.

244. Ciriacy-Wantrup, S.V., & Bishop, R.C. (1975). *Common property as a concept in natural resources policy. Natural Resources Journal*, 15(4): 713-727.

245. Clark, N.L., & Worger, W.H. (2016). *South Africa: The rise and fall of apartheid*. 2nd Edition. London: Routledge.

246. Clayton, S., Litchfield, C., & Geller, E.S. (2013). Psychological science, conservation, and environmental sustainability. *Frontiers in Ecology and the Environment*, 11(7): 377-382.

247. Clayton, S., & Brook, A. (2005). Can psychology help save the world? A model for conservation psychology. *Analyses of Social Issues and Public Policy*, 5(1): 87-102.

248. Cleaver, F. (2017). *Development through bricolage: rethinking institutions for natural resource management*. London: Routledge.

249. Clement, F. (2013). For Critical Social-Ecological System Studies: Integrating Power and Discourses to Move Beyond the Right Institutional Fit. *Environmental Conservation*, 40(1): 1–4.

250. Clement, S., Moore, S. A., Lockwood, M., & Mitchell, M. (2015). Using insights from pragmatism to develop reforms that strengthen institutional competence for conserving biodiversity. *Policy Sciences*, 48(4): 463-489.

251. Clements, H. ., De Vos, A., Bezerra, J.C., Coetzer, K., Maciejewski, K., Mograbi, P.J., & Shackleton, C. (2021). The relevance of ecosystem services to land reform policies: Insights from South Africa. *Land Use Policy*, 100: 104939. Available from: https://doi.org/10.1016/j.landusepol.2020.104939

252. Coad, L., Watson, J. E., Geldmann, J., Burgess, N. D., Leverington, F., Hockings, M., Knights, K., et al. (2019). Widespread shortfalls in protected area resourcing undermine efforts to conserve biodiversity. *Frontiers in Ecology and the Environment*, 17(5): 259–264.

253. Cockayne, J., & Williams, P. (2009). *The Invisible Tide: Towards an International Strategy to Deal with Drug Trafficking Through West Africa*. New York: International Peace Institute. Available from: https://www.ipinst.org/wp-content/uploads/publications/west_africa_drug_trafficking_epub.pdf

254. Cock, J. (2007). *The War Against Ourselves: Nature, Power and Justice*. Johannesburg: Wits University Press.

255. Cock, J., & Fig, D. (2000). From colonial to community based conservation: environmental justice and the national parks of South Africa. *Society in transition*, 31(1): 22-35.

256. Cock, J. & Fig, D. (2002). *From Colonial to Community Based Conservation: Environmental Justice and the Transformation of National Parks (1994–1998)*. In: *Environmental Justice in South Africa*. McDonald, D. (Ed.). Cape Town: University of Cape Town Press.

257. Cockburn, J., Cundill, G., Shackleton, S., & Rouget, M. (2018). Towards place-based research to support social–ecological stewardship. *Sustainability*, 10(5): 1434. Available from: http://dx.doi.org/10.3390/su10051434

258. Cocks, M.L., Dold, T., & Vetter, S. (2012). 'God is my forest': Xhosa cultural values provide untapped opportunities for conservation. *South African Journal of Science*, 108(5-6): 52-59.

259. Coetzee, H., & Nell, W. (2019). The feasibility of national parks in South Africa endorsing a community development agenda: The case of Mokala National Park and two neighbouring rural communities. *Koedoe*: 61(1), 1-13.

260. Colding, J., & Barthel, S. (2019). Exploring the social-ecological systems discourse 20 years later. *Ecology and Society*, 24(1):2. Available from: https://doi.org/10.5751/ES-10598-240102

261. Coleman, J.S. (1990). *Foundations of social theory*. Cambridge, MA: Belknap Press of Harvard University Press.

262. Collen, W., Krause, T., Mundaca, L., & Nicholas, K.A. (2016). Building local institutions for national conservation programs: lessons for developing Reducing Emissions from Deforestation and Forest Degradation (REDD+) programs. *Ecology and Society*, 21(2):4. Available from: http://dx.doi.org/10.5751/ES-08156-210204

263. Colenbrander, D., & Bavinck, M. (2017). Exploring the role of bureaucracy in the production of coastal risks, City of Cape Town, South Africa. *Ocean & Coastal Management*, 150: 35-50.

264. Collier, P. (2019). *Fixing Fragility: The Role of the Private Sector and Local Institutions.* Brookings Institution. Available from: https://www.africaportal.org/publications/fixing-fragility-role-private-sector-and-local-institutions/

265. Collins, S., & Snel, H. (2008). *A perspective on community based tourism from South Africa: The TRANSFORM Programme 1996-2007.* In: *Responsible tourism: critical issues for conservation and development.* Spenceley, A. (Ed). London: Routledge. 85-106.

266. Conand, C., & Mara, E. (2000). *Sea Cucumbers in the Southwest of Madagascar: Problems of the Fishery and Sustainable Management.* In: *Coral Reefs of the Indian Ocean, Their Ecology and Conservation.* McClanahan, T., Sheppard, C.R.C., & Obura, D.O. (Eds.). New York: Oxford University Press. 436-437

267. Connell, J.P., & Kubisch, A.C. (1998). Applying a theory of change approach to the evaluation of comprehensive community initiatives: progress, prospects, and problems. *New approaches to evaluating community initiatives*, 2: 15-44.

268. Connell, J.P., & Klem, A. (2000). You can get there from here: Using a theory of change approach to plan urban education reform. *Journal of Educational and Psychological Consultation*, 11(1): 93-120.

269. Cooke, B. & Kothari, U. (2001). *Participation: The new tyranny.* London: Zed Books.

270. Correia, E., Granadeiro, J. P., Regalla, A., & Catry, P. (2018). Length-weight relationship of fish species from the Bijagós Archipelago, Guinea-Bissau. *Journal of Applied Ichthyology*, 34(1): 177-179.

271. Corson, C. (2012). From rhetoric to practice: how high-profile politics impeded community consultation in Madagascar's new protected areas. *Society & Natural Resources*, 25(4): 336-351.

272. Corson, C. (2017). A history of conservation politics in Madagascar. *Madagascar Conservation & Development,* 12(1): 1-12.

273. Coryn, C.L., Noakes, L.A., Westine, C.D., & Schröter, D.C. (2011). A systematic review of theory-driven evaluation practice from 1990 to 2009. *American journal of Evaluation*, 32(2): 199-226.

274. Costa, S., Casanova, C.C., Sousa, C., & Lee, P.C. (2013). The good, the bad and the ugly: perceptions of wildlife in Tombali (Guinea-Bissau, West Africa). *Journal of Primatology*, 2(1): 110. Available from: http://dx.doi.org/10.4172/2167-6801.1000110

275. Costa, S., Casanova, C., & Lee, P. (2017). What does conservation mean for women? The case of the Cantanhez Forest Natitnal Park. *Conservation & Society*, 15(2): 168-178.

276. Cousins, B. (2016). Land reform in South Africa is failing. Can it be saved?. *Transformation: Critical Perspectives on Southern Africa*, 92(1): 135-157.

277. Cox, R.L., & Underwood, E.C. (2011). The importance of conserving biodiversity outside of protected areas in Mediterranean ecosystems. *PLoS One*, 6(1): e14508. Available from: https://doi.org/10.1371/journal.pone.0014508

278. Cox, M., Arnold, G., & Tomás, S.V. (2010). A review of design principles for community-based natural resource management. *Ecology and Society,* 15(4): 38. Available from: http://www.ecologyandsociety.org/vol15/iss4/art38/

279. Cox, M., Villamayor-Tomas, S., & Arnold, G. (2016). Design principles in commons science: A response to "Ostrom, Hardin and the commons"(Araral). *Environmental Science & Policy*, 61: 238-242.

280. Cripps, G., & Gardner, C.J. (2016). Human migration and marine protected areas: Insights from Vezo fishers in Madagascar. *Geoforum*, 74: 49-62.

281. Crona, B., & Bodin, Ö. (2006). What you know is who you know? Communication patterns among resource users as a prerequisite for comanagement. *Ecology and Society*, 11(2): 7. Available from: http://www.ecologyandsociety.org/vol11/iss2/art7/

282. Crona, B., & Bodin, Ö. (2010). Power asymmetries in small-scale fisheries: a barrier to governance transformability? *Ecology and Society*, 15(4): 32. Available from: http://www.ecologyandsociety.org/vol15/iss4/art32/

283. Crona, B., & Hubacek, K. (2010). The right connections: how do social networks lubricate the machinery of natural resource governance. *Ecology and Society*, 15(4): 18. Available from: http://www.ecologyandsociety.org/vol15/iss4/art18/

284. Crona, B., Ernstson, H., Prell, C., Reed, M., & Hubacek, K. (2011). *Combining social network approaches with social theories to improve understanding of natural resource governance.* In: *Social networks and natural resource management: uncovering the social fabric of environmental governance.* Bodin Ö., & Prell, C. (Eds.). Cambridge: Cambridge University Press. 44–71

285. Crona, B., Gelcich, S., & Bodin, Ö. (2017). The importance of interplay between leadership and social capital in shaping outcomes of rights-based fisheries governance. *World Development*, 91: 70-83.

286. Crook, B.J., & Mann, B.Q. (2002). A critique of and recommendations for a subsistence fishery, Lake St Lucia, South Africa. *Biodiversity and Conservation*, 11: 1223–1235.

287. Cross, H.C. (2014). The Importance of Small-Scale Fishing to Rural Coastal Livelihoods: A Comparative Case-Study in the Bijagós Archipelago Guinea Bissau. PhD book, University College London. Available from: http://discovery.ucl.ac.uk/1417489/2/Helen%20Cross%20Thesis%20edited%20version.pd f

288. Cross, H. (2016). Displacement, disempowerment and corruption: challenges at the interface of fisheries, management and conservation in the Bijagós Archipelago, Guinea-Bissau. *Oryx*, 50(4): 693-701.

289. Cundill, G., & C. Fabricius. (2010). Monitoring the governance dimension of natural resource co-management. *Ecology and Society*, 15(1): 15. Available from: http://www.ecologyandsociety.org/vol15/iss1/art15/

290. Cundill, G., Thondhlana, G., Sisitka, L., Shackleton, S., & Blore, M. (2013). Land claims and the pursuit of co-management on four protected areas in South Africa. *Land use policy*, 35:171–178.

291. Cundill, G., Bezerra, J.C., De Vos, A., & Ntingana, N. (2017). Beyond benefit sharing: Place attachment and the importance of access to protected areas for surrounding communities. *Ecosystem Services*, 28: 140-148.

292. Dart, J.J. (1999). The tale behind the performance story approach. *Evaluation News and Comment*, 8: 12-13.

293. Darwin Initiative. (2017). *Conservation and sustainable use of marine turtles, Southwest Madagascar, Project reference 21-018.* Available from: http://www.darwininitiative.org.uk/documents/21018/24288/21-018%20FR%20-%20Edited.pdf

294. Davidson, J. (2010). Cultivating knowledge: Development, dissemblance, and discursive contradictions among the Diola of Guinea-Bissau. *American Ethnologist*, 37(2): 212-226.

295. Davidson, B. (2017). *No Fist Is Big Enough to Hide the Sky: The Liberation of Guinea-Bissau and Cape Verde, 1963-74.* London: Zed Books Ltd.

296. Davies, R. (2000). *Madikwe Game Reserve: A partnership in conservation.* In: *Wildlife conservation by sustainable use.* Prins, H.H.T., Grootenhuis, J.G., & Dolan, T.T. (Eds.). Dordrecht: Springer. 439-458.

297. Davies, R. (2012). *Criteria for assessing the evaluability of theories of change.* Available from: http://mandenews.blogspot.com/2012/04/criteria-for-assessing-evaluablity-of.html

298. Davies, R., & Dart, J. (2005). *The 'most significant change' (MSC) technique. A guide to its use.* Available from: https://www.betterevaluation.org/en/resources/guides/most_significant_change

299. Davies, T.E., Beanjara, N., & Tregenza, T. (2009). A socio-economic perspective on gear-based management in an artisanal fishery in south-west Madagascar. *Fisheries Management and Ecology*, 16(4): 279-289.

300. Davies, T., Cowley, A., Bennie, J., Leyshon, C., Inger, R., Carter, H., Robinson, B., et al. (2018). Popular interest in vertebrates does not reflect extinction risk and is associated with bias in conservation investment. *PloS one*, 13(9): e0203694. Available from: https://doi.org/10.1371/journal.pone.0203694

301. Davis, A., & Goldman, M. J. (2019). Beyond payments for ecosystem services: considerations of trust, livelihoods and tenure security in community-based conservation projects. *Oryx*, 53(3): 491-496.

302. DEA (Department of Environmental Affairs). (2010) *Conference Report, 4th People and Parks Conference, in Richards Bay, South Africa.* Available from: https://www.environment.gov.za/sites/default/files/docs/4thpeopleandparks_report.pdf

303. DEA (Department of Environmental Affairs). (2019a). *Biodiversity management plans.* Available from: https://www.environment.gov.za/content/management_plans/biodiversity

304. DEA (Department of Environmental Affairs). (2019b). *About People and Parks Programme.* Available from: https://www.environment.gov.za/projectsprogrammes/peopleparks/about

305. De Beer, F. (2012). Community-based natural resource management: living with Alice in Wonderland? *Community Development Journal,* 48(4): 555–570.

306. DeCaro, D.A. (2019). *Understanding the fundamental motivations that drive self-organization and cooperation in commons dilemmas.* In: Hudson, B., Rosenbloom, J., & Cole, D. (Eds.). *Routledge Handbook of the Study of the Commons.* London: Routledge. 117-132

307. DeCaro, D.A., & Stokes, M. (2013). Public participation and institutional fit: a social-psychological perspective. *Ecology and Society*, 18(4): 40. Available from: http://dx.doi.org/10.5751/ES-05837-180440

308. de Koning, M.A.I. (2010). Returning Manyeleti Game Reserve to its rightful owners: land restitution in protected areas in Mpumalanga, South Africa. *Unasylva*, 61: 41-46.

309. DeGeorges, P.A., & Reilly, B.K. (2009). The realities of community based natural resource management and biodiversity conservation in Sub-Saharan Africa. *Sustainability*, 1(3): 734-788.

310. De Koning, M. (2009). Co-management and its options in protected areas of South Africa. *Africanus*, 39(2): 5-17.

311. Delany, S., Scott, D., Dodman, T., & Stroud, D. (2009). An atlas of wader populations in Africa and Western Eurasia. *British Birds*, 102: 639-642.

312. Delgado-Serrano, M.M. (2017). Trade-offs between conservation and development in community-based management initiatives. *International Journal of the Commons*, 11(2): 969-991.

313. Delgado-Serrano, M.D.M., Oteros-Rozas, E., Vanwildemeersch, P., Ortíz-Guerrero, C., London, S. & Escalante, R. (2015). Local perceptions on social-ecological dynamics in Latin America in three community-based natural resource management systems. *Ecology and Society*, 20(4): 24. Available from: http://dx.doi.org/10.5751/ES-07965-200424

314. Denton, G. L., & Harris, J. R. (2019). The Impact of Illegal Fishing on Maritime Piracy: Evidence from West Africa. *Studies in Conflict & Terrorism*, 1-20. Available from: https://doi.org/10.1080/1057610X.2019.1594660

315. Denzin, N. (1979). *The Research Act.* 2nd edition. New York: McGraw-Hill.

316. Desbureaux, S., & Brimont, L. (2015). Between economic loss and social identity: The multi-dimensional cost of avoiding deforestation in eastern Madagascar. *Ecological Economics*, 118: 10-20.

317. de Vente, J., Reed, M.S., Stringer, L.C., Valente, S., & Newig, J. (2016). How does the context and design of participatory decision making processes affect their outcomes? Evidence from sustainable land management in global drylands. *Ecology and Society*, 21(2): 24. Available from: http://dx.doi.org/10.5751/ES-08053-210224

318. de Villiers, P. (2016). *Estuary Management in South Africa–An Overview of the Challenges and Progress Made to Date.* In: *Estuaries: A Lifeline of Ecosystem Services in the Western Indian Ocean.* Diop, S., Scheren, P., & Machiwa, J.F. (Eds.). Springer, Cham. 301-311.

319. Dewar, R.E., & Wright, H.T. (1993). The culture history of Madagascar. *Journal of World Prehistory*, 7(4): 417-466.

320. de Wit, M.J. (2003). Madagascar: heads it's a continent, tails it's an island. *Annual Review of Earth and Planetary Sciences*, 31: 213–248.

321. Díaz, S., Settele, J., Brondízio, E.S., Ngo, H.T., Agard, J., Arneth, A., Balvanera, P., et al. (2019). Pervasive human-driven decline of life on Earth points to the need for transformative change. *Science*, 366(6471): eaax3100. Available from: https://science.sciencemag.org/content/366/6471/eaax3100.full?casa_token=ugF62NZfP5w AAAAA:tbFuWdcgL5bALJk5tLPNFDOywwaU-DaUVYRhFYatwdtDS_H61_rXz8sr6hrP0pt1xK11h16124cYmWs

322. Dietz, T. (2015). *Environmental Value.* In: *Handbook of value: Perspectives from Economics, Philosophy, Psychology, and Sociology.* Brosch, T., & Sander, D. (Eds.). Oxford: Oxford University Press. 329–349.

323. Dietz, T., Ostrom, E., & Stern, P.C. (2003). *The struggle to govern the commons*. Science, vol. 302, no. 5652, p1907-1912.

324. Dietz, J.M., Aviram, R., Bickford, S., Douthwaite, K., Goodstine, A., Izursa, J.L., Kavanaugh, S., et al. (2004). Defining leadership in conservation: a view from the top. *Conservation Biology*, 18(1): 274-278.

325. Diz, D., Johnson, D., Riddell, M., Rees, S., Battle, J., Gjerde, K., Hennige, S., & Roberts, J.M. (2018). Mainstreaming marine biodiversity into the SDGs: the role of other effective area-based conservation measures (SDG 14.5). *Marine Policy*, 93: 251-261.

326. DLA. (Department of Land Affairs). (2007). *Settlement and Implementation Support (SIS) Strategy for Land and Agrarian Reform in South Africa*. Compiled by the Sustainable Development Consortium. Available from: https://www.gov.za/sites/default/files/gcis_document/201411/dlasynthesisreportsept2007.pdf

327. Dobson, A.D., de Lange, E., Keane, A., Ibbett, H. & Milner-Gulland, E.J. (2019). Integrating models of human behaviour between the individual and population levels to inform conservation interventions. *Philosophical Transactions of the Royal Society B*, 374(1781): 20180053. Available from: http://dx.doi.org/10.1098/rstb.2018.0053

328. Dogan, M., & Pelassy, D. (1990). *How to compare nations: Strategies in comparative politics*. Chatham, UK: Chatham House.

329. Dolins, F.L., Jolly, A., Rasamimanana, H., Ratsimbazafy, J., Feistner, A., & Ravoavy, F. (2010). Conservation education in Madagascar: Three case studies in the biologically diverse island-continent. *American Journal of Primatolology*, 72(5): 391–406.

330. Donald, P.F., Buchanan, G.M., Balmford, A., Bingham, H., Couturier, A.R., de la Rosa Jr, G.E., Gacheru, P., et al. (2019). The prevalence, characteristics and effectiveness of Aichi Target 11's "other effective area-based conservation measures"(OECMs) in Key Biodiversity Areas. *Conservation Letters*, 12(5): e12659. Available from: https://doi.org/10.1111/conl.12659

331. Dressel, S., Johansson, M., Ericsson, G., & Sandström, C. (2020). Perceived adaptive capacity within a multi-level governance setting: The role of bonding, bridging, and linking social capital. *Environmental Science & Policy*, 104: 88-97.

332. Dressler, W., Büscher, B., Schoon, M., Brockington, D.A.N., Hayes, T., Kull, C.A., McCarthy, J. & Shrestha, K. (2010). From hope to crisis and back again? A critical history of the global CBNRM narrative. *Environmental conservation*, 37(1): 5-15.

333. Dubow, S. (2014). *Apartheid, 1948-1994*. Oxford: Oxford University Press.

334. Dudley, N. (2008). *Guidelines for Applying Protected Area Management Categories*. Best Practice Protected Area Guidelines Series No. 21. IUCN, Gland, Switzerland. Available from: https://portals.iucn.org/library/sites/library/files/documents/PAG-021.pdf

335. Dudley, N., Higgins-Zogib, L., & Mansourian, S. (2009). The links between protected areas, faiths, and sacred natural sites. *Conservation Biology*, 23: 568–577.

336. Dudley, N., Groves, C., Redford, K. H., & Stolton, S. (2014). Where now for PAs? Setting the stage for the 2014 World Parks Congress. *Oryx*, 48(4): 496-503.

337. Dudley, N., Jonas, H., Nelson, F., Parrish, J., Pyhälä, A., Stolton, S., & Watson, J. E. (2018). The essential role of other effective area-based conservation measures in achieving big bold conservation targets. *Global Ecology and Conservation*, 15: e00424. Available from: https://doi.org/10.1016/j.gecco.2018.e00424

338. Duffy, R. (2006). Non-governmental organisations and governance states: the impact of transnational environmental management networks in Madagascar. *Environmental Politics*, 15: 731–49.

339. Duffy, R., Massé, F., Smidt, E., Marijnen, E., Büscher, B., Verweijen, J., Ramutsindela, M., et al. (2019). Why we must question the militarisation of conservation. *Biological conservation*, 232: 66-73.

340. Ece, M., Murombedzi, J., & Ribot, J. (2017). Disempowering democracy: Local representation in community and carbon forestry in Africa. *Conservation & Society*, 15(4): 357-370. Available from: http://www.conservationandsociety.org/text.asp?2017/15/4/357/221950

341. EcoAfrica. (2013). *Settling the Land Claim. Community development and Land Acquisition Plan (CDLAP) for Ebenhaeser and Papendorp*. Final technical report prepared for the Department of Rural Development and Land Reform. December 2013. Cape Town: EcoAfrica. Available from: https://landportal.org/library/resources/community-development-and-land-acquisition-plan-ebenhaeser-and-papendorp-cdlap

342. Ehrlich, P.R., & Levin, S.A. (2005). The evolution of norms. *PLoS biology*, 3(6): e194. Available from: https://doi.org/10.1371/journal.pbio.0030194

343. Ehrlich, P., & Kennedy, D. (2005). Millennium assessment of human behavior. *Science*, 309: 562–563.

344. Els, H., & Bothma, J.D.P. (2000). Developing partnerships in a paradigm shift to achieve conservation reality in South Africa. *Koedoe*, 43 (1): 19–26.

345. Englefield, E., Black, S.A., Copsey, J.A., & Knight, A.T. (2019). Interpersonal competencies define effective conservation leadership. *Biological Conservation*, 235: 18-26.

346. Epps, M. (2007). *A Socioeconomic Baseline Assessment: Implementing the socioeconomic monitoring guidelines in southwest Madagascar*. Blue Ventures Conservation Report. Available from: http://www.socmon.org/upload/documents/Madagascar_Andavadoaka_2008.pdf

347. Epstein, G., Bennett, A., Gruby, R., Acton, L., & Nenadovic, M. (2014). *Studying Power with the Social-Ecological System Framework*. In: *Understanding Society and Natural Resources*. Manfredo, M.J., Vaske, J.J., Rechkemmer, A., & Duke, E.A. (Eds.). Dordrecht: Springer. 111–135.

348. Epstein, G., Pittman, J., Alexander, S.M., Berdej, S., Dyck, T., Kreitmair, U., Rathwell, K.J. (2015). Institutional fit and the sustainability of social–ecological systems. *Current Opinion in Environmental Sustainability*, 14: 34-40.

349. Equator Initiative. (2019). *Announcing the Equator Prize 2019 Winners*. Available from: https://www.equatorinitiative.org/2019/06/02/ep-2019-meet-the-winners/

350. Ernst, A. (2019). Review of factors influencing social learning within participatory environmental governance. *Ecology and Society* 24(1): 3. Available from: https://doi.org/10.5751/ES-10599-240103

351. Escobar, A. (1998). Whose Knowledge, Whose Nature? Biodiversity, Conservation, and the Political Ecology of Social Movements. *Journal of Political Ecology*, 5: 53-82.

352. Etzioni, A. (1968). *The Active Society: A Theory of Societal and Political Processes*. New York: The Free Press.

353. Fabricius, C. (2004). *The fundamentals of community-based natural resource management*. In: *Rights, Resources and Rural Development: Community-based Natural Resource Management in Southern Africa*. Fabricius, C., Koch, E., Magome, H., & Turner, S. (Eds.). London: Earthscan. 3–43.

354. Fabricius, C., & De Wet, C. (2002). *The Influence of Forced Removals and Land Restitution on Conservation in South Africa*. In: *Conservation and Mobile Indigenous Peoples: Displacement, Forced Settlement and Sustainable Development*. Chatty, D., & Colchester, M. (Ed). Oxford: Berghahn Books. 142–158

355. Fabricius, C., & Collins, S. (2007). Community-based natural resource management: governing the commons. *Water Policy*, 9(2): 83-97.

356. Fabricius, C., & Cundill, G. (2010). *Building Adaptive Capacity in Systems Beyond the Threshold: The Story of Macubeni, South Africa*. In: *Adaptive Capacity and Environmental Governance. Springer Series on Environmental Management*. Armitage, D., & Plummer, R. (Eds.). Berlin: Springer. 43-68.

357. Fabricius, C., Koch, E., & Magome, H. (2001). *Community Wildlife Management in Southern Africa: Challenging the Assumptions of Eden*. Working Paper No. 6. Evaluating Eden Series. London: IIED. Available from: https://pubs.iied.org/7808IIED/

358. Fabricius, C., Cundill, G., & Sisitka, L. (2003). *Guidelines For The Implementation of Community-based Natural Resource Management in South Africa*. Pretoria: DEAT and GTZ Transform. Available from: https://www.westerncape.gov.za/text/2004/5/cbnrm_guidelines_2003.pdf

359. Fabricius, C., Folke, C., Cundill, G., & Schultz, L. (2007). Powerless spectators, coping actors, and adaptive co-managers: a synthesis of the role of communities in ecosystem management. *Ecology and Society*, 12(1): 29. Available from: http://www.ecologyandsociety.org/vol12/iss1/art29/

360. FAO. (1999). *Guidelines for the routine collection of capture fishery data, FAO Fisheries Technical Paper 382, Prepared at the FAO/DANIDA Expert Consultation, Bangkok, Thailand, 18-30 May 1998, FAO, Rome*. Available from: http://www.fao.org/3/x2465e/x2465e0h.htm#ANNEX%205.%20GLOSSARY

361. FAO. (2015). *Voluntary Guidelines for Securing Sustainable Small-Scale Fisheries in the Context of Food Security and Poverty Eradication.* FAO Rome, 2015, 34p. Available from: http://www.fao.org/3/a-i4356en.pdf

362. FAO. (2019a). *Country Profiles.* Available from: http://www.fao.org/countryprofiles/en/

363. FAO. (2019b). *Low-Income Food-Deficit Countries (LIFDCs) - List for 2018.* Available from: http://www.fao.org/countryprofiles/lifdc/en/

364. Feeny, D., Berkes, F., McCay, B.J., & Acheson, J.M. (1990). The tragedy of the commons: twenty-two years later. *Human Ecology*, 18(1): 1-19.

365. Feeny, D., Hanna, S., & McEvoy, A.F. (1996). Questioning the assumptions of the" tragedy of the commons" model of fisheries. *Land Economics*, 72: 187-205.

366. Feka, N.Z., & Morrison, I. (2017). Managing mangroves for coastal ecosystems change: A decade and beyond of conservation experiences and lessons for and from west-central Africa. *Journal of Ecology and The Natural Environment*, 9(6): 99-123.

367. Ferguson, B., & Gardner, C.J. (2010). Looking back and thinking ahead—where next for conservation in Madagascar? *Madagascar Conservation & Development*, 5(2): 75-76.

368. Ferguson, B., Gardner, C.J., Andriamarovolonona, M.M., Healy, T., Muttenzer, F., Smith, S.M., Hockley, N., & Gingembre, M. (2014). *Governing ancestral land in Madagascar. In: For justice and environmental sustainability: Lessons across natural resource sectors in sub-Saharan Africa.* Sowman, M., & Wynberg, R. (Eds.). London: Routledge. 63-92.

369. Fernandes, R. M. (2012). Job satisfaction in the marine and estuarine fisheries of Guinea-Bissau. *Social indicators research*, 109(1): 11-23.

370. Fernández-Llamazares, Á., López-Baucells, A., Rocha, R., Andriamitandrina, S.F., Andriatafika, Z.E., Burgas, D., et al. (2018). Are sacred caves still safe havens for the endemic bats of Madagascar?. *Oryx*, 52(2): 271-275.

371. Ferraro, P., Hanauer, M.M., Miteva, D.A., Canavirebacarreza, J.V., Pattanayak, S.K., & Sims, K.R.E. (2013). More strictly protected areas are not necessarily more protective: evidence from Bolivia, Costa Rica, Indonesia, and Thailand. *Environmental Research Letters*, 8(2): 025011. Available from: http://dx.doi.org/10.1088/1748-9326/8/2/025011

372. Fisher, E., Bavinck, M., & Amsalu, A. (2018). Transforming asymmetrical conflicts over natural resources in the Global South. *Ecology and Society*, 23(4): 28. Available from: https://doi.org/10.5751/ES-10386-230428

373. Fletcher, R. (2010). Neoliberal environmentality: towards a poststructuralist political ecology of the conservation debate. *Conservation & Society*, 8(3): 171-181.

374. Fletcher, R., Dressler, W., Büscher, B., & Anderson, Z. R. (2016). Questioning REDD+ and the future of market-based conservation. *Conservation Biology*, 30(3): 673-675.

375. Flyvbjerg, B. (2006). Five misunderstandings about case-study research. *Qualitative inquiry*, 12(2): 219-245.

376. Folke, C., Hahn, T., Olsson, P., & Norberg, J. (2005). *Adaptive governance of social-ecological systems. Annual Review of Environment and Resources*, 30: 441–473.

377. Foster-Fishman, P.G., Cantillon, D., Pierce, S.J., & Van Egeren, L.A. (2007). Building an active citizenry: the role of neighborhood problems, readiness, and capacity for change. *American Journal of Community Psychology*, 39(1-2): 91-106.

378. Fowler, A. (1997), *Striking a Balance: A Guide to Enhancing Effectiveness of NGOs in International Development.* London: Earthscan Publications.

379. Frank, K. (2011*). Social network models for natural resource use and extraction. In: Social networks and natural resource management: uncovering the social fabric of environmental governance.* Bodin, Ö., & Prell, C. (Eds.). Cambridge: Cambridge University Press. 180–205

380. Fritz-Vietta, N. V., Tahirindraza, H. S., & Stoll-Kleemann, S. (2017). Local people's knowledge with regard to land use activities in southwest Madagascar–Conceptual insights for sustainable land management. *Journal of environmental management*, 199: 126-138.

381. Freudenberger, K. (2010). *Paradise Lost? Lessons from 25 years of USAID Environment Programs in Madagascar.* Washington D.C: International Resources Group. Available from: https://www.usaid.gov/sites/default/files/documents/1860/paradise_lost_25years_env_pro grams.pdf

382. Funnell, S., & Rogers, P. (2011). *Purposeful program theory: Effective use of theories of change and logic models.* San Francisco, CA: Jossey-Bass.

383. Galvin, K.A., Beeton, T.A., & Luizza, M.W. (2018). African community-based conservation: a systematic review of social and ecological outcomes. *Ecology and Society* 23(3): 39. Available from: https://doi.org/10.5751/ES-10217-230339

384. Game, E.T., Meijaard, E., Sheil, D., & McDonald-Madden, E. (2014). Conservation in a wicked complex world; challenges and solutions. *Conservation Letters*, 7(3): 271-277.

385. Ganzhorn, J.U., Wilmé, L., & Mercier, J.L., (2014). *Explaining Madagascar's biodiversity.* In: *Conservation and Environmental Management in Madagascar.* Scales, I.R. (Ed.). London: Routledge. 17-43.

386. Garcia, S.M., Rice, J., & Charles, T. (2014). *Governance of Marine Fisheries and Biodiversity Conservation: Interaction and Co-evolution.* Oxford: Wiley-Blackwell.

387. García-López, G.A. (2019). Rethinking elite persistence in neoliberalism: Foresters and techno-bureaucratic logics in Mexico's community forestry. *World Development,* 120: 169-181.

388. Gardner, C.J. (2011) IUCN management categories fail to represent new, multiple-use protected areas in Madagascar. *Oryx,* 45: 336–346.

389. Gardner, C.J., Nicoll, M.E., Mbohoahy, T., Oleson, K.L.L., Ratsifandrihamanana, A.N., Ratsirarson, J., de Roland, L.A.R., et al. (2013). Protected areas for conservation and poverty alleviation: Experiences from Madagascar. *Journal of Applied Ecology*, 50(6): 1289–1294.

390. Gardner, C.J., Nicoll, M.E., Birkinshaw, C., Harris, A., Lewis, R.E., Rakotomalala, D., & Ratsifandrihamanana, A.N. (2018). The rapid expansion of Madagascar's protected area system. *Biological Conservation*, 220: 29-36.

391. Garnett, S.T., Burgess, N.D., Fa, J.E., Fernández-Llamazares, Á., Molnár, Z., Robinson, C.J., Watson, J.E., et al. (2018). A spatial overview of the global importance of Indigenous lands for conservation. *Nature Sustainability, 1*(7): 369. Available from: https://doi.org/10.1038/s41893-018-0100-6

392. Gavin, M., McCarter, J., Berkes, F., Mead, A., Sterling, E., Tang, R., & Turner, N. (2018). Effective biodiversity conservation requires dynamic, pluralistic, partnership-based approaches. *Sustainability*, 10(6): 1846. Available from: http://dx.doi.org/10.3390/su10061846

393. Gbedomon, RC., Floquet, A., Mongbo, R., Salako, V.K., Fandohan, A.B., Assogbadjo, A.E., & Kakaï, R.G. (2016). Socio-economic and ecological outcomes of community based forest management: A case study from Tobé-Kpobidon forest in Benin, Western Africa. *Forest Policy and Economics*, 64: 46-55.

394. GB-UEMOA (Guinée-Bissau-Union Economique et Monetaire Ouest Africane). (2016). *The Framework Survey of Artisanal Maritime Fisheries in Guinea-Bissau – Year 2014.* Available from: http://atlas.statpeche-uemoa.org/atlas_ecpma/DOCS/gnb.pdf

395. GEF (Global Environmental Facility). (2012). *Contribution to the sustainable management of coral reefs in Ranobe Bay through the creation of community marine protected areas and the promotion of community tourism.* Available from: https://sgp.undp.org/spacial-itemid-projects-landing-page/spacial-itemid-project-search-results/spacial-itemid-project-detailpage.html?view=projectdetail&id=11641

396. GEF (Global Environmental Facility). (2018). *OP6 SGP Guinea - Bissau Country Programme Strategy.* Available from: https://sgp.undp.org//innovation-library/item/download/830_43fde0a25e21c5b876d17d95661ae2c9.html

397. GEF (Global Environmental Facility). (2019). *Community heroes: From Guinea-Bissau to Brazil, Nigeria, and Micronesia, SGP Grantees win Equator Prize.* Available from: https://www.thegef.org/news/community-heroes-guinea-bissau-brazil-nigeria-and-micronesia-sgp-grantees-win-equator-prize

398. Geldmann, J., Barnes, M., Coad, L., Craigie, I.D., Hockings, M., & Burgess, N.D. (2013). Effectiveness of terrestrial protected areas in reducing habitat loss and population declines. *Biological Conservation*, 161: 230-238.

399. Geldmann, J., Coad, L., Barnes, M.D., Craigie, I.D., Woodley, S., Balmford, A., Brooks, T.M., et al. (2018). A global analysis of management capacity and ecological outcomes in terrestrial protected areas. *Conservation Letters*, 11(3): e12434. Available from: https://doi.org/10.1111/conl.12434

400. Gezon, L. (1997). Institutional structure and the effectiveness of integrated conservation and development projects: Case study from Madagascar. *Human Organization*, 56(4): 462-470.

401. Ghimire, K., & Pimbert, M. (1997). *Social Change and Conservation: Environmental Politics and Impacts of National Parks and Protected Areas.* London: Earthscan.

402. Gibson, C.C., Williams, J.T., & Ostrom, E. (2005). The importance of rule enforcement to local-level forest management. *World Development*, 33(2): 273–284.

403. Gifford, R. (2014). Environmental psychology matters. *Psychology*, 65: 541-579.

404. Gill, D.A., Mascia, M.B., Ahmadia, G.N., Glew, L., Lester, S.E., Barnes, M., Craigie, I., Darling, et al. (2017). Capacity shortfalls hinder the performance of marine protected areas globally. *Nature*, 543(7647): 665–669.

405. Gillam, C., & Charles, A. (2018). Fishers in a Brazilian Shantytown: Relational wellbeing supports recovery from environmental disaster. *Marine Policy*, 89: 77–84.

406. Gillibrand, C.J., Harris, A.R. and Mara, E., 2008. Inventory and spatial assemblage study of reef fish in the area of Andavadoaka, South-West Madagascar (Western Indian Ocean). *Western Indian Ocean Journal of Marine Science*, 6(2): 183-197.

407. Glaser, B.G. (1992). *Basics of grounded theory analysis: Emergence vs forcing*. Mill Valley, CA: Sociology Press.

408. Glaser, B.G. (2002). Conceptualization: On theory and theorizing using grounded theory. *International Journal of Qualitative Methods*, 1(2): 23-38.

409. Glaser, B.G., & Strauss, A. (1967). *The discovery of grounded theory: Strategies of qualitative research*. London: Wiedenfeld and Nicholson.

410. Goble, B.J., Lewis, M., Hill, T.R., & Phillips, M.R. (2014). Coastal management in South Africa: Historical perspectives and setting the stage of a new era. *Ocean & coastal management*, 91: 32-40.

411. Golden, A.S., Naisilsisili, W., Ligairi, I., & Drew, J.A. (2014a). Combining Natural History Collections with Fisher Knowledge for Community-Based Conservation in Fiji. *PLoS ONE*, 9(5): e98036. Available from: https://doi.org/10.1371/journal.pone.0098036

412. Golden, C.D., Bonds, M.H., Brashares, J.S., Rodolph Rasolofoniana, B.J., & Kremen, C. (2014b). Economic valuation of subsistence harvest of wildlife in Madagascar. *Conservation Biology*, 28(1): 234-243.

413. Goldman, M., 2003. Partitioned nature, privileged knowledge: community-based conservation in Tanzania. *Development and Change*, 34(5): 833–862.

414. Gollwitzer, L. (2014). *Community-based Micro Grids: A Common Property Resource Problem*. STEPS Working Paper 68, Brighton: STEPS Centre. Available from: http://sro.sussex.ac.uk/53156/1/Rural_Electrification.pdf

415. Gomes, I., Pereira, P. J., Harms, S., Oliveira, A. M., Schneider, P. M., & Brehm, A. (2017). Genetic characterization of Guinea-Bissau using a 12 X-chromosomal STR system: Inferences from a multiethnic population. *Forensic Science International: Genetics*, 31: 89-94.

416. Goodman, S.M., & Benstead, J.P. (2005). Updated estimates of biotic diversity and endemism for Madagascar. *Oryx*, 39: 73-77.

417. Goosen, M., & Blackmore, A.C. (2019). Hitchhikers' guide to the legal context of protected area management plans in South Africa. *Bothalia-African Biodiversity & Conservation*, 49(1): 1-10.

418. Gore, M.L., Ratsimbazafy, J., & Lute, M.L. (2013). Rethinking corruption in conservation crime: insights from Madagascar. *Conservation Letters*, 6(6): 430-438.

419. Gosling, E., & Williams, K. (2010). Connectedness to nature, place attachment and conservation behaviour: Testing connectedness theory among farmers. *Journal of Environmental Psychology*, 30: 298–304.

420. Gough, C., Thomas, T., Humber, F., Harris, A., Cripps, G., & Peabody, S. (2009). *Vezo Fishing: An Introduction to the Methods Used by Fishers in Andavadoaka Southwest Madagascar*. Blue Ventures Conservation Report. Available from: https://blueventures.org/wp-content/uploads/2016/07/Vezo-Fishing.pdf

421. Govan, H. Tawake, A., Tabunakawai, K., Jenkins, A., Lasgorceix, A., Schwarz, A.-M., Aalbersberg, B., Manele, B., et al. (2009). *Status and potential of locally-managed marine areas in the Pacific Island Region: meeting nature conservation and sustainable livelihood targets through wide-spread implementation of LMMAs*. SPREP. WWF. World Fish-Reefbase. CRISP. Munich Personal RePEc Paper No. 23828. Available from: http://mpra.ub.uni-muenchen.de/23828/1/MPRA_paper_23828.pdf

422. Graham, J., Amos, B., & Plumptre, T. (2003). *Governance principles for protected areas in the 21st century, a discussion paper*. Institute on Governance in collaboration with Parks Canada and Canadian International Development Agency, Ottawa. Available from: https://iog.ca/docs/2003_June_pa_governance2.pdf

423. Granek, E.F., & Brown, M.A. (2005). Co-Management Approach to Marine Conservation in Mohéli, Comoros Islands. *Conservation Biology*, 19(6): 1724-1732.

424. Greiner, L.E., & Bhambri, A. (1989). New CEO intervention and dynamics of deliberate strategic change. *Strategic Management Journal*, 10(S1): 67-86.

425. GroundUp. (2018). *Prospect of mining on Olifants River estuary alarms fishermen. By John Yeld, 28 May 2018.* Available from: https://www.groundup.org.za/article/prospect-mining-oiifants-river-estuary-alarms-fishermen/

426. Grove, R.H. (1987). *Early themes in African conservation; the Cape in the nineteenth century.* In: *Conservation in Africa: peoples, policies and practice.* Anderson, D., & Grove, R.H. (Eds.). Cambridge, UK: Cambridge University Press. 22-39.

427. Grundy, I.M., & Michell, N. (2004). Participatory forest management in South Africa. In: *Indigenous forests and woodlands in South Africa: policy, people and practices.* Lawes, M.J., Eeley, H.A.C., Shackleton, C.M., & Geach, B.S. (Eds.). Durban, South Africa: University of KwaZulu-Natal Press. 679-712.

428. Guerrero, A.M., McAllister, R.R.J., Corcoran, J. & Wilson, K.A. (2013) Scale mismatches, conservation planning, and the value of social-network analyses. Conservation Biology, 27: 35-44.

429. Gurney, G.G., Cinner, J., Ban, N.C., Pressey, R.L., Pollnac, R., Campbell, S.J., Tasidjawa, S., & Setiawan, F. (2014). Poverty and protected areas: an evaluation of a marine integrated conservation and development project in Indonesia. *Global Environmental Change*, 26: 98-107.

430. Gurney, G.G., Cinner, J.E., Sartin, J., Pressey, R.L., Ban, N.C., Marshall, N.A., & Prabuning, D. (2016). Participation in devolved commons management: Multiscale socioeconomic factors related to individuals' participation in community-based management of marine protected areas in Indonesia. *Environmental Science & Policy*, 61: 212-220.

431. Habel, J.C., Rasche, L., Schneider, U.A., Engler, J.O., Schmid, E., Rödder, D., Meyer, S.T., et al. (2019). Final countdown for biodiversity hotspots. *Conservation Letters*: e12668. Available from: https://doi.org/10.1111/conl.12668

432. Hackel, J.D. (1999). Community Conservation and the Future of Africa's Wildlife. *Conservation Biology*, 13: 726–734.

433. Hagerman, S., Dowlatabadi, H., Satterfield, T., & McDaniels, T. (2010). Expert views on biodiversity conservation in an era of climate change. *Global environmental change*, 20(1): 192-207.

434. Hall, R., & Kepe, T. (2017). Elite capture and state neglect: new evidence on South Africa's land reform. *Review of African Political Economy*, 44(151): 122-130.

435. Haller, T., Acciaioli, G., & Rist, S. (2016). Constitutionality: Conditions for crafting local ownership of institution-building processes. *Society & Natural Resources*, 29(1): 68-87.

436. Halpern, B.S., Frazier, M., Potapenko, J., Casey, K.S., Koenig, K., Longo, C., Lowndes, J.S., Rockwood, R.C., et al. (2015). Spatial and temporal changes in cumulative human impacts on the world's ocean. *Nature Communications*, 6: 7615. Available from: http://dx.doi.org/10.1038/ncomms8615

437. Hamilton, R.J., Potuku, T., & Montambault, J.R. (2011). Community-based conservation results in the recovery of reef fish spawning aggregations in the Coral Triangle. *Biological Conservation*, 144(6): 1850-1858.

438. Hänke, H., & Barkmann, J. (2017). Insurance function of livestock, Farmers coping capacity with crop failure in southwestern Madagascar. *World Development*, 96: 264-275.

439. Hänke, H., Barkmann, J., Coral, C., Kaustky, E.E., & Marggraf, R. (2017). Social-ecological traps hinder rural development in southwestern Madagascar. *Ecology and Society*, *22*(1): 42. Available from: https://doi.org/10.5751/ES-09130-220142

440. Hanks, J., & Myburgh, W. (2015). *The evolution and progression of transfrontier conservation areas in the Southern African development community.* In: *Institutional arrangements for conservation, development and tourism in Eastern and Southern Africa.* Dordrecht: Springer. 157-179

441. Hardin, G. (1968). The tragedy of the commons. *Science*, 162: 1243–1248.

442. Harries, E., Hodgson, L., & Noble, J. (2014). *Creating your theory of change.* London: New Philanthropy Capital. Available from: http://www.sekonline.org.uk/resources/creatingyourtheoryofchange1.pdf

443. Harris, A. (2007). To live with the sea: Development of the Velondriake community-managed protected area network, Southwest Madagascar. *Madagascar Conservation & Development*, 2(1): 43–49.

444. Harris, A.R. (2011). Out of sight but no longer out of mind: a climate of change for marine conservation in Madagascar. *Madagascar Conservation & Development*, 6(1): 7-14.

445. Harris, A., Mohan, V., Flanagan, M., & Hill, R. (2012). Integrating family planning service provision into community-based marine conservation. *Oryx*, 46(2): 179–186.

446. Harris, J., Branch, G., Sibiya, C., & Bill, C. (2003). *The Sokhulu subsistence mussel-harvesting project: co-management in action.* In: *Waves of change: coastal and fisheries co-management in South Africa.* Hauck, M., & Sowman, M. (Eds.). Cape Town: Juta and Company Ltd. 61-98.

447. Hauck, M., & Sowman, M. (2001). Coastal and fisheries co-management in South Africa: an overview and analysis. *Marine Policy*, 25(3): 173-185.

448. Hauck, M., & Sowman, M. (2003). *Waves of Change: Coastal and fisheries co-management in Southern Africa.* Cape Town: Juta and Company Ltd.

449. Hauzer, M., Dearden, P., & Murray, G. (2013). The effectiveness of community-based governance of small-scale fisheries, Ngazidja Island, Comoros. *Marine Policy*, 38: 346–354.

450. Hayes, T., & Persha, L. (2010). Nesting local forestry initiatives: Revisiting community forest management in a REDD+ world. *Forest Policy and Economics*, 12(8): 545-553.

451. Hearn, J. (2007). Roundtable: African NGOs: the new compradors? *Development and Change*, 38(6): 095-1110.

452. Hearn, A. (2008). The rocky path to sustainable fisheries management and conservation in the Galápagos Marine Reserve. *Ocean & Coastal Management*, 51(8-9): 567-574.

453. Hebinck, P., Kiaka, R. D., & Lubilo, R. (2020). *Chapter 6: Navigating community conservancies and institutional complexities in Namibia.* In: *Natural Resources, Tourism and Community Livelihoods in Southern Africa: Challenges of Sustainable Development.* Stone, M.T., Lenao, M., & Moswete, N. (Eds.). London: Routledge.

454. Herzog, L.M., & Ingold, K. (2019). Threats to common-pool resources and the importance of forums: On the emergence of cooperation in CPR problem settings. *Policy Studies Journal*, 47(1): 77-113.

455. Hockings, K.J., & Sousa, C. (2013). Human-chimpanzee sympatry and interactions in Cantanhez National Park, Guinea-Bissau: current research and future directions. *Primate Conservation*, 26(1): 57-65.

456. Hodgson, G.M. (2006). What Are Institutions? *Journal of Economic Issues*, 40(1): 1-25.

457. Holdgate, M. (1999). *The Green Web: A Union for Conservation.* London: Earthscan.

458. Holling, C.S. (1978). *Adaptive Environmental Assessment and Management.* London: Wiley.

459. Holmes, G., Sandbrook, C., & Fisher, J. A. (2016). Understanding conservationists' perspectives on the new-conservation debate. *Conservation Biology*, 31(2): 353-363.

460. Holmes, G., Smith, T. A., & Ward, C. (2017). Fantastic beasts and why to conserve them: animals, magic and biodiversity conservation. *Oryx*, 52(2): 231-239.

461. Holmes, G., Smith, T. A., & Ward, C. (2018). Fantastic beasts and why to conserve them: animals, magic and biodiversity conservation. *Oryx*, 52(2): 231-239.

462. Holmes-Watts, T., & Watts, S. (2008). Legal frameworks for and the practice of participatory natural resources management in South Africa. Forest Policy and Economics, 10: 435–443.

463. Holt, F.L. (2005). *The catch-22 of conservation: indigenous peoples, biologists and cultural change. Human Ecology*, 33: 199–215.

464. Hoole, A. (2009). Place-power-prognosis: Community-based conservation, partnerships, and ecotourism enterprises in Namibia. *International Journal of the Commons*, 4(1): 78-99.

465. Horwich, R.H., Lyon, J., & Bose, A. (2011). *What Belize can teach us about grassroots conservation. Solutions*, 2(3): 51-58.

466. Horwich, R.H., Lyon, J., Bose, A., & Jones CB. (2012). *Preserving biodiversity and ecosystems: catalyzing conservation contagion.* In: *Deforestation Around the World.* Moutinho, P. (Ed.). Rijeky, Croatia: InTech. 283-318.

467. Humber, F., Godley, B.J., Ramahery, V., & Broderick, A.C. (2011). Using community members to assess artisanal fisheries: the marine turtle fishery in Madagascar. *Animal Conservation*, 14: 175–85.

468. Humber, F., Andriamahefazafy, M., Godley, B.J., & Broderick, A.C. (2015). Endangered, essential and exploited: How extant laws are not enough to protect marine megafauna in Madagascar. *Marine Policy*, 60: 70-83.

469. Hume, D.W. (2006). Swidden agriculture and conservation in eastern Madagascar: stakeholder perspectives and cultural belief systems. *Conservation & Society*, 4(2): 287-303.

470. Hushlak, A.A. (2012). *Integrating traditional ecological knowledge in South Africa's small-scale fisheries: the Olifants Estuary Gillnet Fishery*. Masters book, University of Cape Town. Available from: https://open.uct.ac.za/bitstream/handle/11427/11608/book_sci_2012_hushlak_a_a.pdf?sequence=1&isAllowed=y

471. Hutton, J.M, & Leader-Williams, N. (2003). *Sustainable use and incentive-driven conservation: Realigning human and conservation interests*. *Oryx*, 37: 215-226.

472. Hutton, J., Adams, W.M., & Murombedzi, J.C. (2005). *Back to the barriers? Changing narratives in biodiversity conservation*. *Forum for development studies*, 32(2): 341-370.

473. Ibarra, J.T., Barreau, A., Campo, C.D., Camacho, C.I., Martin, G.J., & McCandless, S.R. (2011). When formal and market-based conservation mechanisms disrupt food sovereignty: impacts of community conservation and payments for environmental services on an indigenous community of Oaxaca, Mexico. *International Forestry Review*, 13(3): 318-337.

474. Igoe, J. (2005). Global Indigenism and Spaceship Earth. *Globalization*, 2(3): 377-390.

475. Igoe, J., & Brockington, D. (2007). Neoliberal Conservation: A Brief Introduction. Conservation and Society, 5: 432-49. Available from: http://www.conservationandsociety.org/text.asp?2007/5/4/432/49249

476. Iida, T. (2005). The Past and Present of the Coral Reef Fishing Economy in Madagascar: Implications for Self-Determination in Resource Use. *Senri Ethnological Studies*, 67: 237-258.

477. IMVP (Instituto Marquês de Valle Flôr) (2017). *Urok Osheni! Conservation, development and sovereignty in the Urok Islands*. Available from: https://www.imvf.org/wp-content/uploads/2017/12/brochuraurok_osheni_final.pdf

478. Indjai, B. (2017). O saber local sobre a utilização das plantas medicinais na Área Marinha Protegida Comunitária da Ilhas Urok (Reserva da Biosfera do Arquipélago Bolama Bijagós, Guiné-Bissau). PhD book, Universidade Nova de Lisboa. Avaiable from: https://run.unl.pt/bitstream/10362/23211/1/Disserta%C3%A7%C3%A3o_Mestrado_ANC_Final_Indjai_2017.pdf

479. Infield, M., Entwistle, A., Anthem, H., Mugisha, A., & Phillips, K. (2017). Reflections on cultural values approaches to conservation: lessons from 20 years of implementation. *Oryx*, 52(2): 220-230.

480. Ingold, T. (2000). *The Perception of the Environment: Essays on Livelihood, Dwelling and Skill*. London: Routledge.

481. Intchama, J.F., Belhabib, D., Jumpe, T., & Joaquim, R. (2018). Assessing Guinea Bissau's Legal and Illegal Unreported and Unregulated Fisheries and the Surveillance Efforts to Tackle Them. *Frontiers in Marine Science*, 5: 79. Available from: https://doi.org/10.3389/fmars.2018.00079

482. Irland, L.C. (2008). State failure, corruption, and warfare: challenges for forest policy. *Journal of Sustainable Forestry*, 27(3): 189-223.

483. Isaacs, M. (2011). Individual transferable quotas, poverty alleviation and challenges for small-country fisheries policy in South Africa. *MAST*, 10(2): 63-84.

484. Isaacs, M., & Witbooi, E. (2019). Fisheries crime, human rights and small-scale fisheries in South Africa: A case of bigger fish to fry. *Marine Policy*, 105: 158-168.

485. IUCN. (2005). Benefits Beyond Boundaries: Proceedings of the Vth IUCN World Parks Congress: Durban, South Africa 8-17 September 2003. World Conservation Union, Cambridge UK. Available from: https://portals.iucn.org/library/node/8662

486. IUCN (2014a). The Promise of Sydney: Innovative Approaches for Change. Sydney: IUCN World Parks Congress. Available from: http://worldparkscongress.org/about/promise_of_sydney_innovative_approaches.html

487. IUCN (2014b). IUCN, World Parks congress 2014 bulletin, Summary Report, vol. 89, no. 16, 22 November 2014. Available from: http://enb.iisd.org/download/pdf/sd/crsvol89num16e.pdf

488. Jackson, S., Sowman, M., & J. Cox, J. (2013). *Fishers' Proposals for Fishery Management in the Olifants River*. Unpublished report.

489. Jenkins, C.N., & Van Houtan, K.S. (2016). Global and regional priorities for marine biodiversity protection. *Biological Conservation*, 204: 333-339.

490. Jentoft, S. (2000). The community: a missing link in fisheries management. *Marine Policy*, 24: 53-59.

491. Jentoft, S. (2007a). The limits of governability: Institutional implications for fisheries and coastal governance. *Marine Policy*, 31: 360–370.
492. Jentoft, S. (2007b). In the power of power: the understated aspect of fisheries and coastal management. *Human Organization*, 66(4): 426-437.
493. Jentoft, S. (2017). Small-scale fisheries within maritime spatial planning: knowledge integration and power. *Journal of Environmental Policy & Planning*, 19(3): 266-278.
494. Jick, T.D. (1979). Mixing qualitative and quantitative methods: Triangulation in action. *Administrative Science Quarterly*, 24: 602–611.
495. Johnson, R.B., & Onwuegbuzie, A.J. (2004). Mixed-methods research: a research paradigm whose time has come. *Educational Researcher*, 33(7): 14-26.
496. Johnson, R.B., Onwuegbuzie, A.J., & Turner, L.A. (2007). Toward a definition of mixed methods research. *Journal of mixed methods research*, 1(2): 112-133.
497. Johnson, M.F., Karanth, K.K., & Weinthal, E. (2018). Compensation as a Policy for Mitigating Human-wildlife Conflict Around Four Protected Areas in Rajasthan, India. *Conservation & Society*, 16(3): 305-319.
498. Jonas, H., & MacKinnon, K. (2016). *Advancing Guidance on Other Effective Area-based Conservation Measures: Report of the Second Meeting of the IUCN-WCPA Task Force on Other Effective Area-based Conservation Measures.* Bundesamt für Naturschutz: Bonn. 87-89. Available from: http://www.bfn.eu/fileadmin/BfN/ina/Dokumente/Tagungsdoku/2016/Task_Force_on_OEC Ms_Rp2016.pdf
499. Jones, K.N. (2012). Examining trends in taste preference, market demand, and annual catch in an indigenous marine turtle fishery in southwest Madagascar. Honors College book, Washington State University. Available from: https://digitalcollections.sit.edu/isp_collection/999/
500. Jones, J.P.G., Andriamarovololona, M.M., & Hockley, N. (2008) The importance of taboos and social norms to conservation in Madagascar. *Conservation Biology*, 22: 976-986.
501. Jones, K.R., Venter, O., Fuller, R.A., Allan, J.R., Maxwell, S.L., Negret, P.J., & Watson, J.E. (2018a). One-third of global protected land is under intense human pressure. *Science*, 360(6390): 788-791.
502. Jones, K.R., Klein, C.J., Halpern, B.S., Venter, O., Grantham, H., Kuempel, C.D., Shumway, N., et al. (2018b). The location and protection status of Earth's diminishing marine wilderness. *Current Biology*, 28(15): 2506-2512.
503. Jones, P.J.S., Murray, R. H., & Vestergaard, O. (2019a). Enabling Effective and Equitable Marine Protected Areas: Guidance on Combining Governance Approaches. Ecosystems Division, UN Environment. Available from: https://wedocs.unep.org/bitstream/handle/20.500.11822/27790/MPA.pdf?sequence=1
504. Jones, J.P., Ratsimbazafy, J., Ratsifandrihamanana, A.N., Watson, J.E., Andrianandrasana, H.T., Cabeza, M., Cinner, J.E., et al. (2019b). Madagascar: Crime threatens biodiversity. *Science,* 363(6429): 825.
505. Julia, P., Jones, G., Mijasoa, M., Andriamarovolona, M., Hockley, N. (2008). The importance of taboos and social norms to conservation in Madagascar. *Conservation Biology*, 22(4): 976–86.
506. Kaczynski, V.M., Fluharty, D.L. (2002). European policies in West Africa: Who benefits from fisheries agreements? *Marine Policy*, 26: 75–93.
507. Kamat, S. (2004). The privatization of public interest: theorizing NGO discourse in a neoliberal era. *Review of International Political Economy*, 11(1): 155-176.
508. Kalamandeen, M., & Gillson, L. (2007). Demything "wilderness": implications for protected area designation and management. *Biodiversity and Conservation*, 16(1): 165-182.
509. Kaplan-Hallam, M., & Bennett, N. J. (2018). Adaptive social impact management for conservation and environmental management. *Conservation Biology*, 32(2): 304-314.
510. Kashwan, P. (2013). The politics of rights-based approaches in conservation. *Land Use Policy*, 31: 613-626.
511. Kashwan, P (2016). Integrating Power in Institutional Analysis: A Micro-Foundation Perspective. *Journal of Theoretical Politics,* 28(1): 5-26.
512. Kashwan, P., MacLean, L.M., & García-López, G.A. (2019). Rethinking power and institutions in the shadows of neoliberalism: An introduction to a special issue of World Development. *World Development,* 120: 133-146.

513. Kaufman, J.C. (2014). *Contrasting visions of nature and landscapes*. In: *Conservation and Environmental Management in Madagascar*. Scales, I. (Ed.). New York: Routledge. 329-330.

514. Kautto, P., & Silila, J. (2005). Recently introduced policy instruments and intervention theories. *Evaluation*, 11 (1): 55 – 68.

515. Kawaka, J.A., Samoilys, M.A., Murunga, M., Church, J., Abunge, C., & Maina, G.W. (2017). Developing locally managed marine areas: lessons learnt from Kenya. *Ocean & coastal management*, 135: 1-10.

516. Keen, M., Brown, V., & Dyball, R. (2005). *Social Learning in Environmental Management*. London: EarthScan.

517. Kellert, S.R., Mehta, J.N., Ebbin, S.A., & Lichtenfeld, L.L. (2000). Community natural resource management: promise, rhetoric, and reality. *Society & Natural Resources*, 13(8): 705-715.

518. Kelly, M. (2008). *Anything new under the sun? Lessons learned from community based natural resource management*. In: *Community Development and Ecology: engaging ecological sustainability through community development*. Geelong, Australia: Centre for Citizenship, Development and Human Rights. 184-197

519. Kemp, R., Parto, S., & Gibson, R.B. (2005). Governance for sustainable development: moving from theory to practice. *International Journal of Sustainable Development*, 8(1-2): 12-30.

520. Kepe, T. (2008). Land claims and co-management of protected areas in South Africa: exploring the challenges. *Environmental management*, 41: 311–321.

521. Kepe, T. (2009). Shaped by race. *Local Environment*, 14(9): 871–878.

522. Kepe, T. (2012). Land and justice in South Africa: Exploring the ambiguous role of the state in the land claims process. *African and Asian Studies*, 11: 391–409.

523. Kepe, T. (2018). *Meanings, alliances and the state in tensions over land rights and conservation in South Africa*. In: *Land Rights, Biodiversity Conservation and Justice: Rethinking Parks and People*. Mollett, S., & Kepe, T. (Eds.). London: Routledge. 17-30.

524. Kepe, T., & Hall, R. (2018). Land redistribution in South Africa: Towards decolonisation or recolonisation?. *Politikon*, 45(1): 128-137.

525. Kepe, T., Wynberg, R., Ellis, W. (2005). Land reform and biodiversity conservation in South Africa: complementary or in conflict? *The International Journal of Biodiversity Science and Management*, 1(1): 3-16.

526. Kiefer, I., Lopez, P., Ramiarison, C., Barthlott, W. and Ibish, P.L. (2010). *Development, biodiversity conservation and global change in Madagascar*. In: *Interdependence of Biodiversity and Development Under Global Change*. Ibisch, P.L., Vega, E., & Hermann, T.M. (Eds.). Montreal, Canada: Secretariat of the Convention on Biological Diversity. 58–83.

527. King, B.H. (2007). Conservation and community in the new South Africa: A case study of the Mahushe Shongwe Game Reserve. *Geoforum*, 38(1): 207-219.

528. King, B., & Peralvo, M. (2010). Coupling community heterogeneity and perceptions of conservation in rural South Africa. *Human Ecology*, 38(2): 265-281.

529. Kinzig, A.P., Ehrlich, P.R., Alston, L.J., Arrow, K., Barrett, S., Buchman, T.G., Daily, G.C., et al. (2013). Social norms and global environmental challenges: the complex interaction of behaviors, values, and policy. *BioScience*, 63(3): 164-175.

530. Klein, J., Re'au, B., & Edwards, M. (2008). *Zebu landscapes: Conservation and cattle in Madagascar*. In: *Greening the Great Red Island: Madagascar in nature and culture*. Kaufmann, J.C. (Ed.). Pretoria: Africa Institute of South Africa. 157–178.

531. Klute, G., & Fernandes, R. (2014). Global Challenges and the (Re) Development of Neo-traditional Land Rights. Research in Legal Anthropology in Guinea-Bissau. *Modern Africa: Politics, History and Society*, 1(2): 60-86.

532. Knight, A.T., Cowling R.M., & Campbell B.M. (2006). An operational model for implementing conservation action. *Conservation Biology*, 20(2): 408–419.

533. Knight, A.T., Cowling R.M., Rouget M., Balmford A., Lombard A.T., & Campbell B.M. (2008). Knowing but not doing: selecting priority conservation areas and the research-implementation gap. *Conservation Biology*, 22(3): 610–617.

534. Kooiman, J. (2005). *Principles for fisheries governance: Introduction*. In: *Fish for life: interactive governance for fisheries*. Kooiman, J., Bavinck, M., Jentoft, S., & Pullin, R. (Eds.). Amsterdam: Amsterdam University Press. 241-244.

535. Kooiman, J., & Bavinck, M. (2005). *The governance perspective*. In: *Fish for life: interactive governance for fisheries*. Kooiman, J., Bavinck, M., Jentoft, S., & Pullin, R. (Eds.). Amsterdam: Amsterdam University Press. 11-24.

536. Kooiman, J., Bavinck, M., Jentoft, S., & Pullin, R. (2005a). *Fish for life: interactive governance for fisheries*. Amsterdam: Amsterdam University Press.

537. Kooiman, J., Jentoft, S., Bavinck, M., Chuenpagdee, R., & Sumaila, U.R. (2005b). *Meta-Principles*. In: *Fish for life: interactive governance for fisheries*. Kooiman, J., Bavinck, M., Jentoft, S., & Pullin, R. (Eds.). Amsterdam: Amsterdam University Press. 266-284.

538. Koot, S. (2019). The limits of economic benefits: Adding social affordances to the analysis of trophy hunting of the Khwe and Ju/'hoansi in Namibian community-based natural resource management. *Society & Natural Resources*, 32(4):417–33.

539. Koot, S., & Büscher, B. (2019). Giving Land (Back)? The Meaning of Land in the Indigenous Politics of the South Kalahari Bushmen Land Claim, South Africa. *Journal of Southern African Studies*, 1-18. Available from: https://doi.org/10.1080/03057070.2019.1605119

540. Koot, S., Hebinck, P., & Sullivan, S. (2020). Science for Success—A Conflict of Interest? Researcher Position and Reflexivity in Socio-Ecological Research for CBNRM in Namibia. *Society & Natural Resources*, 1-18. Available from: https://doi.org/10.1080/08941920.2020.1762953

541. Kothari, A., Camill, P., & Brown, J. (2013). Conservation as if people also mattered: Policy and practice of community-based conservation. *Conservation & Society*, 11: 1-15. Available from: http://www.conservationandsociety.org/text.asp?2013/11/1/1/110937

542. Krause, T., & Nielsen, T.D. (2014). The legitimacy of incentive-based conservation and a critical account of social safeguards. *Environmental Science & Policy*, 41: 44-51.

543. Krosnick, J.A. (1991). Cognitive demands of attitude measures. *Applied Cognitive Psychology*, 5: 213-236.

544. Kreuter, U., Peel, M., & Warner, E. (2010). Wildlife conservation and community-based natural resource management in southern Africa's private nature reserves. *Society & Natural Resources*, 23(6): 507-524.

545. Krüger, R., Cundill, G., & Thondhlana, G. (2016). A case study of the opportunities and trade-offs associated with deproclamation of a protected area following a land claim in South Africa. *Local Environment*, 21(9): 1047-1062.

546. Kull, C. (2002). Empowering pyromaniacs in Madagascar: Ideology and legitimacy in community-based natural resource management. *Development and Change*, 33: 57-78.

547. Lahiff, E. (2008). *Land reform in South Africa: a status report 2008*. Programme for Land and Agrarian Studies research report no. 38. Available from: https://www.africaportal.org/publications/land-reform-in-south-africa-a-status-report-2008/

548. Lammers, P.L., Richter, T., Lux, M., Ratsimbazafy, J., & Mantilla-Contreras, J. (2017). The challenges of community-based conservation in developing countries—A case study from Lake Alaotra, Madagascar. *Journal for Nature Conservation*, 40: 100-112.

549. Lane, M.B., & Corbett, T. (2005). The Tyranny of localism: indigenous participation in community-based environmental management. *Journal of Environmental Policy and Planning*, 7: 141–159

550. Langholz, L.J., & Lassoie, J.P. (2001). Perils and Promise of Privately Owned Protected Areas", *BioScience*, 51(12): 1079-1085.

551. Langholz, J., & Krug, W. (2004). New forms of biodiversity governance: Non-State actors and the private protected area action plan. *Journal of International Wildlife Law and Policy*, 7: 1-21.

552. Langley, J.M. (2006). *Vezo knowledge: Traditional ecological knowledge in Andavadoaka, southwest Madagascar*. Blue Ventures Conservation: London, UK. Available from: http://blueventures.org/wp-content/uploads/2015/03/vezo-knowledge-traditional-ecological-knowledge-andavadoaka-english.pdf

553. Laroche, J., & Ramananarivo, N. (1995). A preliminary survey of the artisanal fishery in the coral reefs of the Tulear region (southwest Madagascar). *Coral Reefs*, 14: 193-200.

554. Larrosa, C., Carrasco, L. R., & Milner-Gulland, E. J. (2016). Unintended feedbacks: challenges and opportunities for improving conservation effectiveness. *Conservation Letters*, 9(5): 316-326.

555. Larson, A.M. (2003). Decentralisation and forest management in Latin America: towards a working model. *Public Administration and Development*, 23: 211–226.

556. Lauber, T.B., Decker, D.J., & Knuth, B.A. (2008). Social networks and community-based natural resource management. *Environmental Management*, 42(4): 677-687.

557. Lauer, M., & Aswani, S. (2009). Indigenous ecological knowledge as situated practices: understanding fishers' knowledge in the Western Solomon Islands. *American Anthropologist*, 111(3), 317-329.

558. Leach, M., Scoones, I., & Stirling, A. (2011). *Dynamic sustainabilities: technology, environment, social justice.* London: Earthscan.

559. Lele, S., Wilshusen, P., Brockington, D., Seidler, R., & Bawa, K. (2010). Beyond exclusion: alternative approaches to biodiversity conservation in the developing tropics. *Current Opinion in Environmental Sustainability*, 2(1-2): 94-100.

560. Le Manach, F., Gough, C., Harris, A., Humber, F., Harper S., & Zeller, D. (2012). Unreported fishing, hungry people and political turmoil: The recipe for a food security crisis in Madagascar? *Marine Policy*, 36(1): 218-225.

561. Le Manach, F., Andrianaivojaona, C., Oleson, K., Clausen, A., & Lange, G-M. (2013). *Natural Capital Accounting and Management of the Malagasy Fisheries Sector. A technical case study for the WAVES Global Partnership in Madagascar.* Available from: https://www.wavespartnership.org/sites/waves/files/images/NCA%20and%20mgt%20of% 20the%20Malagasy%20fisheries%20sector%20FINAL%2008 03 13.docx .

562. Lemieux, C.J., Gray, P.A., Devillers, R., Wright, P.A., Dearden, P., Halpenny, E.A., Groulx, M., et al. (2019). How the race to achieve Aichi Target 11 could jeopardize the effective conservation of biodiversity in Canada and beyond. *Marine Policy*, 99: 312-323.

563. L'Haridon, L. (2006). *Evolution de la collecte de poulpe sur la côte Sud-Ouest de Madagascar: éléments de reflexion pour une meilleure gestion des ressources.* Blue Ventures Conservation Report. http://blueventures.org/wp-content/uploads/2015/03/evolution-de-la-collecte-de-poulpes.pdf

564. Leisher, C., Booker, F., Agarwal, B., Day, M., Matthews, E., Prosnitz, D., Roe, D., et al. (2018). A preliminary theory of change detailing how women's participation can improve the management of local forests and fisheries. *SocArXiv, July 2018.* Available from: https://osf.io/rgakw/download/?format=pdf

565. Lejano, R., Araral, E., Agrawal, A. (2014). Special Issue: Interrogating The Commons. *Environmental Science & Policy*, 36: 1-92. D

566. Levine, A.S., & Richmond, L.S. (2014). Examining enabling conditions for community-based fisheries comanagement: comparing efforts in Hawai'i and American Samoa. *Ecology and Society*, 19(1): 24. Available from: http://dx.doi.org/10.5751/ES-06191-190124

567. Levine, A.S., & Richmond, L.S. (2015). Using common-pool resource design principles to assess the viability of community-based fisheries co-management systems in American Samoa and Hawai'i. *Marine Policy*, 62: 9-17.

568. Lewin, K. (1952). *Psychological ecology.* In: *Field theory in social science.* Cartwright, D. (Ed.). London: Social Science Paperbacks. 170-187.

569. Lewis, D. (1998). Development NGOs and the challenge of partnership: changing relations between North and South. *Social policy & administration*, 32(5): 501-512.

570. Lippitt, R., Watson, J., & Westley, B. (1958). *The Dynamics of Planned Change.* New York: Harcourt, Brace and World.

571. Locke, H., & Deardon, P. (2005). Rethinking protected area categories and the new paradigm. *Environmental Conservation*, 32(1): 1-10.

572. Lockwood, M., Davidson, J., Curtis, A., Stratford, E., & Griffith, R. (2010). Governance principles for natural resource management. *Society & Natural Resources*, 23(10): 986-1001.

573. Long, S. (2017). Short-term impacts and value of a periodic no take zone (NTZ) in a community-managed small-scale lobster fishery, Madagascar. *PloS one*, 12(5): e0177858. Available from: https://doi.org/10.1371/journal.pone.0177858

574. Lourenço, P.M., Catry, T., & Granadeiro, J. P. (2017). Diet and feeding ecology of the wintering shorebird assemblage in the Bijagós archipelago, Guinea-Bissau. *Journal of Sea Research*, 128: 52-60.

575. Ludwig, D. (2001). The era of management is over. *Ecosystems*, 4(8): 758-764.

576. Ludwig, D., Hilborn, R., & Walters, C. (1993). Uncertainty, resource exploitation, and conservation: lessons from history. *Science*, 260(5104): 17&18.

577. Lukes, S. (1982). *Power: A Radical View.* London: The MacMillan Press Ltd.

578. Lund, J.F., & Trueu, T. (2008). Are we getting there? Evidence of decentralized forest management from the Tanzanian Miombo Woodlands. *World Development*, 36(12): 2780–2800.

579. Lund, J.F., & Saito-Jensen, M. (2013). Revisiting the issue of elite capture of participatory initiatives. *World development*, 46: 104-112.
580. Lundy, B.D. (2012). Playing the market: How cashew "Comodityscape" is redefining Guinea-Bissau's countryside. *Culture, Agriculture, Food and Environment*, 34: 33-52.
581. Lyons, I., & Cavaye, J. (2016) Community-Led Engagement With Government and the Role of Community Brokers in East New Britain, Papua New Guinea. *Society & Natural Resources*, 29(4): 462-478.
582. Mackelworth, P., Holcer, D., & Fortuna, C.M. (2008). Multiple use marine protected areas as complex commons. *Proceedings of the 12th Biennial Conference of the International Association for the Study of Commons, Cheltenham, England, University of Gloucestershire.* https://dlc.dlib.indiana.edu/dlc/bitstream/handle/10535/1756/Mackelworth_144701.pdf?sequence=1
583. Madeira, J.P. (2015). GUINEA-BISSAU: the role of local and international NGOs in environmental preservation and sustainability of the Bijagos Archipelago. *InterEspaço: Revista de Geografia e Interdisciplinaridade, 1*(3), 191-202. Available from: https://core.ac.uk/download/pdf/38682990.pdf
584. Madeira, João Paulo. (2016). BIJAGOS ARCHIPELAGO: impacts and challenges for environmental sustainability. *InterEspaço: Revista de Geografia e Interdisciplinaridade, 2*(5): 291-305. Available from: https://core.ac.uk/download/pdf/80950549.pdf
585. Maffi, L., & Woodley, E. (2012). *Biocultural diversity conservation: a global sourcebook.* London: Routledge.
586. Magome, H., & Murombedzi, J. (2003). *Sharing South African national parks: Community land and conservation in a democratic South Africa.* In: *Decolonizing nature: Strategies for conservation in a post-colonial era.* Adams, W., & Mulligan, M. (Eds.). London: Taylor & Francis Group.108-134.
587. Magome, H., & Fabricius, C. (2004). *Reconciling biodiversity conservation with rural development: the Holy Grail of CBNRM.* In: *Rights, Resources and Rural Development: Community-based Natural Resource Management in Southern Africa.* Fabricius, C., Koch, E., Magome, H., & Turner, S. (Eds.). London: Earthscan. 93–114.
588. Mahajan, S.L., & Daw, T. (2016). Perceptions of ecosystem services and benefits to human well-being from community-based marine protected areas in Kenya. *Marine Policy*, 74: 108-119.
589. Mallarach, J.M., Morrison, J., Kothari, A., Sarmiento, F., Atauri, J.A., & Wishitemi, B. (2008). *In defence of protected landscapes: a reply to some criticisms of Category V protected areas and suggestions for improvement.* In: *Defining Protected Areas: An International Conference in Almeria, Spain, May 2007.* Dudley, N., & Stolton, S. (Eds.). IUCN World Commission on Protected Areas, Gland, Switzerland. 31–37.
590. Mandondo, A. (2001). Use of woodland resources within and across villages in a Zimbabwean communal area. *Agriculture and Human Values*, 18(2):177-194.
591. Manfredo, M.J., Bruskotter, J.T., Teel, T.L., Fulton, D., Schwartz, S.H., Arlinghaus, R., Oishi, S., et al. (2016). Why social values cannot be changed for the sake of conservation. *Conservation Biology*, 31(4): 772-780.
592. Mann, B.Q. (2003). *The St Lucia subsistence gillnet fishery.* In: *Waves of Change: Coastal and Fisheries Co-management in South Africa.* Hauck, M., & Sowman, M. (Eds.). Cape Town: University of Cape Town Press. 100-122.
593. Margoluis, R., Stem, C., Swaminathan, V., Brown, M., Johnson, A., Placci, G., Salafsky, N., & Tilders, I. (2013). Results chains: a tool for conservation action design, management, and evaluation. *Ecology and Society*, 18(3): 22.
594. Marie, C. N., Sibelet, N., Dulcire, M., Rafalimaro, M., Danthu, P., & Carrière, S. M. (2009). Taking into account local practices and indigenous knowledge in an emergency conservation context in Madagascar. *Biodiversity and Conservation*, 18(10): 2759-2777.
595. Marikandia, M. (2001). The Vezo of the Fiherena coast, southwest Madagascar: yesterday and today. *Ethnohistory*, 48(1-2): 157-170.
596. Marks, F., Rabehanta, N., Baker, S., Panzner, U., Park, S.E., Fobil, J.N., Meyer, C.G. & Rakotozandrindrainy, R., (2016). A way forward for healthcare in Madagascar?. *Clinical Infectious Diseases*, 62(1): S76-S79. Available from: https://doi.org/10.1080/16549716.2017.1329961
597. Marshall, G. (2008). Nesting, subsidiarity, and community-based environmental governance beyond the local scale. *International Journal of the Commons*, 2(1): 75-97.

598. Marvier, M. (2014). New conservation is true conservation. *Conservation Biology* 28(1): 1-3.
599. Mascia, M.B., & Mills, M. (2018). When conservation goes viral: the diffusion of innovative biodiversity conservation policies and practices. *Conservation Letters*, 11(3): e12442. Available from: https://doi.org/10.1111/conl.12461
600. Mascia, M.B., Brosius, J.P., Dobson, T.A., Forbes, B.C., Horowitz, L., McKean, M.A., & Turner, N.J. (2003). Conservation and the social sciences. *Conservation biology*, 17(3): 649-650.
601. Mascia, M.B., Pailler, S., Thieme, M.L., Rowe, A., Bottrill, M.C., Danielsen, F., Geldmann, J., et al. (2014). Commonalities and complementarities among approaches to conservation monitoring and evaluation. *Biological Conservation*, 169: 258-267.
602. Masuku Van Damme, L.S., & Meskell, L.M. (2009). Producing conservation and community. *Ethics, Place & Identity,* 12: 69-89.
603. Matenga, C. (2015). *Community-based Wildlife Management Programmes and Livelihoods in Zambia: Empowering or Disempowering Local Communities? In: Contemporary Concerns in Development Studies: Perspectives from Tanzania and Zambia.* Kilonzo, R., & Kontinen, T. (Eds.). Finland: Publications of the Department of Political and Economic Studies, number 23 (2015), Development Studies, University of Helsinki. 141-155.
604. Matose, F. (2006). Co-management options for reserved forests in Zimbabwe and beyond: policy implications of forest management strategies. *Forest Policy and Economics*, 8(4): 363-374.
605. Matose, F., & Watts, S. (2010). Towards community-based forest management in Southern Africa: do decentralization experiments work for local livelihoods?. *Environmental Conservation*, 37(3): 310-319.
606. Matzikama (Matzikama Municipality). (2016). Matzikama Municipality Ward 2 Report, March 2016.
607. Mayne, J. (2015). Useful Theory of Change Models. *Canadian Journal of Program Evaluation*, 30(2): 119-142.
608. Mayne, J. (2017). Theory of Change Analysis: Building Robust Theories of Change. *Canadian Journal of Program Evaluation*, 32(2): 155-173.
609. Mayne, J., & Johnson, N. (2015). Using theories of change in the Agriculture for Nutrition and Health CGIAR research program. *Evaluation*, 21(4): 407–428.
610. Mayol, T.L. (2013). Madagascar's nascent locally managed marine area network. *Madagascar Conservation & Development*, 8(2): 91-95.
611. Mbaru, E.K., & Barnes, M.L. (2017). Key players in conservation diffusion: Using social network analysis to identify critical injection points. *Biological Conservation*, 210: 222-232.
612. Mbatha, N.P. (2018). The influence of plural governance systems on rural coastal livelihoods: the case of Kosi Bay. PhD book, University of Cape Town. Available from: https://open.uct.ac.za/handle/11427/29768
613. McCann, K., Kloppers, R., & Venter, A. (2015). *Using biodiversity stewardship as a means to secure the natural wild values on communal land in South Africa. In: Science and stewardship to protect and sustain wilderness values: Tenth World Wilderness Congress symposium; 2013, 4-10 October, Salamanca, Spain.* Watson, A., Carver, S., Krenova, Z., McBride, B. (Eds.). 133-
140. Available from: https://www.fs.usda.gov/treesearch/pubs/49607
614. McCauley, D.J., Pinsky, M.L., Palumbi, S.R., Estes, J.A., Joyce, F.H., & Warner, R.R. (2015). Marine defaunation: Animal loss in the global ocean. *Science*, 347(6219): 1255641.
615. McCay, B.J., & Acheson, J. (1989). *The Question of the Commons: The Culture and Ecology of Communal Resources.* Tucson: University of Arizona Press.
616. McCay, B.J., Micheli, F., Ponce-Díaz, G., Murray, G., Shester, G., Ramirez-Sanchez, S., & Weisman, W. (2014). Cooperatives, concessions, and co-management on the Pacific coast of Mexico. *Marine Policy*, 44: 49-59.
617. McClanahan, T.R., & Abunge, C.A. (2016). Perceptions of fishing access restrictions and the disparity of benefits among stakeholder communities and nations of south-eastern Africa. *Fish and Fisheries*, 17(2): 417-437.
618. McClanahan, T.R., Allison, E.H., & Cinner, J.E. (2015). Managing fisheries for human and food security. *Fish and Fisheries*, 16(1): 78-103.
619. McClanahan, T.R., Muthiga, N.A., & Abunge, C.A. (2016). Establishment of Community Managed Fisheries' Closures in Kenya: Early Evolution of the Tengefu Movement. *Coastal Management*, 44(1): 1-20.

620. McCusker, B., Moseley, W.G., & Ramutsindela, M. (2016). *Land reform in South Africa: an uneven transformation*. Maryland, USA: Rowman & Littlefield.

621. McKean, M.A. (2000). *Common property: what is it, what is it good for, and what makes it work*. In: *People and forests: Communities, institutions, and governance*. Gibson, C.C., McKean, M.A., & Ostrom, E. (Eds.). International Conference on Chinese Rural Collectives and Voluntary Organizations: Between State Organization and Private Interest, University of Lieden, The Netherlands. 27-55.

622. McNeely, J.A. (1993). *Parks for Life: Report of the Fourth World Congress on National Parks and Protected Areas, 10–21 February 1992*. IUCN, Protected Areas Programme, WWF, IUCN, Gland. Available from: https://www.iucn.org/content/parks-life-report-fourth-world-congress-national-parks-and-protected-areas-10-21-february-1992

623. McNeely, J.A., & Miller K.R. (1984). *National Parks, Conservation, and Development: The Role of Protected Areas in Sustaining Society*. In: *Proceedings of the World Congress on National Parks, Bali, Indonesia, 11-22 October 1982*. Washington: Smithsonian Institution Press. Available from: https://portals.iucn.org/library/node/5846

624. McShane, T.O., & Wells, M.P. (2004). *Getting biodiversity projects to work: towards better conservation and development*. New York: Columbia University Press.

625. Meer, T., & Schnurr, M.A. (2013). The community versus community-based natural resource management: the case of Ndumo game reserve, South Africa. *Canadian Journal of Development Studies/Revue canadienne d'études du développement*, 34(4): 482-497.

626. Meinzen-Dick, R., Raju, K. V., and Gulati, A. (2002). *What affects organization and collective action for managing resources? Evidence from canal irrigation systems in India*. World Development, vol. 30, no. 4, p649-666.

627. Mercier, J.R. (2006). The preparation of the National Environmental Action Plan (NEAP): Was it a false start?. *Madagascar Conservation & Development*, 1(1): 50-54.

628. Mermet, L. (2018). Knowledge that is actionable by whom? Underlying models of organized action for conservation. *Environmental Science & Policy*. Available from: https://doi.org/10.1016/j.envsci.2018.04.004

629. Meissner, R., Funke, N., Nienaber, S., & Ntombela, C. (2013). The status quo of research on South Africa's water resource management institutions. *Water SA, 39*(5): 721-732.

630. Metcalfe, K., Ffrench-Constant, R., & Gordon, I. (2010). Sacred sites as hotspots for biodiversity: the Three Sisters Cave complex in coastal Kenya. *Oryx*, 44(01): 118-123.

631. Mezirow, J. (1993). *A transformation theory of adult learning*. In: *Adult Education Research Annual Conference Proceedings*. Kleiber, P., & Tisdell, L. (Eds.). 141-146. Available from: https://files.eric.ed.gov/fulltext/ED357160.pdf#page=153

632. Michie, S., van Stralen, M.M., & West, R. (2011). The behaviour change wheel: A new method for characterising and designing behaviour change interventions. *Implementation Science*, 6(42): 11. Available from: http://www.implementationscience.com/content/6/1/42

633. Michler, L.M., Treydte, A.C., Hayat, H., & Lemke, S. (2019). Marginalised herders: Social dynamics and natural resource use in the fragile environment of the Richtersveld National Park, South Africa. *Environmental Development*, 29: 29-43.

634. MIHARI (2015). *Madagascar's locally managed marine area network*. Available from: http://bjyv3zhj902bwxa8106gk8x5-wpengine.netdna-ssl.com/wp-content/uploads/2015/05/Mihari-Leaflet-2015-ENGLISH-online.pdf

635. Miller, D.C. (2014). Explaining global patterns of international aid for linked biodiversity conservation and development. *World Development*, 59: 341-359.

636. Miller, B.W., Caplow, S.C., & Leslie, P.W. (2012). Feedbacks between conservation and social-ecological systems. *Conservation Biology*, 26(2): 218-227.

637. Miller, D.C., Agrawal, A., & Roberts, J.T. (2013). Biodiversity, governance, and the allocation of international aid for conservation. *Conservation Letters*, 6(1): 12-20.

638. Miller, T.R., Minteer, B.A., & Malan, L.C. (2011). *The new conservation debate: the view from practical ethics. Biological Conservation*, 144(3): 948-957.

639. Milner-Gulland, E.J. (2012). Interactions between human behaviour and ecological systems. *Philosophical Transactions of the Royal Society B: Biological Sciences*, 367(1586): 270-278.

640. Mills, M., Pressey, R.L., Ban, N.C., Foale, S., Aswani, S., & Knight, A.T. (2013). Understanding characteristics that define the feasibility of conservation actions in a common pool marine resource governance system. *Conservation Letters*, 6(6): 418-429.

641. Mittermeier, R.A., Hoffmann, M., Pi Lgrim, J.D., Brooks, T.M., Mittermeier, C.G., Lamoreux, J.F., & da Fonseca, G.A.B. (2004). *Hotspots Revisited: Earth's Biologically Richest and Most Endangered Ecoregions*. Mexico City: Cemex Mexico.

642. Molinos, J.G., Halpern, B.S., Schoeman, D.S., Brown, C.J., Kiessling, W., Moore, P.J., Pandolfi, et al. (2016). Climate velocity and the future global redistribution of marine biodiversity. *Nature Climate Change*, 6(1): 83-88.

643. Monteiro, F., Catarino, L., Batista, D., Indjai, B., Duarte, M.C., & Romeiras, M.M. (2017). Cashew as a high agricultural commodity in West Africa: insights towards sustainable production in Guinea-Bissau. *Sustainability*, 9(9): 1666. Available from: https://doi.org/10.3390/su9091666

644. Montoya, J.M., Donohue, I., & Pimm, S.L. (2018). Planetary boundaries for biodiversity: implausible science, pernicious policies. *Trends in ecology & evolution*, 33(2): 71-73.

645. Moon, K., Blackman, D.A., Adams, V.M., Colvin, R.M., Davila, F., Evans, M.C., Januchowski-Hartley, S.R., et al. (2019a). Expanding the role of social science in conservation through an engagement with philosophy, methodology, and methods. *Methods in Ecology and Evolution*, 10(3): 294-302.

646. Moon, K., Adams, V. M., & Cooke, B. (2019). Shared personal reflections on the need to broaden the scope of conservation social science. *People and Nature*, 1(4): 426-434.

647. Moore, S.A., & Rodger, K. (2010). Wildlife tourism as a common pool resource issue: enabling conditions for sustainability governance. *Journal of Sustainable Tourism*, 18(7): 831-844.

648. Moore, M.L., & Westley, F. (2011). Surmountable chasms: networks and social innovation for resilient systems. *Ecology and Society*, 16(1): 5. Available from: http://www.ecologyandsociety.org/vol16/iss1/art5/

649. Morin, J.F., & Orsini, A. (Eds.). (2020). *Essential concepts of global environmental governance*. London: Routledge.

650. Morrison, S.A. (2015). A framework for conservation in a human-dominated world. *Conservation Biology*, 29(3): 960-964.

651. Morrison, T.H., Adger, W.N., Brown, K., Lemos, M.C., Huitema, D., Phelps, J., ... & Quinn, T. (2019). The black box of power in polycentric environmental governance. *Global Environmental Change*, 57: 101934. Available from: https://doi.org/10.1016/j.gloenvcha.2019.101934

652. Mountjoy, N.J., Seekamp, E., Davenport, M.A., & Whiles, M.R. (2014). Identifying capacity indicators for community-based natural resource management initiatives: focus group results from conservation practitioners across Illinois. *Journal of Environmental Planning and Management*, 57(3): 329-348.

653. Mudliar, P., & Koontz, T. (2018). The muting and unmuting of caste across inter-linked action arenas: Inequality and collective action in a community-based watershed group. *International Journal of the Commons*, 12(1): 225–248.

654. Mufune, P. (2015). Community based natural resource management (CBNRM) and sustainable development in Namibia. *Journal of land and rural studies*, 3(1): 121-138.

655. Mulder, M.B., & Coppolillo, P. (2005). *Conservation: linking ecology, economics, and culture*. New Jersey, USA: Princeton University Press.

656. Mulrennan, M.E., Mark, R., & Scott, C.H. (2012). Revamping community-based conservation through participatory research. *The Canadian Geographer/Le Géographe canadien*, 56(2): 243-259.

657. Murphree, M.W. (2000). *Community–based conservation: old ways, new myths and enduring challenges*. Paper presented at the conference African Wildlife Management in the New Millennium Mweka, Tanzania, 13-15 December 2000. Available from: https://rmportal.net/framelib/marshalmurphree-mweka2000.pdf

658. Murphree, M.W. (2002). *Protected areas and the commons. Common Property Resource Digest*, 60: 1-3.

659. Murphree, M.W. (2009). The strategic pillars of communal natural resource management: Benefit, empowerment and conservation. *Biodiversity and Conservation*, 18(10): 2551-2562.

660. Murray, W.E., & Overton, J.D. (2011). Neoliberalism is dead, long live neoliberalism? Neostructuralism and the international aid regime of the 2000s. *Progress in Development Studies*, 11(4): 307-319.

661. Musgrave, M.K., & Wong, S. (2016). Towards a more nuanced theory of elite capture in development projects. The importance of context and theories of power. *Journal of Sustainable Development*, 9(3): 87-103.

662. Nagendra, H., & Gokhale, Y. (2008). Management regimes, property rights, and forest biodiversity in Nepal and India. *Environmental Management*, 41(5), 719.

663. Nagendra, H., Karna, B., & Karmacharya, M. (2005). Examining institutional change: Social conflict in Nepal's leasehold forestry programme. *Conservation & Society*, 3(1): 72-91.

664. Napier, V.R., Branch, G.M., & Harris, J.M. (2005). Evaluating conditions for successful co-management of subsistence fisheries in KwaZulu-Natal, South Africa. *Environmental Conservation*, 32(2): 165-177.

665. Nash, R.F. (1967). *Wilderness and the American Mind.* New Haven: Yale University Press.

666. Nelson, F. (2012). *Community rights, conservation and contested land: the politics of natural resource governance in Africa.* London: Routledge.

667. Neumann, R.P. (1996). *Dukes, Earls and ersatz Edens: Aristocratic nature preservationists in colonial Africa. Environment and Planning D: Society and Space*, 14(1): 79-98.

668. Neumann, R.P. (1998). *Imposing Wilderness: Struggles Over Livelihood and Nature Preservation in Africa.* Berkeley, USA: University of California Press.

669. Neumann, R.P. (2001). Africa's 'last wilderness': Reordering space for political and economic control in colonial Tanzania. *Africa*, 71(4): 641-665.

670. Neumann, R.P. (2002). The postwar conservation boom in British colonial Africa. *Environmental History*, 7(1): 22-47.

671. Neumann, R.P. (2004). *Nature-state-territory: Towards a critical theorization of conservation enclosures.* In: *Liberation ecologies: environment, development, social movements.* Peet, R., & Watts, M. (Eds.). London: Routledge. 195-217

672. Nilsson, D., Baxter, G., Butler, J.R., & McAlpine, C.A. (2016). How do community-based conservation programs in developing countries change human behaviour? A realist synthesis. *Biological Conservation*, 200: 93-103.

673. Nissanke, M., & Ndulo, M. (2017). *Poverty Reduction in the Course of African Development.* Oxford: Oxford University Press.

674. Noble, M., & Wright, G. (2013). Using Indicators of Multiple Deprivation to Demonstrate the Spatial Legacy of Apartheid in South Africa. *Social Indicators Research*, 112 (1): 187–201.

675. Nordgren, L. (2014). *Opportunity costs of growth-overfishing: socioeconomic evaluation of the beach-seine fishery, Bay of Ranobe, Madagascar.* PhD book, University of Akureyri. Available from: https://www.semanticscholar.org/paper/Opportunity-costs-of-growth-overfishing-%3A-of-the-of-Nordgren/3110088c25707bc2b2c508033201cf7f02f67cdc

676. Normann, A.K. (2006). *Troubled Waters, Troubled Times: Fisheries Policy Reforms in the Transition to Democracy in South Africa and Mozambique.* PhD book, University of Tromso, Norway. Available from: https://munin.uit.no/bitstream/handle/10037/988/book.pdf?sequence=1&isAllowed=y

677. North, D.C. (1990). *Institutions, Institutional Change and Economic Performance.* Cambridge: Cambridge University Press

678. North, D.C. (1993). Institutions and credible commitment. *Journal of Institutional and Theoretical Economics (JITE)/Zeitschrift für die gesamte Staatswissenschaft*, 149(1): 11-23. Available from: https://www.jstor.org/stable/40751576

679. Novellie, P., Biggs, H., & Roux, D. (2016). National laws and policies can enable or confound adaptive governance: examples from South African national parks. *Environmental Science & Policy*, 66: 40-46.

680. Nuesiri, E.O. (2018). *Strengths and Limitations of Conservation NGOs in Meeting Local Needs.* In: *The Anthropology of Conservation NGOs. Palgrave Studies in Anthropology of Sustainability.* Larsen P., & Brockington D. (Eds.). Cham: Palgrave Macmillan. 203-225.

681. Nuggehalli, R.K., & Prokopy, L.S. (2009). Motivating factors and facilitating conditions explaining women's participation in co-management of Sri Lankan forests. *Forest Policy and Economics*, 11(4): 288-293.

682. Nyumba, T.O., Wilson, K., Derrick, C.J., & Mukherjee, N. (2018). The use of focus group discussion methodology: Insights from two decades of application in conservation. *Methods in Ecology and Evolution*, 9(1): 20-32.

683. Ntshona, Z., Kraai, M., Kepe, T., & Saliwa, P. (2010). From land rights to environmental entitlements: Community discontent in the 'successful' Dwesa-Cwebe land claim in South Africa. *Development Southern Africa*, 27(3): 353-361.

684. Oates, J.F. (1995). The dangers of conservation by rural development: A case study from the forests of Nigeria. *Oryx*, 29(2): 115-122.

685. Oates, J.F. (2006). *Conservation, development and poverty alleviation: time for a change in attitudes.* In: *Gaining Ground: In Pursuit of Ecological Sustainability.* Lavigne, D. (Ed.). Guelph, Canada: International Fund for Animal Welfare. 277–284.

686. Obiri, J.F., & Lawes, M.J. (2002). Challenges Facing New Forest Policies in South Africa: Attitudes of Forest Users Toward Management of the Coastal Forests of the Eastern Cape Province. *Environmental Conservation*, 29, 519-529.

687. O'Connell, M.J., Nasirwa, O., Carter, M., Farmer, K.H., Appleton, M., Arinaitwe, J., ... & Duthie, A. (2019). Capacity building for conservation: problems and potential solutions for sub-Saharan Africa. *Oryx*, 53(2): 273-283.

688. OECD. (Organisation for Economic Co-operation and Development). (2007). *Fragile States: Policy Commitment and Principles for Good International Engagement in Fragile States and Situations.* Paris: DCD/DAC. Available from: https://www.oecd.org/dac/conflict-fragility-resilience/docs/38368714.pdf

689. OECD (Organisation for Economic Co-operation and Development). (2018a). *States of Fragility 2018. Available* from: http://www.oecd.org/dac/states-of-fragility-2018-9789264302075-en.htm

690. OECD (Organisation for Economic Co-operation and Development). (2018b). *Sustainable financing for marine ecosystem services in Mauritania and Guinea-Bissau.* Country Study, OECD Environment Policy Paper No. 10. Available from: https://www.oecd.org/countries/mauritania/Policy-Paper-Sustainable-financing-for-marine-ecosystem-services-in-Mauritania-Guinea-Bissau.pdf

691. Ojha, H.R., Ford, R., Keenan, R.J., Race, D., Vega, D.C., Baral, H., & Sapkota, P. (2016). Delocalizing communities: Changing forms of community engagement in natural resources governance. *World Development*, 87: 274-290.

692. Okafor-Yarwood, I. (2019). Illegal, unreported and unregulated fishing, and the complexities of the sustainable development goals (SDGs) for countries in the Gulf of Guinea. *Marine Policy,* 99: 414-422.

693. Oldekop, J. A., Holmes, G., Harris, W. E., & Evans, K. L. (2016). A global assessment of the social and conservation outcomes of protected areas. *Conservation Biology*, *30*(1), 133-141.

694. Oliver, T. A., Oleson, K. L., Ratsimbazafy, H., Raberinary, D., Benbow, S., & Harris, A. (2015). Positive Catch & Economic Benefits of Periodic Octopus Fishery Closures: Do Effective, Narrowly Targeted Actions 'Catalyze' Broader Management?. *PloS one*, 10(6), e0129075. Available from: https://doi.org/10.1371/journal.pone.0129075

695. Olmedo, A., Sharif, V., & Milner-Gulland, E. J. (2018). Evaluating the design of behavior change interventions: a case study of rhino horn in Vietnam. *Conservation Letters*, *11*(1), e12365.

696. Olsen, W. (2004). *Triangulation in social research: Qualitative and quantitative methods can really be mixed.* In: *Developments in Sociology.* Holborn, M. (Ed.). Causeway Press: Ormskirk. 1–30.

697. Olsen, S., & Christie, P. (2000). What are we learning from tropical coastal management experiences?. *Coastal Management*, 28(1): 5-18.

698. Olsson, P., & Galaz, V. (2012). *Social-ecological innovation and transformation.* In: *Social innovation: blurring boundaries to reconfigure markets.* Nicholls, A., & Murdock, A. (Eds.). Basingstoke, UK: Palgrave Macmillan. 223-243.

699. Olsson, P., Folke, C., & T. Hahn, T. (2004). Social-ecological transformation for ecosystem management: The development of adaptive co-management of a wetland landscape in southern Sweden. Ecology and Society 9(4): 2. Available from: http://www.ecologyandsociety.org/vol9/iss4/art2/

700. Olsson, P., Folke, C., Galaz, V., Hahn, T., & Schultz, L. (2007). Enhancing the fit through adaptive co-management: creating and maintaining bridging functions for matching scales in the Kristianstads Vattenrike Biosphere Reserve, Sweden. *Ecology and society*, *12*(1): 28. Available from: http://www.ecologyandsociety.org/vol12/iss1/art28/

701. O'Regan, D., & Thompson, P. (2013). *Advancing Stability and Reconciliation in Guinea-Bissau: Lessons from Africa's First Narco-State.* Available from: https://africacenter.org/wp-content/uploads/2016/06/ASR02EN-Advancing-Stability-and-Reconciliation-in-Guinea-Bissau-Lessons-from-Africa%E2%80%99s-First-Narco-State.pdf

702. Osborne, T.K., Belghith, N.B.H., Bi, C., Thiebaud, A., Mcbride, L., & Jodlowski, M.C. (2016). *Shifting fortunes and enduring poverty in Madagascar: recent findings.* Washington, D.C.: World Bank Group. Available from:

http://documents.worldbank.org/curated/en/413071489776943644/Shifting-fortunes-and-enduring-poverty-in-Madagascar-recent-findings

703. Osterhoudt, S.R. (2018). Community Conservation and the (Mis) appropriation of Taboo. *Development and Change*, 49(5): 1248-1267.

704. Ostrom, E. (1990). *Governing the Commons: The Evolution of Institutions for Collective Action.* Cambridge: Cambridge University Press.

705. Ostrom, E. (1992). *Crafting Institutions for Self-Governing Irrigation Systems.* San Francisco: Institute for Contemporary Studies Press.

706. Ostrom, E. (2000). *Social Capital: A Fad or Fundamental Concept?* In: *Social capital: A multifaceted perspective.* Dasgupta, P. & Seragilden, I. (Eds.). Washington, DC: World Bank. 195-198.

707. Ostrom, E. (2009). A general framework for analyzing sustainability of Social-Ecological Systems. *Science*, 325(5939):419-422.

708. Ostrom, E. (2010a). Institutional analysis and development framework and the commons. *Cornell Law Review*, 95(4): 807–815.

709. Ostrom, E. (2010b). Analyzing collective action. *Agricultural Economics*, 41: 155-166.

710. Ostrom, E., & Nagendra, H. (2006). Insights on linking forests, trees, and people from the air, on the ground, and in the laboratory. *Proceedings of the National Academy of Sciences*, 103: 19224–19231.

711. Ostrom, E., & Ahn, T. (2009). *The meaning of social capital and its link to collective action.* In: *Handbook of Social Capital: The Troika of Sociology, Political Science and Economics.* Svendsen, G.T., & Svendsen, G.L.H. (Eds.). Cheltenham, UK: Edward Elgar.

712. Ostrom, E., Burger, J., Field, C.B., Norgaard, R.B., & Policansky, D. (1999). Revisiting the commons: local lessons, global challenges. *Science*, 284(5412): 278-282.

713. Ostrom, E., Dietz, T., Dolsak, N., Stern, P., Stonich, S., and EU, W. (2002). *The Drama of the Commons.* National Research Council.

714. Padgee, A., Kim, Y.-S., & Daugherty, P.J. (2006). What makes community forestry management successful: a meta-study from community forests throughout the world. *Society & Natural Resources,* 19(1): 33-52.

715. Paterson, A.R. (2011). *Bridging the gap between conservation and land reform: communally-conserved areas as a tool for managing South Africa's natural commons.* PhD book, University of Cape Town, South Africa. Available from: https://open.uct.ac.za/handle/11427/11498

716. Paterson, A.R. (2015). *Protected Areas and community based conservation.* In: *Environmental Law in South Africa.* Glazewski, J., & du Toit, L. (Eds.). South Africa: LexisNexis. Loose-Leaf Edition (Issue 3), 11(1)- 11(50). Available from: https://store.lexisnexis.co.za/products/environmental-law-in-south-africa-skuZASKUPG1566

717. Paterson, A. (2018a). A critical review of South Africa's forestry legislation in promoting participatory forest management. *South African Law Journal*, 135(1): 121-158.

718. Paterson, A. (2018b). Maintaining the ecological flows of estuaries: a critical reflection on the application and interpretation of the relevant legal framework through the lens of the Klein River Estuary. *Potchefstroom Electronic Law Journal/Potchefstroomse Elektroniese Regsblad*, 21(1). Available from: http://dx.doi.org/10.17159/1727-3781/2018/v21i0a2781

719. Paterson, A., & Mkhulisi, M. (2014). Traversing South Africa's conservation and land reform objectives-lessons from the Dwesa-Cwebe Nature Reserve. *South African Law Journal*, 131(2): 365-407.

720. Paulson, S., Gezon, L.L., & Watts, M. (2003). Locating the political in political ecology: An introduction. *Human organization*, 62(3): 205-217.

721. Pawson, R. (2013). *The science of evaluation: a realist manifesto.* California: Sage Publications, Inc.

722. Pawson, R., & Tilley, N. (2004). *Realist evaluation.* Available from: http://www.communitymatters.com.au/RE_chapter.pdf .

723. Persha, L., Agrawal, A., & Chhatre, A. (2011). Social and ecological synergy: local rulemaking, forest livelihoods, and biodiversity conservation. *Science*, 331(6024): 1606-1608.

724. Persha, L., & Andersson, K. (2014). Elite capture risk and mitigation in decentralized forest governance regimes. *Global Environmental Change*, 24: 265-276.

725. Peterson, A.M., & Stead, S.M. (2011). Rule breaking and livelihood options in marine protected areas. *Environmental Conservation*, 38(3): 342-352.

726. Peterson, R.B., Russell, D., West, P., & Brosius, J.P. (2010). Seeing (and doing) conservation through cultural lenses. *Environmental Management*, 45(1): 5-18.

727. Petty, R.E., Wegener, D.T., & Fabrigar, L.R. (1997). Attitudes and attitude change. *Annual review of psychology*, 48(1): 609-647.

728. Phillips, A. (2003). *Turning ideas on their heads: a new paradigm for protected areas. George Wright Forum*, 20(2): 8–32.

729. Pimm, S.L., Jenkins, C.N., Abell, R., Brooks, T.M., Gittleman, J.L., Joppa, L.N., Raven, P.H., et al. (2014). The biodiversity of species and their rates of extinction, distribution, and protection. *Science*, 344(6187): 1246752. Available from: https://doi.org/10.1126/science.1246752

730. Pittman, J., Armitage, D., Alexander, S., Campbell, D., & Alleyne, M. (2015). Governance fit for climate change in a Caribbean coastal-marine context. *Marine Policy*, 51: 486-498.

731. Platteau, J.P. (2004). Monitoring elite capture in community-driven development. *Development and Change*, 35(2): 223–246.

732. Pollini, J., & Lassoie, J. P. (2011). Trapping farmer communities within global environmental regimes: the case of the GELOSE legislation in Madagascar. *Society & Natural Resources*, 24(8): 814-830.

733. Pollini, J., Hockley, N., Muttenzer, F.D., & Ramamonjisoa, B.S. (2014). *The transfer of natural resource management rights to local communities*. In: *Conservation and Environmental Management in Madagascar*. Scales, IR. (Ed.). 172-192.

734. Pollnac, R.B., Crawford, B.R., & Gorospe, M.L. (2001). Discovering factors that influence the success of community-based marine protected areas in the Visayas, Philippines. *Ocean & Coastal Management*, 44(11-12): 683-710.

735. Pomeroy, R.S., Katon, B.M., & Harkes, I. (2001). *Conditions affecting the success of fisheries co-management: lessons from Asia. Marine Policy*, 25(3): 197-208.

736. Pool-Stanvliet, R., Stoll-Kleemann, S., & Giliomee, J. H. (2018). Criteria for selection and evaluation of biosphere reserves in support of the UNESCO MAB programme in South Africa. *Land use policy*, 76: 654-663.

737. PRCM. (Regional Programme for Coastal and Marine Conservation in West Africa). (2012). *Case studies on best practices for a cherished and protected biodiversity*. PRCM, Nouakchott, Mauritania. Available from: http://www.prcmarine.org/sites/prcmarine.org/files/9B Case studies on best pratices fo r protected biodiversity.pdf

738. Prell, C., Hubacek, K., & Reed, M. (2009). Stakeholder Analysis and Social Network Analysis in Natural Resource Management. *Society & Natural Resources*, 22(6): 501-518.

739. Pressman, J.F., & Wildavsky, A. (1983). *Implementation*. Berkeley: University of California Press.

740. Pretty, J. (2003). Social capital and the collective management of resources. *Science*, 302: 1912–1914.

741. Pretty, J., & Smith, D. (2004). Social capital in biodiversity conservation and management. *Conservation Biology*, 18(3): 631-638.

742. Pretty, J., Adams, W., Berkes, F., Ferreira de Athayde, S., Dudley, N., Hunn, E., Maffi, L., et al. (2010). The intersections of biological diversity and cultural diversity: towards integration. *Conservation & Society*, 7(2): 100-112.

743. Prinsen, G., & Nijhof, S. (2015). Between logframes and theory of change: reviewing debates and a practical experience. *Development in Practice*, 25(2): 234-246.

744. Prochaska, J.O., & DiClemente, C.C. (1982). Transtheoretical therapy: Toward a more integrative model of change. *Psychotherapy: theory, research and practice*, 19: 276-288.

745. Prochaska, J.O., & DiClemente, C.C. (1986). *Toward a comprehensive model of change*. In: *Treating addictive behaviors*. Miller, W.R., & Heather, N. (Eds.). Boston: Springer. 3-27.

746. Pyle, R.M. (2003). Nature matrix: reconnecting people and nature. *Oryx*, 37(2): 206-214.

747. Quinn, C.H., Huby, M., Kiwasila, H., & Lovett, J.C. (2007). Design principles and common pool resource management: An institutional approach to evaluating community management in semi-arid Tanzania. *Journal of environmental management*, 84(1): 100-113.

748. Rabearivony, J., Thorstrom, R., de Roland, L.R., Rakotondratsima, M., Andriamalala, T.R., Sam, S.T., Razafimanjato, G., et al. (2010). Protected area surface extension in Madagascar: Do

endemism and threatened species remain useful criteria for site selection? *Madagascar Conservation & Development*, 5(1): 35-47.

749. Radel, C.A. (2012). Outcomes of conservation alliances with women's community-based organizations in southern Mexico. *Society & Natural Resources*, 25(1): 52-70.

750. Raik, D.B., & Decker, D.J. (2007). A multisector framework for assessing community-based forest management: lessons from Madagascar. *Ecology and Society*, 12(1): 14. Available from: http://www.ecologyandsociety.org/vol12/iss1/art14/

751. Raik, D.B., Wilson, A.L., & Decker, D.J. (2008). Power in natural resources management: an application of theory. *Society & Natural Resources*, 21(8): 729-739.

752. Rakotomalala, R. (2018). *State capture and the exploitation of natural resources: The 'Rosewood Scandal' in Madagascar.* In: *State Capture in Africa: Old Threats, New Packaging.* Meirotti, M., & Masterson, G. (Eds.). Johannesburg, South Africa: EISA. Available from: https://www.eisa.org.za/pdf/sym2017papers.pdf

753. Rakotosamimanana, V.R., Arvisenet, G., & Valentin, D. (2014). Studying the nutritional beliefs and food practices of Malagasy school children parents. A contribution to the understanding of malnutrition in Madagascar. *Appetite*, 81: 67-75.

754. Rakotoson, L. R., & Tanner, K. (2006). Community-based governance of coastal zone and marine resources in Madagascar. *Ocean & Coastal Management*, 49(11): 855-872.

755. RAMSAR (2012). *AWARD FOR WETLAND MANAGEMENT 2012.* Available from: https://www.ramsar.org/activities/award-for-wetland-management-2012

756. RAMSAR. (2014). *Guinea-Bissau designates Bijagós Archipelago.* Available from: https://www.ramsar.org/news/guinea-bissau-designates-bijagos-archipelago

757. Randriamboarison, R., Rasoamanajara, F., & Solonandrasana, B. (2013). Tourism return frequency demand in Madagascar. *Tourism Economics*, 19(4): 943-958.

758. Ramutsindela, M., & Shabangu, M. (2013). Conditioned by neoliberalism: a reassessment of land claim resolutions in the Kruger National Park. *Journal of Contemporary African Studies*, 31(3): 441-456.

759. Ramutsindela, M., & Shabangu, M. (2018). *The promise and limit of environmental justice through land restitution in protected areas in South Africa.* In: *Land Rights, Biodiversity Conservation and Justice: Rethinking Parks and People.* Mollett, S., & Kepe, T. (Eds.). London: Routledge. 31-48.

760. Rands, M.R., Adams, W.M., Bennun, L., Butchart, S.H., Clements, A., Coomes, D., Entwistle, A., et al. (2010). Biodiversity conservation: challenges beyond 2010. *Science*, 329(5997): 1298-1303.

761. RARE (n.d.). *Theory of Change for community-based conservation.* Available from: https://www.europarc.org/wp-content/uploads/2015/05/2014-Theory-of-Change-Theory-of-Change.pdf

762. RARE, & BIT (The Behavioural Insights Team). (2019). *Behavior Change for Nature: A Behavioral Science Toolkit for Practitioners.* Arlington, VA: Rare. Available from: https://www.bi.team/wp-content/uploads/2019/04/2019-BIT-Rare-Behavior-Change-for-Nature-digital.pdf

763. Ratner, B.D., Cohen, P., Barman, B., Mam, K., Nagoli, J,. & Allison, E.H. (2013). Governance of aquatic agricultural systems: analyzing representation, power, and accountability. *Ecology and Society,* 18(4): 59. Available from: http://dx.doi.org/10.5751/ES-06043-180459

764. Ratsimbazafy, H., Lavitra, T., Kochzius, M., & Hugé, J. (2019). Emergence and diversity of marine protected areas in Madagascar. *Marine Policy*, 105: 91-108.

765. Raycraft, J. (2019). Conserving Poverty. *Conservation & Society*, 17(3): 297-309.

766. Razafindrakoto, M., Roubaud, F., & Wachsberger, J.M. (2015). *The Puzzle of Structural Fragility in Madagascar: A Political Economy Approach.* Berkley: Center for Effective Global Action, University of California. Available from: http://cega.berkeley.edu/assets/miscellaneous_files/79_-ABCDA_2015_-_Madagascar_Structural_Fragility.pdf

767. Reddy, S.M., Montambault, J., Masuda, Y.J., Keenan, E., Butler, W., Fisher, J.R., Asah, S.T., & Gneezy, A. (2017). Advancing conservation by understanding and influencing human behavior. *Conservation Letters*, 10(2): 248-256.

768. Redmond, W.H. (2005). A framework for the analysis of stability and change in formal institutions. *Journal of Economic Issues*, 39(3): 665-681.

769. Redmore, L., Stronza, A., Songhurst, A., & McCulloch, G. (2018). *Which Way Forward? Past and New Perspectives on Community-Based Conservation in the Anthropocene.* In:

Encyclopaedia of the Anthropocene. Volume 3. DellaSala, D.A., & Goldstein, M.I. (Eds.). 453-460. Available form: https://doi.org/10.1016/B978-0-12-809665-9.09838-4

770. Redpath, S.M., Young, J., Evely, A., Adams, W.M., Sutherland, W.J., Whitehouse, A., Amar, A., et al. (2013). Understanding and managing conservation conflicts. *Trends in ecology & evolution, 28*(2): 100-109.

771. Reed, J., Barlow, J., Carmenta, R., van Vianen, J., & Sunderland, T. (2020). Engaging multiple stakeholders to reconcile climate, conservation and development objectives in tropical landscapes. *Biological Conservation, 238,* 108229. Available from: 10.1016/j.biocon.2019.108229

772. Reef Doctor. (2012). *The Marine Reserves of the Bay of Ranobe: 2012 Report.* Available from: http://www.reefdoctor.org/wp-content/uploads/Bay-of-Ranobe-Marine-Reserves-Report-2012-English-Version.pdf

773. Reef Doctor. (2019). *New Marine Reserve and Artificial Reef Site in the Bay of Ranobe.* 5 February 2019. Available from: https://www.reefdoctor.org/new-marine-reserve-and-artificial-reef-site-in-the-bay-of-ranobe/

774. Relly, P. (2008). *Madikwe game reserve, South Africa–investment and employment.* In: *Responsible tourism: Critical issues for conservation and development.* Spenceley, A. (Ed.). London: Earthscan. 267-284.

775. Renard, Y., & Touré, O. (2012). *Exploring Pathways of creating Marine Protected Areas in West Africa – Experiences and Lessons Learnt.* FIBA/RAMPAO/PRCM, Dakar, Senegal. Available from: http://en.mava-foundation.org/wp-content/uploads/2015/07/AMP_AN_web.pdf

776. RGB (Republic of Guinea-Bissau). (2014). *Fifth National Report to the Convention on Biological Diversity.* Available from: https://www.cbd.int/doc/world/gw/gw-nr-05-en.pdf

777. Rhodes, R. (2007). Understanding Governance: Ten years on. *Organization Studies,* 28(08): 1-20.

778. Ribot, J.C. (2004). *Waiting for Democracy: The Politics of Choice in Natural Resource Decentralization.* Washington, DC, USA: World Resources Institute.

779. Ribot, J.C. (2007). Representation, citizenship and the public domain in democratic decentralization. *Development,* 50(1): 43–49.

780. Ribot, J.C., Agrawal, A., & Larson, A.M. (2006). Recentralizing while decentralizing: how national governments reappropriate forest resources. *World Development,* 34(11): 1864–1886.

781. Ribot, J.C., Lund, J.F., & Treue, T. (2010). Democratic decentralization in sub-Saharan Africa: its contribution to forest management, livelihoods, and enfranchisement. *Environmental Conservation,* 37(1): 35-44.

782. Rice, W.S. (2015). *Contextualising the bycatch 'problem' in the Olifants Estuary Small-Scale Gillnet Fishery using an Ecosystem Approach to Fisheries.* Masters book, University of Cape Town. Available from: http://open.uct.ac.za/bitstream/handle/11427/19987/book_sci_2015_rice_wayne_stanley.pdf?sequence=1&isAllowed=y

783. Rice, W.S., Serge, J.P.R., & Sowman, M.R. (2017). Understanding bycatch using an EAF approach: The case of the Olifants estuary small-scale gillnet-fishery, South Africa. *Ocean & coastal management,* 149: 22-32.

784. Richerson, P.J., Boyd, R., & Paciotti, B. (2002). *An evolutionary theory of commons management.* In: *The Drama of the Commons.* Ostrom, E., Dietz, T., Dolšak, N., Stern, P.C., Stonich, S., & Weber, E.U. (Eds.). Washington, D.C.: National Academy Press. 403-442.

785. Robertson, J., & Lawes, M.J. (2005). User perceptions of conservation and participatory management of iGxalingenwa forest, South Africa. *Environmental Conservation,* 32(1): 64-75.

786. Robbins, S.P., & Judge, T. (2009). *Essentials of organizational behaviour.* 10th edition. Upper Saddle River, New Jersey: Prentice Hall.

787. Robbins, P., McSweeney, K., Chhangani, A.K., & Rice, J.L. (2009). Conservation as it is: illicit resource use in a wildlife reserve in India. *Human Ecology,* 37(5), 559. Available from: https://doi.org/10.1007/s10745-009-9233-6

788. Robson, L., & Rakotozafy, F. (2015). The freedom to choose: integrating community-based reproductive health services with locally-led marine conservation initiatives in southwest Madagascar. *Madagascar Conservation & Development,* 10(1): 6-12.

789. Rodriguez-Izquierdo, E., Gavin, M.C., & Macedo-Bravo, M.O. (2010). Barriers and triggers to community participation across different stages of conservation management. *Environmental Conservation,* 37(3): 239-249.

790. Roe, D., Mayers, J., Grieg-Gran, M., Kothari, A., Fabricius, C., & Hughes, R. (2000). *Evaluating Eden*. London: IIED. Available from: http://pubs.iied.org/pdfs/7810IIED.pdf.

791. Roe, D., Mohammed, E.Y., Porras, I., & Giuliani, A. (2013). Linking biodiversity conservation and poverty reduction: de-polarizing the conservation-poverty debate. *Conservation Letters*, 6(3): 162-171.

792. Roe, D., Booker, F., Day, M., Zhou, W., Allebone-Webb, S., Hill, N.A., Kumpel, N., et al. (2015). Are alternative livelihood projects effective at reducing local threats to specified elements of biodiversity and/or improving or maintaining the conservation status of those elements?. *Environmental Evidence*, 4(1): 22.

793. Rogers, P. (2007). Theory-based evaluations: Reflections ten years on. *New Directions for Evaluation*, 114: 63–67.

794. Rogers, P. (2014a). *Overview of Impact Evaluation, Methodological Briefs: Impact Evaluation 1*. Florence: UNICEF Office of Research. Available from: https://www.unicef-irc.org/publications/pdf/brief_1_overview_eng.pdf

795. Rogers, P. (2014b). *Theory of Change, Methodological Briefs: Impact Evaluation 2*. Florence: UNICEF Office of Research. Available from: https://www.unicef-irc.org/publications/pdf/brief_2_theoryofchange_eng.pdf

796. Rolston, H, III. (1996). *Feeding People versus Saving Nature?* In: Aiken, W., and LaFollette, H. *World Hunger and Morality*. 2nd Edition. Upper Saddle River, New Jersey: Prentice-Hall.

797. Romero, C., & Putz, F. (2018). Theory-of-change development for the evaluation of forest stewardship council certification of sustained timber yields from natural forests in Indonesia. *Forests*, 9(9), 547. Available from: https://doi.org/10.3390/f9090547

798. Rousseau, Y., Watson, R. A., Blanchard, J. L., & Fulton, E. A. (2019). Defining global artisanal fisheries. *Marine Policy*, 108: 103634. Available from: https://doi.org/10.1016/j.marpol.2019.103634

799. RSA. (Government of the Republic of South Africa). (1994). *National Restitution of Land Rights Act No. 22 of 1994*. Government Gazette No. 2011, 22 November 1994. Available from: http://www.justice.gov.za/lcc/docs/1994-022.pdf

800. RSA. (Government of the Republic of South Africa). (1996). *National Communal Property Associations Act 28 of 1996*. Government Gazette No. 17205, 22 May 1996. Available from: https://www.gov.za/sites/default/files/gcis_document/201409/act28of1996.pdf

801. RSA. (Government of the Republic of South Africa). (1998a). *National Environmental Management Act No. 107 of 1998*. Government Gazette No. 19519, 27 November 1998. Available from: https://www.environment.co.za/documents/legislation/NEMA-National-Environmental-Management-Act-107-1998-G-19519.pdf

802. RSA. (Government of the Republic of South Africa). (1998b). *National Marine Living Resources Act 18 of 1998*. Government Gazette, No. 18930, 27 May 1998. Available from: https://www.environment.gov.za/sites/default/files/legislations/marine_livingresources_act18_0.pdf

803. RSA. (Government of the Republic of South Africa). (1998c). *National Forests Act No. 84 of 1998*. Government Gazette No. 19408, 30 October 1998. Available from: https://www.gov.za/sites/default/files/gcis_document/201409/a84-98.pdf

804. RSA. (Government of the Republic of South Africa). (2004a). *National Environmental Management: Biodiversity Act No. 10 of 2004*. Government Gazette No. 26436, 7 June 2004. Available from: https://www.environment.gov.za/sites/default/files/legislations/nema_amendment_act10.pdf

805. RSA. (Government of the Republic of South Africa). (2004b). *National Environmental Management: Protected Areas Act No. 57 of 2003*. Government Gazette No. 26025, 18 February 2004. Available from: https://www.environment.gov.za/sites/default/files/legislations/nema_amendment_act57.pdf

806. RSA. (Government of the Republic of South Africa). (2009). *National Environmental Management: Integrated Coastal Management Act No. 24 of 2008*. Government Gazette No. 31884, 7 February 2009. Available from: https://www.gov.za/sites/default/files/gcis_document/201409/31884138.pdf

807. RSA. (Government of the Republic of South Africa). (2010). *National Protected Area Expansion Strategy for South Africa 2008: Priorities for expanding the protected area network for ecological sustainability and climate change adaptation*. Pretoria: Government of South Africa. Available

from:
https://www.environment.gov.za/sites/default/files/docs/nationalprotected_areasexpansio
n_strategy.pdf

808. RSA. (Government of the Republic of South Africa). (2012). *National Policy for the Small Scale Fisheries Sector in South Africa.* Government Gazette No. 35455, 20 June 2012. Available from: https://www.nda.agric.za/docs/policy/policysmallscalefishe.pdf

809. RSA. (Government of the Republic of South Africa). (2014a). *National Restitution of Land Rights Amendment Act No. 15 of 2014.* Government Gazette No. 37791, 1 July 2014. Available from: https://www.gov.za/sites/default/files/gcis_document/201409/37791act-15-2014-restitution-land-rights-amendment-acta.pdf

810. RSA. (Government of the Republic of South Africa). (2014b). *National Environmental Management: Integrated Coastal Management Amendment Act No. 36 of 2014.* Government Gazette No. 38171, 31 October 2014. Available from: https://www.gov.za/sites/default/files/gcis_document/201501/3817131-10act36of2014integratedcoastalmanagema.pdf

811. RSA. (Government of the Republic of South Africa). (2014c). *National Marine Living Resources Amendment Act 5 of 2014.* Government Gazette, No. 37659, 19 May 2014. Available from: https://www.gov.za/sites/default/files/gcis_document/201409/37659act5of2014marinelivin gres19may2014.pdf

812. RSA. (Government of the Republic of South Africa). (2014d). *National Environmental Biodiversity Act (10/2004): Norms and Standards for Biodiversity Management Plans.* Government Gazette, No. 37302, 7th February 2014. Available from: https://cer.org.za/wp-content/uploads/2004/09/Norms-and-Standards-for-Biodiversity-Management-Plans-for-Ecosystems-gazetted-7-Feb-2014.pdf

813. RSA. (Government of the Republic of South Africa). (2016). *South Africa's National Biodiversity Strategy and Action Plan 2015-2025.* Available from: https://www.environment.gov.za/sites/default/files/docs/publications/SAsnationalbiodiver sity_strategyandactionplan2015_2025.pdf

814. RSA (Government of the Republic of South Africa). (2019a). *President Cyril Ramaphosa: Ebenhaeser land claim settlement ceremony.* Presidential Speech, 22 March 2019. Available from: https://www.gov.za/speeches/president-cyril-ramaphosa-ebenhaeser-land-claim-settlement-ceremony-23-mar-2019-0000

815. RSA (Government of the Republic of South Africa). (2019b). *Operation Phakisa.* Available from: https://www.operationphakisa.gov.za/pages/home.aspx

816. RSA (Government of the Republic of South Africa). (2019c). *OCEANS ECONOMY SUMMARY PROGRESS REPORT: JUNE 2019. Operation Phakisa.* Available from: https://www.environment.gov.za/sites/default/files/docs/publications/oceans-economy-summary-progress-report-June2019.pdf

817. Ruiters, G. (2001). Environmental racism and justice in South Africa's transition. *Politikon: South African Journal of Political Studies*, 28(1): 95-103.

818. Ruiz-Mallen, I., Newing, H., Porter-Bolland, L., Pritchard, D.J., Garcia-Frapolli, E., Méndez-López, M.E., et al. (2014). Cognisance, participation and protected areas in the Yucatan Peninsula. *Environmental Conservation*, 41(3): 265-275.

819. Ruiz-Mallén, I., Schunko, C., Corbera, E., Rös, M. and Reyes-García, V. (2015). Meanings, drivers, and motivations for community-based conservation in Latin America. *Ecology and Society,* 20(3): 33.

820. Runte, A. (1987). *National Parks: The American Experience.* Lincoln: University of Nebraska Press.

821. Runte, A. (1990). *Yosemite: The Embattled Wilderness.* Lincoln: University of Nebraska Press.

822. Saito-Jensen, M., Nathan, I., & Treue, T. (2010). Beyond elite capture? Community-based natural resource management and power in Mohammed Nagar village, Andhra Pradesh, India. *Environmental Conservation*, 37(3): 327-335.

823. Sanderson, S., & Redford, K.H. (2003). Contested relationships between biodiversity conservation and poverty alleviation. *Oryx,* 37(4): 389–390.

824. Sandstrom, A., & Carlsson, L. (2008). The performance of policy networks: the relation between network structure and network performance. *Policy Studies Journal*, 36(4): 497-524.

825. Salafsky, N., & Wollenberg, E. (2000). Linking livelihoods and conservation: a conceptual framework and scale for assessing the integration of human needs and biodiversity. *World Development*, 28(8): 1421–1438.

826. Salazar, G., Mills, M., & Veríssimo, D. (2018). Qualitative impact evaluation of a social marketing campaign for conservation. *Conservation Biology*, 33(3): 634-644.
827. Samoilys, M., Osuka, K., Muthiga, N., & Harris, A. (2017). *Locally managed fisheries in the Western Indian Ocean: a review of past and present initiatives.* Available from: http://cordioea.net/wp-content/uploads/2018/04/Samolys-et-al-2017-Locally-managed-fisheries-WIO-Final_English.pdf
828. Sandbrook, C. (2015). What is conservation?. *Oryx*, 49(4): 565-566.
829. Sandbrook, C., Fisher, J. A., Holmes, G., Luque-Lora, R., & Keane, A. (2019). The global conservation movement is diverse but not divided. *Nature Sustainability*, 2(4): 316-323.
830. Sanders, M.J., Miller, L., Bhagwat, S.A., van der Grient, J.M.A., & Rogers, A.D. (2019). Practitioner insights as a means of setting a context for conservation. *Conservation Biology*, 34(1): 113-124.
831. Sangreman, C., Delgado, F., & Martins L. V. (2018). Guinea-Bissau (2014 - 2016). An empirical study of economic and social human rights in a fragile state. *Advances in Social Sciences Research Journal*, 5(3): 66-84.
832. SAR (Situational Assessment Report). (2017). *Olifants Estuarine Management Plan: Revised Situation Assessment Report.* Draft, June 2017.
833. Sarkki, S., & Acosta García, N. (2019). Merging social equity and conservation goals in IPBES. *Conservation Biology*.
834. Sarrasin, B. (2013) Ecotourism, Poverty and Resources Management in Ranomafana, Madagascar. *Tourism Geographies*, 15(1): 3-24.
835. Saunders, F.P. (2014). The promise of common pool resource theory and the reality of commons projects. *International Journal of the Commons*, 8(2): 636-656.
836. Scales, I.R. (2014). *Conservation and environmental management in Madagascar.* London: Routledge.
837. Scanlon, L.J., & Kull, C.A. (2009). Untangling the links between wildlife benefits and community-based conservation at Torra Conservancy, Namibia. *Development Southern Africa*, 26(1), 75-93.
838. Schaefer, G.F. (1991). The rise and fall of subsidiarity. *Futures*, 23(7): 681-694.
839. Schill, C., Anderies, J.M., Lindahl, T., Folke, C., Polasky, S., Cárdenas, J.C., Crépin, A.S., et al. (2019). A more dynamic understanding of human behaviour for the Anthropocene. *Nature Sustainability*, 1-8. Available from: https://doi.org/10.1038/s41893-019-0419-7
840. Schober, M.F., & Conrad, F.G. (1997). Does conversational interviewing reduce survey measurement error?. *Public opinion quarterly*, 61: 576-602.
841. Schober, M.F., & Conrad, F.G. (2015). Improving social measurement by understanding interaction in survey interviews. *Policy Insights from the Behavioral and Brain Sciences*, 2(1): 211-219.
842. Schreiber, L. (2017). *Putting Justice into Practice: Communal Land Tenure in Ebenhaeser, South Africa, 2012 – 2017.* Innovations for Successful Societies, Princeton University. Available from: https://successfulsocieties.princeton.edu/publications/communal-land-tenure-ebenhaeser-south-africa
843. Schultz, P.W. (2011). Conservation means behavior. *Conservation Biology*, 25(6): 1080-1083.
844. Schultz, P.W. (2014). Strategies for promoting proenvironmental behavior. *European Psychologist*, 19: 107-117.
845. Schultz, P.W., Nolan, J.M., Cialdini, R.B., Goldstein, N.J., & Griskevicius, V. (2007). The constructive, destructive, and reconstructive power of social norms. *Psychological Science*, 18(5): 429–434.
846. Schultz, L., Folke, C., Österblom, H., & Olsson, P. (2015). Adaptive governance, ecosystem management, and natural capital. *Proceedings of the National Academy of Sciences*, 112(24): 7369-7374.
847. Schwitzer, C., Mittermeier, R.A., Johnson, S.E., Donati, G., Irwin, M., Peacock, H., Ratsimbazafy, J., et al. (2014). Averting lemur extinctions amid Madagascar's political crisis. *Science*, 343(6173): 842-843.
848. Scott, W.R. (2014). *Institutions and organizations: Ideas, interests, and identities.* 4th edition. Thousand Oaks, CA: Sage.
849. Scott, W.R., & Davis, G.F. (2015). *Organizations and organizing: Rational, natural and open systems perspectives.* London: Routledge.

850. Seixas, C.S., & Davy, B. (2008). Self-organization in integrated conservation and development initiatives. *International Journal of the Commons*, 2(1): 99-125.

851. Selig, E.R., Turner, W.R., Troëng, S., Wallace, B.P., Halpern, B.S., Kaschner, K., Lascelles, B.G., et al. (2014). Global priorities for marine biodiversity conservation. *PloS one*, 9(1): e82898. Available from: https://doi.org/10.1371/journal.pone.0082898

852. Selman, P. (2004). Community participation in the planning and management of cultural landscape. *Journal of Environmental Planning and Management* 47(3): 365–92.

853. Shackleton, C. (2009). Will the real custodian of natural resource management please stand up. *South African Journal of Science*, 105(3-4): 91-93.

854. Shackleton, C., Shackleton, S. (2004). The importance of non-timber forest products in rural livelihood security and as safety nets: a review of evidence from South Africa. *South African Journal of Science*, 100(11): 658-664.

855. Shove, E., and Walker, G. (2007). Commentary. CAUTION! Transitions ahead: politics, practice, and sustainable transition management. *Environment and Planning A*, 39(4): 763-770.

856. Shukla, S.R., & Sinclair, A.J. (2010). Strategies for self-organization: learning from a village-level community-based conservation initiative in India. *Human Ecology*, 38(2): 205-215.

857. Sim, J., Saunders, B., Waterfield, J., & Kingstone, T. (2018). Can sample size in qualitative research be determined a priori?. *International Journal of Social Research Methodology*, 21(5): 619-634.

858. Sidaway, J. (2000). Recontextualising positionality: Geographical research and academic fields of power. *Antipode*, 32 (3):260–70.

859. Simier, M., Ecoutin, J., & Tito de Morais, L. (2019). The PPEAO experimental fishing dataset: Fish from West African estuaries, lagoons and reservoirs. *Biodiversity Data Journal*, 7: e31374. Available from: https://doi.org/10.3897/BDJ.7.e31374

860. Simon, B., & Oakes, P. (2006). Beyond dependence: An identity approach to social power and domination. *Human Relations*, 59(1): 105–139.

861. Singleton, S. (2009). Native people and planning for marine protected areas: how "Stakeholder" processes fail to address conflicts in complex, real-world environments. Coastal Management, 37(5): 421–440.

862. Singleton, R.L., Allison, E.H., Le Billon, P., & Sumaila, U.R. (2017). Conservation and the right to fish: International conservation NGOs and the implementation of the Voluntary Guidelines for securing Sustainable Small-Scale Fisheries. *Marine Policy*, 84: 22-32.

863. Skutsch, M.M., & Ba, L. (2010). Crediting carbon in dry forests: The potential for community forest management in West Africa. *Forest Policy and Economics*, 12(4): 264-270.

864. Só, B., Franco, E.F., Carvalho, H.C., Santos, J.R.D., & Armenia, S. (2018). Nobody deserves this fate: the vicious cycle of low human development in Guinea-Bissau. *Kybernetes*, 47(2): 392-408.

865. Song, A.M., Chuenpagdee, R., & Jentoft, S. (2013). Values, images, and principles: What they represent and how they may improve fisheries governance. *Marine Policy*, 40: 167-175.

866. Soulé, M. (2013). The "New Conservation." *Conservation Biology*, 27(5): 895–897.

867. Southall, R. (2016). *The new black middle class in South Africa*. New York: Boydell & Brewer.

868. Sousa, J. (2014). Shape-shifting nature in a contested landscape of southern Guinea-Bissau. Unpublished PhD book, Oxford Brookes University.

869. Sousa, C., Gippoliti, S., & Akhlas, M. (2005). *Republic of Guinea-Bissau*. In: *World Atlas of Great Apes and Their Conservation. UNEP World Conservation*. Caldecott, J., & Miles, L. (Eds.). Berkeley, USA: University of California Press. 362–365

870. Sousa, J., Hill, C. M., & Ainslie, A. (2017). Chimpanzees, sorcery and contestation in a protected area in Guinea-Bissau. *Social Anthropology*, 25(3): 364-379.

871. Soutschka, N. (2014). *Community-based resource use monitoring at the Olifants River Estuary*. Masters book, University of Cape Town. Available from: https://open.uct.ac.za/bitstream/handle/11427/13276/book_sci_2014_soutschka_n.pdf?sequence=1&isAllowed=y

872. Sowards, S.K., Tarin, C.A., & Upton, S.D. (2017). Place-Based Dialogics: adaptive cultural and interpersonal approaches to environmental conservation. *Frontiers in Communication*, 2: 9. Available from: https://doi.org/10.3389/fcomm.2017.00009

873. Sowman, M. (2003). *Co-management of the Olifants River harder fishery*. In: *Waves of Change: Coastal and Fisheries Co-management in Southern Africa*. Hauck, M., & Sowman, M. (Eds.). Cape Town: University of Cape Town Press. 269-298.

874. Sowman, M. (2009). An Evolving Partnership: Collaboration between university 'experts' and net-fishers. *Gateways: International Journal of Community Research and Engagement, 2:* 119-143.

875. Sowman, M. (2011). New perspectives in small-scale fisheries management: challenges and prospects for implementation in South Africa. *African Journal of Marine Science,* 33(2): 297-311.

876. Sowman, M. (2017). *Turning the tide: Strategies, innovation and transformative learning at the Olifants estuary, South Africa.* In: *Governing the Coastal Commons: Communities, Resilience and Transformation.* Armitage, D., Berkes, F. and Charles, T. (Eds.). London: Earthscan/Routledge, 39-56.

877. Sowman, M., Sunde, J., Raemaekers, S., & Schultz, O. (2014). Fishing for equality: Policy for poverty alleviation for South Africa's small-scale fisheries. *Marine Policy,* 46: 31-42.

878. Spinola, C., Bruges-Armas, J., Brehm, A., & Spinola, H. (2008). HLA-A polymorphisms in four ethnic groups from Guinea-Bissau (West Africa) inferred from sequence-based typing. *Tissue antigens,* 72(6): 593-598.

879. SSA (Statistics South Africa). (2019). Available from: http://www.statssa.gov.za/

880. Standing, A. (2008). Corruption and industrial fishing in Africa. *U4 Issue 2008:7.* U4 AntiCorruption Resource Centre. Available from: https://open.cmi.no/cmi-xmlui/bitstream/handle/11250/2474539/Corruption%20and%20Industrial%20Fishing%20in%20Africa?sequence=1

881. Steenbergen, D.J. (2016). Strategic customary village leadership in the context of marine conservation and development in Southeast Maluku, Indonesia. *Human Ecology,* 44(3): 311-327.

882. Steenbergen, D.J., & Warren, C. (2018). Implementing strategies to overcome social-ecological traps: the role of community brokers and institutional bricolage in a locally managed marine area. *Ecology and Society,* 23(3): 10.

883. Stein, D., & Valters, C. (2012). *Understanding theory of change in international development.* Available from: http://eprints.lse.ac.uk/56359/1/JSRP_Paper1_Understanding_theory_of_change_in_international_development_Stein_Valters_2012.pdf

884. Steinberg, P.F. (2009). Institutional resilience amid political change: the case of biodiversity conservation. *Global Environmental Politics,* 9(3): 61-81.

885. Steins, N., Edwards, V., & Roling, N. (2000). Re-designed principles for CPR theory. *The Common Property Resource Digest,* 53: 1-5.

886. Sterling, E., Ticktin, T., Morgan, T.K.K., Cullman, G., Alvira, D., Andrade, P., Bergamini, N., et al. (2017). Culturally grounded indicators of resilience in socialecological systems. *Environment and Society: Advances in Research,* 8(1): 63–95.

887. Stern, P.C. (2000). Toward a coherent theory of environmentally significant behavior. *Journal of Social Issues,* 56(3): 407–424.

888. Stern, P.C. (2011). Design principles for global commons: Natural resources and emerging technologies. *International Journal of the Commons,* 5(2): 213-232.

889. Stern, P.C., Dietz, T., & Ostrom, E. (2002). Research on the Commons: lessons for Environmental Resource Managers. *Environmental Practice,* 4(2): 61-64.

890. Stone, M.T., Lenao, M., & Moswete, N. (2020). *Natural Resources, Tourism and Community Livelihoods in Southern Africa: Challenges of Sustainable Development.* London: Routledge.

891. Streak, J.C. (2004). The Gear legacy: did Gear fail or move South Africa forward in development?. *Development Southern Africa,* 21(2): 271-288.

892. Strydom, H. (2015). *Introduction to regional environmental law of the African Union.* In: *Regional Environmental Law.* Scholtz, W., & Verschuuren, J. (Eds.). Elgaronline: Edward Elgar Publishing. 21-50.

893. Suich, H., Child, B., & Spenceley, A. (2012). *Evolution and innovation in wildlife conservation: parks and game ranches to transfrontier conservation areas.* London: Earthscan.

894. Sunde, J. (2014). *Customary Governance and Expressions of Living Customary Law at Dwesa-Cwebe: Contributions to Small-Scale Fisheries Governance in South Africa.* PhD book, University of Cape Town, South Africa. Available from: https://open.uct.ac.za/handle/11427/13275

895. Sunde, J., Sowman, M., Smith, H., & Wicomb, W. (2013). Emerging proposals for governance of tenure in small-scale fisheries in South Africa. *Land Tenure Journal,* 1: 117-144.

896. Sundnes, F. (2013). Scrubs and squatters: the coming of the Dukuduku forest, an indigenous forest in KwaZulu-Natal, South Africa. *Environmental History*, 18(2): 277-308.

897. Sunseri, T. (2014). *Wielding the Ax: State forestry and social conflict in Tanzania, 1820–2000*. Ohio: Ohio University Press.

898. Sundström, A. (2016). Corruption and violations of conservation rules: A survey experiment with resource users. *World Development, 85:* 73-83.

899. Sutton, A.M., & Rudd, M.A. (2014). Deciphering contextual influences on local leadership in community-based fisheries management. *Marine Policy*, 50: 261-269.

900. Temple, B., & Young, A. (2004). Qualitative research and translation dilemmas. *Qualitative research*, 4(2): 161-178.

901. Temudo, M.P. (2011). Planting knowledge, Harvesting Agro-Biodiversity: A Case Study of Southern Guinea-Bissau Rice Farming. *Human Ecology*, 39(3): 309-321.

902. Temudo, M.P. (2012). "The white men bought the forests": conservation and contestation in Guinea-Bissau, Western Africa. *Conservation & Society*, 10(4): 354–66.

903. Temudo, M.P., & Abrantes, M. (2014). The cashew frontier in Guinea-Bissau, West Africa: changing landscapes and livelihoods. *Human Ecology*, 42(2), 217-230.

904. Temudo, M.P., & Abrantes, M. (2015). The Pen and the Plough: Balanta Young Men in Guinea-Bissau. *Development and Change*, 46(3): 464-485.

905. Tengö, M., Brondizio, E.S., Elmqvist, T., Malmer, P., & Spierenburg, M. (2014). Connecting diverse knowledge systems for enhanced ecosystem governance: the multiple evidence base approach. *Ambio*, 43(5): 579-591.

906. Terborgh, J. (1999). *Requiem for Nature*. Washington DC: Island Pres/Shearwater Books.

907. Terborgh, J. (2000). The fate of tropical forests: a matter of stewardship. *Conservation Biology* 14(5): 1358–1361.

908. Terborgh, J. (2004). *Reflections of a Scientist on the World Parks Congress. Conservation Biology*, 18(3): 619–620.

910. Terborgh, J., & Peres, C.A. (2017). Do Community-Managed Forests Work? A Biodiversity Perspective. *Land*, 6(2), 22.

911. Thiault, L., Marshall, P., Gelcich, S., Collin, A., Chlous, F., & Claudet, J. (2018). Mapping social–ecological vulnerability to inform local decision making. *Conservation biology*, 32(2): 447-456.

912. Thompson, J. (1995). Participatory approaches in government bureaucracies: Facilitating the process of institutional change. *World development*, 23(9): 1521-1554.

913. Thondhlana, G., & Shackleton, S. (2015). Cultural values of natural resources among the San people neighbouring Kgalagadi Transfrontier Park, South Africa. *Local Environment*, 20(1): 18-33.

914. Thondhlana, G., & Muchapondwa, E. (2014). Dependence on environmental resources and implications for household welfare: Evidence from the Kalahari drylands, South Africa. *Ecological Economics*, 108: 59-67.

915. Thondhlana, G., & Cundill, G. (2017). Local people and conservation officials' perceptions on relationships and conflicts in South African protected areas. *International Journal of Biodiversity Science, Ecosystem Services & Management*, 13(1): 204-215.

916. Thondhlana, G., Shackleton, S., & Muchapondwa, E. (2011). Kgalagadi Transfrontier Park and its land claimants: a pre- and post-land claim conservation and development history. *Environmental Research Letters*, 6(2): 1–12.

917. Thondhlana, G., Shackleton, S., & Blignaut, J. (2015). Local institutions, actors, and natural resource governance in Kgalagadi Transfrontier Park and surrounds, South Africa. *Land Use Policy*, 47: 121–129.

918. Thondhlana, G., Cundill, G., & Kepe, T. (2016). Co-management, land rights, and conflicts around South Africa's Silaka Nature Reserve. *Society & Natural Resources*, 29(4): 403–417

919. Tiniguena. (2019). *The Urok Islands.* Available from: http://www.tiniguenagb.org/biodivercidade/

920. Titz, A., Cannon, T., & Krüger, F. (2018). Uncovering 'Community': Challenging an Elusive Concept in Development and Disaster Related Work. *Societies*, 8(3): 71.

921. Toillier, A., Lardon, S., & Herve, D. (2008). An environmental governance support tool: community-based forest management contracts (Madagascar). *International Journal of Sustainable Development*, 11(2-4): 187-205.

922. Toillier, A., Serpantié, G., Hervé, D., & Lardon, S. (2011). Livelihood strategies and land use changes in response to conservation: Pitfalls of community-based forest management in Madagascar. *Journal of Sustainable Forestry*, 30(1-2): 20-56.

923. Tole, L. (2010). Reforms from the ground up: a review of community-based forest management in tropical developing countries. *Environmental Management*, 45(6): 1312–1331.

924. Tool, M. (1979). *The Discretionary Economy: A Normative Theory of Political Economy*. Santa Monica, California: Goodyear Publishing Company.

925. Transparency International. (2019). *Corruption Perception Index 2018.* Available from: https://www.transparency.org/cpi2018

926. Traynor, C.H., & Hill, T. (2008). Mangrove Utilisation and Implications for Participatory Forest Management, South Africa. *Conservation & Society,* 6(2): 109–116.

927. Tumusiime, D.M., & Vedeld, P. (2012). False promise or false premise? Using tourism revenue sharing to promote conservation and poverty reduction in Uganda. *Conservation & Society*, 10(1): 15–28.

928. Turnbull, D. (1997). *Reframing Science and Other Local Knowledge Traditions.* Futures, 29(6): 551-562.

929. Turner, R.A., Forster, J., Fitzsimmons, C., Gill, D., Mahon, R., Peterson, A., & Stead, S. (2018). Social fit of coral reef governance varies among individuals. *Conservation Letters*, 11(3): e12422.

930. Turpie, J.K., Adams, J.B., Joubert, A., Harrison, T.D., Colloty, B.M., Maree, R.C., Whitfield, A.K., et al. (2002). Assessment of the conservation priority status of South African estuaries for use in management and water allocation. *Water SA*, 28(2): 191-206. Available from: https://www.ajol.info/index.php/wsa/article/viewFile/4885/12529

931. Turpie, J.K., Forsythe, K.J., Knowles, A., Blignaut, J., & Letley, G. (2017). Mapping and valuation of South Africa's ecosystem services: A local perspective. *Ecosystem services*, 27: 179-192.

932. Turreira-García, N., Lund, J., Domínguez, P., Carrillo-Anglés, E., Brummer, M., Duenn, P., & Reyes-García, V. (2018). What's in a name? Unpacking "participatory" environmental monitoring. *Ecology and Society*, 23(2): 24.

933. Tvedten, I. (1990). The difficult transition from subsistence to commercial fishing. The Case of the Bijagós of Guinea-Bissau. *Maritime Anthropological Studies,* 3(1): 119-130.

934. Uberhuaga, P., Larsen, H.O., & Treue, T. (2011). Indigenous forest management in Bolivia: potentials for livelihood improvement. *The International Forestry Review*, 13(1): 80-95.

935. Ulambayar, T., Fernández-Giménez, M.E., Baival, B., & Batjav, B. (2017). Social Outcomes of Community-based Rangeland Management in Mongolian Steppe Ecosystems. *Conservation Letters*, 10(3): 317-327.

936. UNDP. (United Nations Development Programme). (2015). *Briefing Note for Countries on the 2015 Human Development Report: Guinea-Bissau.* Available from: http://hdr.undp.org/sites/all/themes/hdr_theme/country-notes/GNB.pdf

937. UNDP. (United Nations Development Programme). (2016). *Human Development Report 2016: Human Development for Everyone. Briefing note for countries on the 2016 Human Development Report - Madagascar.* Available from: http://hdr.undp.org/sites/all/themes/hdr_theme/country-notes/MDG.pdf

938. UNESCO. (United Nations Educational, Scientific and Cultural Organization). (2013). Decision: 37 COM 8B.17 Bijagós Archipelago – Motom Moranghajogo (Guinea Bissau). Available from: https://whc.unesco.org/en/decisions/5132

939. UNESCO. (United Nations Educational, Scientific and Cultural Organization). (2019). Guinea-Bissau. Available from: http://uis.unesco.org/country/GW

940. UNICEF, WHO, & World Bank (United Nations Children's Fund, World Health Organization, and World Bank). (2016). *Joint Child Malnutrition Estimates. Global Database on Child Growth and Malnutrition.* Available from: http://www.who.int/nutgrowthdb/estimates2014/en/

941. UNIOGBIS (United Nations Integrated Peacebuilding Office in Guinea-Bissau). (2017). WFD: GUINEA-BISSAU FORESTS STILL AT RISK DESPITE GOOD PROTECTION SYSTEM, 5 April 2017. Available from: https://uniogbis.unmissions.org/en/wfd-guinea-bissau-forests-still-risk-despite-good-protection-system

942. UNEP-WCMC. (United Nations Environmental Programme-World Conservation Monitoring Centre). (2020). *Aichi Target 11 Dashboard.* Cambridge, UK; Gland, Switzerland; and Washington, DC: USA. Available from: https://www.protectedplanet.net/target-11-dashboard

943. UNEP-WCMC. (United Nations Environmental Programme-World Conservation Monitoring Centre). (2019a). *Protected Planet: The World Database on Protected Areas (WDPA). Country Profile, Guinea-Bissau.* Available from: https://www.protectedplanet.net/country/GW

944. UNEP-WCMC. (United Nations Environmental Programme-World Conservation Monitoring Centre). (2019b). *Protected Planet: The World Database on Protected Areas (WDPA). Available* from: https://www.protectedplanet.net/ilhas-formosa-nago-tchedia-urok-marine-community-protected-area

945. Valters, C. (2015). *Theories of change: time for a radical approach to learning in development.* London: Overseas Development Institute. Available from: https://www.odi.org/sites/odi.org.uk/files/odi-assets/publications-opinion-files/9835.pdf

946. Vance-Borland, K., & Holley, J. (2011). Conservation stakeholder network mapping, analysis, and weaving. *Conservation Letters*, 4(4): 278-288.

947. van den Broek, K. (2018). Illuminating divergence in perceptions in natural resource management: A case for the investigation of the heterogeneity in mental models. *Journal of Dynamic Decision Making*, 4(1). Available from: https://doi.org/10.11588/jddm.2018.1.51316

948. van der Linde, M., & Feris, L. (2010). *Compendium of South African Environmental Legislation* 2nd edition. Pretoria, South Africa: Pretoria University Law Press (PULP). Available from: http://www.pulp.up.ac.za/component/edocman/compendium-of-south-african-environmental-legislation-second-edition

949. van Es, M., Guijt, I., & Vogel, I. (2015). *Theory of Change Thinking in Practice.* The Hague: Hivos. Available from: http://www.theoryofchange.nl/sites/default/files/resource/hivos_toc_guidelines_final_nov_2015.pdf

950. van Eeden, D.G., van Rensburg, B.J., De Wijn, M., & du P. Bothma, J. (2006). The value of community-based conservation in a heterogeneous landscape: an avian case study from sand forest in Maputaland, South Africa. *South African Journal of Wildlife Research*, 36(2): 153–157.

951. Van Laerhoven, F., & Ostrom, E. (2007). *Traditions and Trends in the Study of the Commons. International Journal of the Commons*, 1(1): 3-28.

952. Van Niekerk, L., Taljaard, S., Ramjukadh, C-L., Adams, J.B., Lamberth, S.J., Weerts, S., Petersen, C., et al. (2017). A multi-sectoral Resource Planning Platform for South Africa's estuaries. Water Research Commission Report No K5/2464. Available from: http://www.wrc.org.za/mdocs-posts/tt748-web/

953. Van Niekerk, L., Adams, J.B., Allan, D.G., Taljaard, S., Weerts, S.P., Louw, D., Talanda, C., & Van Rooyen, P. (2019). Assessing and planning future estuarine resource use: A scenario-based regional-scale freshwater allocation approach. *Science of The Total Environment*, 657: 1000-1013.

954. Van Sittert, L. (1998). Keeping the Enemy at Bay: The Extermination of Wild Carnivora in the Cape Colony, 1889 - 1910. *Environmental History*, 3(3): 333-356.

955. Vasconcelos, M.J.P., Biai, J.M., Araujo, A., & Diniz, M.A. (2002). Land cover change in two protected areas of Guinea-Bissau (1956-1998). *Applied Geography*, 22(2): 139-156.

956. Vasconcelos, M.J., Cabral, A.I., Melo, J.B., Pearson, T.R., Pereira, H.D.A., Cassamá, V., & Yudelman, T. (2015). Can blue carbon contribute to clean development in West-Africa? The case of Guinea-Bissau. *Mitigation and Adaptation Strategies for Global Change*, 20(8): 1361-1383.

957. Veldkornet, D.A., Adams, J.B., & Van Niekerk, L. (2015). Characteristics and landcover of estuarine boundaries: implications for the delineation of the South African estuarine functional zone. *African journal of marine science*, 37(3): 313-323.

958. Veríssimo, D. (2013). Influencing human behaviour: an underutilised tool for biodiversity management. *Conservation Evidence*, 10(1): 29-31.

959. Virah-Sawmy, M., Gardner, C.J., & Ratsifandrihamanana, A.N. (2014). *The Durban Vision in practice: experiences in the participatory governance of Madagascar's new protected areas.* In: *Conservation and environmental management in Madagascar.* Scales, I.R. (Ed.). Oxon and New York: Routledge. 216–251.

960. Virdin, J., Kobayashi, M., Akester, S., Vegh, T., & Cunningham, S. (2019). West Africa's coastal bottom trawl fishery: Initial examination of a trade in fishing services. *Marine Policy*, 100: 288-297.

961. Visconti, P., Butchart, S.H., Brooks, T.M., Langhammer, P.F., Marnewick, D., Vergara, S., Yanosky, A., & Watson, J.E. (2019). Protected area targets post-2020. *Science*, 364(6437): 239-241. Available from:

962. Vogel, I. (2012). *Review of the use of 'Theory of Change' in international development.* Report commissioned by the United Kingdom Department for International Development. Available from: https://assets.publishing.service.gov.uk/media/57a08a5ded915d3cfd00071a/DFID_ToC_Review_VogelV7.pdf

963. Vonk, G., Geertman, S., & Schot, P. (2007). A SWOT analysis of planning support systems. *Environment and Planning A*, 39(7): 1699–1714.

964. Vucetich, J.A., Burnham, D., Macdonald, E.A., Bruskotter, J.T., Marchini, S., Zimmermann, A., & Macdonald, D.W. (2018). Just conservation: What is it and should we pursue it?. *Biological Conservation*, 221: 23-33.

965. Wabnitz, C., Karibuhoye, C., & Fall, M. (2008). *West African marine protected areas network.* The Sea Around Us Project, Newsletter Issue 48 – July/August 2008, p1-3. Available from: http://www.seaaroundus.org/newsletter/Issue48.pdf

966. Wade, R. (1987). *Village Republics: Economic Conditions for Collective Action.* Cambridge: Cambridge University Press.

967. Waeber, P.O., Wilmé, L., Mercier, J.R., Camara, C., & Lowry II, P.P. (2016). How effective have thirty years of internationally driven conservation and development efforts been in Madagascar?. *PloS one*, *11*(8), e0161115. Available from: https://doi.org/10.1371/journal.pone.0161115

968. Waldron, A., Miller, D.C., Redding, D., Mooers, A., Kuhn, T.S., Nibbelink, N., Roberts, J.T., et al. (2017). Reductions in global biodiversity loss predicted from conservation spending. *Nature*, 551(7680): 364-367.

969. Wals, A.E.J. (2007). Learning in a changing world and changing in a learning world: reflexively fumbling towards sustainability. *Southern African Journal of Environmental Education*, 24: 35–45.

970. Walsh, A. (2002). Responsibility, taboos and 'the freedom to do otherwise' in Ankarana, northern Madagascar. *Journal of the Royal Anthropological Institute*, 8(3): 451-468.

971. Walter, R.K., & Hamilton, R.J. (2014). A cultural landscape approach to community-based conservation in Solomon Islands. *Ecology and Society,* 19(4): 41.

972. Walters, G., Schleicher, J., Hymas, O., & Coad, L. (2015). Evolving hunting practices in Gabon: lessons for community-based conservation interventions. *Ecology and Society*, 20(4): 31.

973. Warren, C., & Visser, L. (2016). The local turn: an introductory essay revisiting leadership, elite capture and good governance in Indonesian conservation and development programs. *Human Ecology*, 44(3): 277-286.

974. Wasserman, S., & Faust, K. (1994). *Social Network Analysis—Methods and Applications.* Cambridge: Cambridge University Press.

975. Watson, J.E., Dudley, N., Segan, D.B., & Hockings, M. (2014). The performance and potential of protected areas. *Nature*, 515(7525): 67-73.

976. WAVES (Wealth Accounting and Valuation of Ecosystem Services). (2015). *Madagascar Country Report 2015. Priority Policy Linkages and Workplan: An Update of Progress June 2015.* Available from: https://www.wavespartnership.org/sites/waves/files/images/Country%20Report%20Madagascar.pdf

977. Waylen, K.A., Fischer, A., McGowan, P.J.K., Thirgood, S.J., & Milner-Gulland, E.J. (2010). Effect of local cultural context on the success of community-based conservation interventions. *Conservation Biology*, 24(4): 1119-1129.

978. Waylen, K.A., Fischer, A., McGowan, P.J., & Milner-Gulland, E.J. (2013). Deconstructing community for conservation: why simple assumptions are not sufficient. *Human Ecology*, 41(4): 575-585.

979. Westlund, L., Charles, A., Garcia, S., Sanders, j. (2017). *Marine protected areas: Interactions with fishery livelihoods and food security. FAO* Technical paper no. 603. Rome: FAO.

980. WCPA (IUCN-World Commission on Protected Areas). (2010). *50 Years of Working for Protected Areas-A brief history of IUCN World Commission on Protected Areas. Gland, Switzerland.* Available from https://www.iucn.org/sites/dev/files/import/downloads/history_wcpa_15july_web_version_1.pdf

981. Weber, M. (2017). *Methodology of Social Sciences.* London: Routledge.

982. Weihrich H. (1982). The TOWS matrix – a tool for situational analysis. *Long Range Planning*, 15(2): 54–66.

983. Weiss, C.H. (1997). How can Theory-Based Evaluation Make Greater Headway? *Evaluation Review*, 21(4): 501–524.

984. Welch-Devine, M., & Campbell, L.M. (2010). Sorting out roles and defining divides: social sciences at the World Conservation Congress. *Conservation & Society*, 8(4): 339-348.

985. Wells, M., & Brandon, K. (1992). *People and Parks: Linking Protected Areas with Local Communities.* Washington, DC: World Bank.

986. Wells, S., Samoilys, M., Makoloweka, S., & Kalombo, H. (2010). Lessons learnt from a collaborative management programme in coastal Tanzania. *Ocean & Coastal Management*, 53(4): 161-168.

987. Wepener, V., & Degger, N. (2019). *South Africa.* In: *World Seas: An Environmental Evaluation. 2nd edition. Volume II: The Indian Ocean to the Pacific.* Sheppard, C. (Ed.). New York: Elsevier Academic Press. 101-119

988. West, P., Igoe, J., & Brockington, D. (2006). Parks and peoples: the social impact of protected areas. *Annual Review of Anthropology.*, *35*: 251-277.

989. Western, D., & Wright, R.M. (1994). *The background to community- based conservation.* In: *Natural connections: perspectives in community-based conservation.* Western, D., Wright, R.M., & Strum, S. (Eds.). Washington, D.C.: Island Press. 1-12

990. Western, D., Waithaka, J., & Kamanga, J. (2015). Finding space for wildlife beyond national parks and reducing conflict through community-based conservation: The Kenya experience. *Parks, 21*: 51-62.

991. Westerman, K., & Benbow, S. (2013). The role of women in community-based small-scale fisheries management: the case of the southern Madagascar octopus fishery. *Western Indian Ocean Journal of Marine Science*, 12(2): 119-132.

992. Westerman, K., & Gardner, C.J. (2013). Adoption of socio-cultural norms to increase community compliance in permanent marine reserves in southwest Madagascar. *Conservation Evidence*, 10: 4-9.

993. Westerman, K., Olesen, K.L.L., & Harris, A.R. (2012). Building socio-ecological resilience to climate change through community-based coastal conservation and development: experiences in southern Madagascar. *Western Indian Ocean Journal of Marine Science,* 11(1): 87–97.

994. WFP (World Food Programme). (2019). *Guinea-Bissau.* Available from: https://www1.wfp.org/countries/guinea-bissau

995. Whitfield, A.K. (2016). Biomass and productivity of fishes in estuaries: A South African case study. *Journal of Fish Biology*, 89(4): 1917-1930.

996. Wight, D., Wimbush, E., Jepson, R., & Doi, L. (2016). Six steps in quality intervention development (6SQuID). *Journal of Epidemiology and Community Health*, 70(5): 520-525.

997. Wijen, F., & Ansari, S. (2007). Overcoming inaction through collective institutional entrepreneurship: insights from regime theory. *Organization Studies*, 28(7): 1079–1100.

998. Williams, S. (2013). *Beyond rights: Developing a conceptual framework for understanding access to coastal resources at Ebenhaeser and Covie, Western Cape, South Africa.* PhD book, University of Cape Town. Available from: https://open.uct.ac.za/bitstream/handle/11427/4819/book_sci_2013_williams_samantha.pdf?sequence=1&isAllowed=y

999. Williams, B., & Imam, I. (2007). *Systems concepts in evaluation: An expert anthology.* Point Reyes, CA: EdgePress of Inverness.

1000. Williams, A., & Le Billon, P. (2017). *Corruption, Natural Resources and Development: From Resource Curse to Political Ecology.* Cheltenham, UK: Edward Elgar Publishing.

1001. Wilson, D.C. (2003). *Fisheries co-management and the knowledge base for management decisions.* In: *The Fisheries Co-management Experience.* Wilson, D.C., Nielsen, P., & Degnbol, P. (Eds.). Dordrecht: Springer. 265-279

1002. Wilson, M., Pavlowich, T., & Cox, M. (2016). Studying common-pool resources over time: A longitudinal case study of the Buen Hombre fishery in the Dominican Republic. *Ambio*, 45(2): 215-229.

1003. Wintle, B.A., Kujala, H., Whitehead, A., Cameron, A., Veloz, S., Kukkala, A., Moilanen, A., et al. (2019). Global synthesis of conservation studies reveals the importance of small habitat patches for biodiversity. *Proceedings of the National Academy of Sciences*, 116(3): 909-914.

1004. Wissink, H. (2019). *The Struggle for Land Restitution and Reform in Post-Apartheid South Africa.* In: *Trajectory of Land Reform in Post-Colonial African States. Advances in African Economic, Social and Political Development.* Akinola A., & Wissink H. (Eds.). Cham: Springer. 57-73.

1005. Wolmer, W. (2005). Wilderness gained, wilderness lost: Wildlife management and land occupations in Zimbabwe's southeast lowveld. *Journal of Historical Geography*, 31(2): 260-280.

1006. World Bank. (2014). *Face of poverty in Madagascar: Poverty, Gender, and Inequality Assessment.* Washington, D.C.: World Bank Group. Available from: https://openknowledge.worldbank.org/handle/10986/18250

1007. World Bank. (2015a). *Madagascar - Systematic country diagnostic.* Washington, D.C.: World Bank Group. Available from: http://documents.worldbank.org/curated/en/743291468188936832/Madagascar-Systematic-country-diagnostic

1008. World Bank. (2015b). *Analysis of Community Forest Management (CFM) in Madagascar Report no. 10113424, September 2015.* Washington, D.C.: World Bank Group. Available from: https://openknowledge.worldbank.org/bitstream/handle/10986/23348/Analysis0of0co000 CFM00in0Madagascar.pdf?sequence=1&isAllowed=y

1009. World Bank. (2015c). *Guinea-Bissau's Green Development Takes Root, Starting with Biodiversity Conservation.* Washington, D.C.: World Bank Group. Available from: http://documents.worldbank.org/curated/en/347181468036532328/Guinea-Bissau-Country-economic-memorandum-Terra-ranca-A-Fresh-Start

1010. World Bank. (2016). Guinea Bissau Biodiversity Conservation (P122047). Implementation Completion Report (ICR), Report Number: ICRR0020566. Available from: http://documents.worldbank.org/curated/en/898631487646853577/pdf/ICRR-Disclosable-P122047-02-20-2017-1487646829547.pdf

1011. World Bank. (2017). *SCALING UP NUTRITION IN GUINEA-BISSAU: What Will It Cost? World Bank Health, Nutrition, and Population Discussion Paper, January 2017.* The World Bank Group, Washington, DC. Available from: https://openknowledge-worldbank-org.ezproxy.uct.ac.za/bitstream/handle/10986/26417/113817-WP-PUBLIC-GHNGE-HNP-Discussion-Paper-ScalingUpNutritionGuineaBissau.pdf?sequence=1

1012. World Bank. (2019). *DataBank. World Development Indicators.* Washington, D.C.: World Bank Group. Available from: https://databank.worldbank.org/home

1013. World Justice Project. (2019). *Rule of Law Index 2019.* Washington, DC.: The World Justice Project, Available from: https://worldjusticeproject.org/our-work/research-and-data/wjp-rule-law-index-2019

1014. Wright, V.C. (2017). Turbulent terrains: The contradictions and politics of decentralised conservation. *Conservation & Society*, 15(2): 157-167.

1015. Wright, S.C., Taylor, D.M., & Moghaddam, F.M. (1990). Responding to membership in a disadvantaged group: from acceptance to collective protest. *Journal of Personality and Social Psychology*, 58: 994-1003.

1016. Wright, G., Andersson, K., Gibson, C., & Evans, T. (2015). What incentivizes local forest conservation efforts? Evidence from Bolivia. *International Journal of the Commons*, 9(1): 322-346.

1017. Wright, J.H., Hill, N.A., Roe, D., Rowcliffe, J.M., Kümpel, N.F., Day, M., Booker, F., & Milner-Gulland, E.J. (2016). Reframing the concept of alternative livelihoods. *Conservation Biology*, 30(1): 7-13.

1018. Wright, D.R., Stevens, C.M., Marnewick, D., & Mortimer, G. (2018). Privately Protected Areas and Biodiversity Stewardship in South Africa: Challenges and Opportunities for Implementation Agencies. *PARKS*, 24(2): 45-62.

1019. WWF (Worldwide Fund for Nature). (2017). *The Nature of Change. The science influencing behaviour change.* Available from: https://www.worldwildlife.org/pages/the-nature-of-change

1020. WWF (Worldwide Fund for Nature). (2019). *Meet the guardians of the grasslands, Angus Burns, 7 May 2019.* Available from: https://www.wwf.org.za/our_news/blog/meet_the_guardians_of_the_grasslands/

1021. Wynberg, R. (2002). A decade of biodiversity conservation and use in South Africa: tracking progress from the Rio Earth Summit to the Johannesburg World Summit on Sustainable Development. *South African Journal of Science*, 98(5): 233–243.

1022. Young, O. (2002) *The Institutional Dimensions of Environmental Change: Fit, Interplay, and Scale.* Cambridge, MA, USA: MIT Press.

1023. Young, J.C., Searle, K., Butler, A., Simmons, P., Watt, A.D., & Jordan, A. (2016). The role of trust in the resolution of conservation conflicts. *Biological Conservation*, 195: 196–202.

1024. Young, J.C., Rose, D.C., Mumby, H.S., Benitez-Capistros, F., Derrick, C.J., Finch, T., Garcia, C., et al. (2018). A methodological guide to using and reporting on interviews in conservation science research. *Methods in Ecology and Evolution*, 9(1): 10-19.

1025. Zanotti, L. (2013). Resistance and the politics of negotiation: women, place and space among the Kayapó in Amazonia, Brazil. *Gender, Place & Culture*, 20(3): 346-362.

1026. Jensen, S., & Zenker, O. (2018). *South African Homelands as Frontiers: Apartheid's Loose Ends in the Postcolonial Era–An Introduction.* In: *South African Homelands as Frontiers: Apartheid's Loose Ends in the Postcolonial Era.* Jensen, S., & Zenker, O. (Eds.). London: Routledge. 9-24

1027. Zinner, D., Wygoda, C., Razafimanantsoa, L., Rasoloarison, R., Andrianandrasana, H.T., Ganzhorn, J.U., & Torkler, F. (2014). Analysis of deforestation patterns in the central Menabe, Madagascar, between 1973 and 2010. *Regional Environmental Change*, 14(1): 157–166.

1028. Zulu, L.C. (2008). Community forest management in southern Malawi: solution or part of the problem? *Society & Natural Resources*, 21(8): 687–703.

1029. Zulu, L.C. (2012). Neoliberalization, decentralization and community-based natural resources management in Malawi: The first sixteen years and looking ahead. *Progress in Development Studies*, 12(2-3): 193-212.

Appendices:

Appendix 1: Phase One: Literature Review Search String

("community conservation" OR "community-based conservation" OR "CBC" or "community based conservation" OR "community-based natural resource management" OR "collaborative management" OR "co-management" OR "co management" OR "Indigenous and Locally Conserved Areas" OR "Indigenous and Community Conserved Areas" OR "Indigenous Peoples' and Community Conserved Territories and Areas" OR "ICCAs" OR "Indigenous Protected Areas" OR "IPAs" OR "Locally Managed Marine Areas" OR "LMMAs" OR "community based natural resource management" OR "CBNRM" OR "Community-Based Wildlife Management" OR "Community-Based Forestry Management" OR "Community-Based Fisheries Management") AND "Conservation" AND ("Community" OR "indigenous people*") AND ("developing nation*" OR "Africa" OR "Namibia" OR "Botswana" OR "Zambia" OR " Zimbabwe" OR "Mozambique" OR "Kenya" OR "Tanzania" OR "Madagascar" OR "South Africa" OR "Asia" OR "India" OR "Sri Lanka" OR "Mongolia" OR "Philippines" OR "Indonesia" OR "Australia" OR "South America" OR "Brazil" OR "Chile" OR "Peru" OR "Latin America" OR "Mexico" OR "Emerging nation*" OR "emerging countr*" OR "developing world" OR "developing countr*" OR "Pacific Island*" OR "Polynesia*" OR "Fiji" OR "Solomon Islands")

Appendix 2: South African National CBC Conservation Actor Semi-Structured Interview Questions:

1. What do you understand by the concept of *community-based conservation* (CBC)?
2. In your own words, how would you generally describe the current landscape of CBC in South Africa? To what extent are communities involved and have powers been devolved to communities, etc...?
3. The current South African environmental legislation can be considered progressive and enabling for CBC.

   Given this enabling legislative environment please highlight the main institutional/ governance barriers or obstacles you have encountered when attempting to implement CBC initiatives.
4. Please elaborate on any outstanding constraining or enabling factors and conditions in general which you have experienced within CBC initiatives you have been associated with and their causes.
5. How would you generally rate levels of active (i.e. participate & have decision-making power) and passive (i.e. consulted) involvement of local communities in CBC initiatives you have been involved in, across the following (based on a scale of 1 [very low] – 5 [very high]):

	Active	Passive
a. Planning Phase: decision-making	①②③④⑤	①②③④⑤
b. Planning Phase: design of management plans	①②③④⑤	①②③④⑤
c. Monitoring of natural resource usage	①②③④⑤	①②③④⑤
d. Enforcement of natural resource usage	①②③④⑤	①②③④⑤
e. Tangible benefit-sharing of natural resource use	①②③④⑤	①②③④⑤

6. In CBC initiatives you have been associated with, how would you generally rate levels of ecological sustainability (based on a scale of 1 [very low] – 5 [very high])?　　①②③④⑤
7. In CBC initiatives you have been associated with, how would you generally rate levels of socio-economic sustainability (based on a scale of 1 [very low] – 5 [very high])?　　①②③④⑤
8. How would you generally describe local community member's attitudes toward CBC initiatives you have been associated with (based on a scale of 1 [completely 'pro'] – 5 [completely 'against'])? ①②③④⑤
9. Please briefly describe the relations between local communities and other stakeholders before and after implementation of a CBC initiative you have been associated with. And, what were some of the key causal factors that led to an improvement or deterioration of the relationship?
10. In your opinion, do you think that engaging local communities in conservation efforts can be productive and promote socially and ecologically sustainable resource use? Yes/ No. Please provide reasons.
11. What are the key changes required, in terms of governance, legislation, capacity, attitudes, behaviour, etc... to better enable the implementation and governance of CBC initiatives in South Africa?

Appendix 3: Example of Case Study National Partner Organization Semi-Structured Interview Questions:

Guinea-Bissau CBC Interview Questions: Partner Organisations (e.g. NGO; government; academic; private sector)

1. What is your understanding of the term *community-based conservation (CBC)*?
2. Please name any CBC initiatives you were/ are involved in.
3. What led to your organisations involvement in establishing/ supporting the CBC initiative(s)? Who initiated the partnership, i.e. you or the community? What institutions/ laws were required/ used?
4. What are the governance arrangements of the CBC initiative(s)? How does your organisation fit in?
5. What are the roles and responsibilities of the other actors? E.g. community members, village committee representatives, traditional authorities, and partner organizations such as a conservation agency and /or government ministry/ department, local government, NGOs, academic institutions, private sector, etc....
6. Who has ultimate decision-making power in the CBC initiative(s)?
7. How would you rate the nature of community involvement in the CBC initiative(s), across the following categories? (Please cross one of the following: L - Low; M - Medium; H - High)

	L M H
a. Participation at meetings	L M H
b. Decision-making about access and use of natural resources	L M H
c. Design of management plans	L M H
d. Monitoring of natural resource access and usage L M H	
e. Formulation of rules of natural resource access, use and management	L M H
f. Enforcement of rules of natural resource access, use and management	L M H
g. Distribution of benefits from natural resource use	L M H

8. How is the community represented in the CBC initiative(s)? And in your opinion, how effectively are community interests being represented?
9. Does the community make use of any cultural practices to manage the natural resources? Are these effective and generally respected by the community?
10. In your experience, what have been some of the difficulties experienced in implementing the CBC initiative(s)?
11. What have been some of the benefits of the CBC initiative(s)? What tangible benefits (e.g. monetary), capacity-building and skills development, and employment opportunities have arisen?
12. From your own experience, please describe the attitudes and perceptions of local communities toward the CBC initiative(s).
13. How would you describe the relations between community members, local-level village committees, and partner organisations, including your organisation in the CBC initiative(s)? And have these relations changed since its establishment? If so, how and why?
14. How would you generally rate local community member's willingness to engage in conservation management activities? E.g. participating in meetings, designing management plans, monitoring activities, etc....
    (Please cross one of the following: L - Low; M - Medium; H - High)   L M H
15. Please list any factors and/ or conditions which you have experienced to be challenging/ favourable in participating in the management of the CBC initiative(s). And what factors/ conditions are important to increase its functionality moving forward?
16. In your opinion, does the community aspire to take greater control of the CBC initiative(s)? And if so, are there any plans to afford greater responsibility to the community in the future?
17. What support do communities need, and from whom, to manage the CBC initiative(s) on their own? (E.g. capacity-building; financial capital; legal rights; etc....)
18. In your opinion, what are the key changes required, in terms of governance, institutions, capacity, attitudes, behaviour, relations etc... to better enable the implementation and management of CBC initiatives?
    Any additional comments by respondent?

Appendix 4: Example of Case Study Partner Organization Semi-Structured Interview Questions:

Case Study: The Urok CMPA, Guinea-Bissau

1. What is your understanding of the term *community-based conservation (CBC)*?
2. What led to your organisations involvement in establishing or supporting the Urok Community-Managed Marine Protected Area (CMPA)? Who initiated the partnership, i.e. you or the community? What institutions/ laws were required/ used?
3. What are the governance arrangements of the Urok CMPA? How does your organisation fit in?
4. What are the roles and responsibilities of the other actors in the Urok CMPA? E.g. community members, village committee representatives, traditional authorities, and partner organizations such as a conservation agency and /or government ministry (e.g. IBAP), local government, NGOs (e.g. Tiniguena), academic institutions, private sector partners, etc....
5. Who has ultimate decision-making power in the Urok CMPA?
6. How would you rate the nature of community involvement in the Urok CMPA, across the following categories? (Please cross one of the following: L - Low; M - Medium; H - High)

	(L)(M)(H)
a. Participation at meetings	(L)(M)(H)
b. Decision-making about access and use of natural resources	(L)(M)(H)
c. Design of management plans	(L)(M)(H)
d. Monitoring of natural resource access and usage	(L)(M)(H)
e. Formulation of rules of natural resource access, use and management	(L)(M)(H)
f. Enforcement of rules of natural resource access, use and management	(L)(M)(H)
g. Distribution of benefits from natural resource use	(L)(M)(H)

7. How is the community represented in the Urok CMPA? And in your opinion, how effectively are community interests being represented?
8. Does the community make use of any cultural practices to manage the CMPA? And are these effective and generally respected by the community?
9. In your experience, what have been some of the difficulties experienced in implementing the Urok CMPA?
10. What have been some of the benefits of the Urok CMPA? What tangible benefits, capacity-building and skills development, and employment opportunities have arisen?
11. From your own experience, please describe the attitudes and perceptions of local communities toward the Urok CMPA.
12. How would you describe the relations between community members, local-level village associations/ committees, and partner organisations, including your organisation in managing the Urok CMPA? And have these relations changed since its establishment? If so, how and why?
13. How would you generally rate local community member's willingness to engage in conservation management activities? E.g. participating in meetings, designing management plans, monitoring activities, etc.... (L)(M)(H)
(Please cross one of the following: L - Low; M - Medium; H - High)
14. Please list any factors and/ or conditions which you have experienced to be challenging/ favourable in participating in the management of the Urok CMPA. And what factors/ conditions are important to increase its functionality moving forward?
15. In the Urok CMPA, does the community aspire to take greater control? And if so, are there any plans to afford greater responsibility to the local community in the future?
16. What support does the community need, and from whom, to manage the Urok CMPA moving forward? (E.g. capacity-building; financial capital; legal rights; etc....)
17. In your opinion, what are the key changes required, in terms of governance, institutions, capacity, attitudes, behaviour, relations etc... to better enable the implementation and management of CBC initiatives?

Any additional comments by interviewee:

Appendix 5: Example of Case Study Local Representative Semi-Structured Interview Questions:

Case Study: The Bay of Ranobe, Madagascar

1. Please briefly describe your position and responsibilities within your Village Association.
2. What natural resources are harvested in this area? Who has access to these natural resources?
3. What led to this community conservation area being established? And how has your association participated in establishing it? What is the purpose of your association?
4. How has access to, and use of, these natural resources changed since initiating the community conservation area?
5. How were natural resources managed historically in this area? Who made decisions about access and use? Who made the rules? Who was responsible for managing the resources? And in what way/s has this changed?
6. Are there any norms, rules or customary practices that determine *what, where, when* and *how* much natural resources can be harvested at present? If so, can you describe the main customary practices? And are these rules and practices respected by community members?
7. What are the roles and responsibilities of the different actors in this community conservation area, e.g. community members, community representatives, traditional authorities, conservation agency/ government ministry, local government, NGO partners like Reef doctor, academic institutions, private sector partners like Hotels and Dive Operators, etc....?
8. Who has ultimate decision-making powers and over what issues for this community conservation area?
9. How would you rate the nature of community involvement in this community conservation area, across the following categories?
   (Please cross one of the following: L - Low; M - Medium; H - High):

   a. Participation at meetings   (L)(M)(H)

   b. Decision-making about allocation of resources   (L)(M)(H)

   c. Design of management plans   (L)(M)(H)

   d. Monitoring of natural resource usage   (L)(M)(H)

   e. Formulation of rules of natural resource access, use and management   (L)(M)(H)

   f. Enforcement of rules of natural resource access, use and management   (L)(M)(H)

   g. Distribution of benefits from natural resource use   (L)(M)(H)

10. How is the community represented? And how effectively are community representatives representing community interests?
11. In your opinion, what do you think is the best way of interacting with your community to improve the management of natural resources? Give reasons.
12. In your experience, what have been some of the difficulties experienced in establishing this community conservation area?
13. What have been some of the benefits of this community conservation area? What tangible benefits, capacity-building and skills development, and employment opportunities have arisen?
14. From your own experience, please describe the perceptions and attitudes of the community toward this community conservation area.
15. How would you describe the relations between community members and the partner organisation(s) of this community conservation area? And how have these relations changed since establishing the community conservation area.
16. How would you describe the relations between the community members and your Village Association? And how have these relations changed since the implementation of your community conservation area.
17. How would you describe the relations between the community members and FIMIHARA? And how have these relations changed since the implementation of your community conservation area.
18. Please list any conditions which you have experienced to be challenging/ favourable in participating in the management of this community conservation area. What factors/ conditions are important to increase the functionality of this community conservation area?
19. What if any plans do the community have for taking greater responsibility over this community conservation area from partner organisations in the future?
20. What support does your community and your authority need, and from whom, to manage this community conservation area? (e.g. capacity-building; financial capital; legal rights; etc....)

Any additional comments from the interviewee:

Appendix 6: Example of Case Study Community Member Semi-Structured Interview Questions:

Case Study: The Bay of Ranobe, Madagascar

1. What natural resources can you access, and which do you use in this community conservation area?
2. For what purposes do you use these natural resources? And to what extent do they contribute to your livelihood/ household income?
3. Are you engaged in any other income generating activities? Yes/No (Circle one). If yes, please list & very briefly discuss (e.g. seasonal harvesting; road works when opportunity arises, etc.)
4. What value does the *Bay of Ranobe* hold for you besides being a source of resources? How is this value expressed?
5. Did you live in this area before the community conservation area was established? Yes/No (Circle one). If yes, how were natural resources managed historically in this area? Who made decisions about access and use? Who made the rules? Who was responsible for managing the resources?
6. Are there any norms, rules or customary practices that determine *what*, *where*, *when* and *how* much natural resources can be harvested at present, e.g. Dina or fady? If so, can you describe the main customary practices? And are these rules and practices respected by community members?
7. How has access to, and use of, these natural resources changed since establishing the community conservation area?
8. What are the roles and responsibilities of the different actors in this community conservation area, e.g. community members, village representatives, traditional authorities, conservation agency/ government ministry, local government, NGO partners like Reef Doctor, academic institutions, private sector partners, etc....?
9. Who has ultimate decision-making powers and over what issues for this community conservation area?
10. How would you rate the nature of the community involvement in this community conservation area, across the following categories?

    (Please cross one of the following: L - Low; M - Medium; H - High):

    a. Participation at meetings   Ⓛ Ⓜ Ⓗ

    b. Decision-making about allocation of resources   Ⓛ Ⓜ Ⓗ

    c. Design of management plans   Ⓛ Ⓜ Ⓗ

    d. Monitoring of natural resource usage   Ⓛ Ⓜ Ⓗ

    e. Formulation of rules of natural resource access, use and management   Ⓛ Ⓜ Ⓗ

    f. Enforcement of rules of natural resource access, use and management   Ⓛ Ⓜ Ⓗ

    g. Distribution of benefits from natural resource use   Ⓛ Ⓜ Ⓗ

11. How is the community represented in this community conservation area? And how effectively are community representatives (e.g. village council members) at representing community interests?
12. In your opinion, what do you think is the best way of getting local communities involved in managing natural resources? Give reasons.
13. In your experience, what have been some of the difficulties experienced in establishing this community conservation area?
14. What have been some of the benefits of this community conservation area? What tangible benefits, capacity-building and skills development, and employment opportunities have arisen?
15. From your own experience, please describe the perceptions and attitudes of the community toward this community conservation area.
16. How would you describe the relations between community members and partner organisations? Have these relations changed since the establishment of this community conservation area? If so, how and why?
17. How would you describe the relations between community members and the village representatives? Have these relations changed since the establishment of this community conservation area? If so, how and why?
18. How would you describe the relations between community members and FIMIHARA? Have these relations changed since the establishment of this community conservation area? If so, how and why?
19. Please list any conditions which you have experienced to be challenging/ favourable in participating in the management of this community conservation area. What factors/ conditions are important to increase the functionality of this community conservation area? (E.g. coastal migration of non-Vezo).
20. Does your community aspire to taking over more control of this community conservation area from partner organisations in the future? And if so, what support does your community need, and from whom, to manage this community conservation area? (E.g. capacity-building; financial capital; legal rights; etc....)

Appendix 7: Example of Case Study Social Relations & Network Appraisal

Case Study: *The Bay of Ranobe*, Madagascar

Note: Questions require naming a specific actor within the social network, however, should answers be dependent upon specific actions or examples, please specify in brackets next to the actor mentioned. For example: question 14 is about the power to make decisions about 'day to day conservation practices'? i.e. monitoring catches; enforcing area restrictions, etc.... E.g. answer could then be: 1. Reef Doctor (monitoring) 2. FIMIHARA (enforcing area restrictions) etc....

Social Network Attribute	Question	Answer(s)
Interactional Support	1. Whom do you find the most approachable (i.e. you trust the most) to work with over your natural resource use concerns? **List up to 3** in order from most to least approachable (e.g. answers: other fishers, other community members, Reef Doctor, village council, village president, FIMIHARA, Local government, Ministry of Fisheries, academic institutions, etc....)	1.________________ 2.________________ 3.________________
	2. How often do you discuss these concerns with them? (tick for each of the 3 mentioned above)	1. ☐ once a week ☐ once a month ☐ once a year ☐ when issues arise 2. ☐ once a week ☐ once a month ☐ once a year ☐ when issues arise 3. once a week once a month once a year when issues arise
Knowledge Acquisition	**From whom do you acquire knowledge or information about the following (List up to 3 - e.g. answer: Ministry of Fisheries, Local government, FIMIHARA, Reef Doctor, other fishers, other community members, village council, etc....):**	
	3. ...legal issues & rights about natural resource access & use?	1.________________ 2.________________ 3.________________
	4. ...sources of financial support for the community conserved area? i.e. from loans; donors, sources of income financing operating costs, etc....	1.________________ 2.________________ 3.________________
	5. ...sources of non-monetary resources? i.e. fishing gear, boats, etc....	1.________________ 2.________________ 3.________________
	6. ...ecological aspects of conservation? i.e. status of fish & the marine environment, such as decreasing fish stocks	1.________________ 2.________________ 3.________________

<table>
<tr><td rowspan="4" style="vertical-align:middle">Knowledge Diffusion</td><td colspan="2">With whom do you share knowledge or information about the following (List up to 3 - e.g. answer: Ministry of Fisheries, Local government, FIMIHARA, Reef Doctor, other fishers, other community members, village council, etc....):</td></tr>
<tr><td>7. ...legal issues & rights about natural resource access & use?</td><td>1.______________
2.______________
3.______________</td></tr>
<tr><td>8. ...sources of financial support for the community conserved area? i.e. from loans; donors, sources of income financing operating costs, etc....</td><td>1.______________
2.______________
3.______________</td></tr>
<tr><td>9. ...sources of non-monetary resources? i.e. fishing gear, boats, etc....</td><td>1.______________
2.______________
3.______________</td></tr>
<tr><td>10. ...ecological aspects of conservation? i.e. status of fish & the marine environment, such as decreasing fish stocks</td><td>1.______________
2.______________
3.______________</td></tr>
</table>

<table>
<tr><td rowspan="8" style="vertical-align:middle">Power & Politics</td><td colspan="2">With whom rests the power to make decisions about the following (List up to 3 - e.g. answer: Ministry of Fisheries, Local government, FIMIHARA, Reef Doctor, other fishers, other community members, village council, etc....):</td></tr>
<tr><td>11. ...legal rights associated with natural resource access & use? E.g. Fisher permits, catch & size limits, prohibited species & areas, etc.... (Where applicable)</td><td>1.______________
2.______________
3.______________</td></tr>
<tr><td>12. ...local conservation practices associated with natural resource access & use? E.g. *Dina* or *fady* about establishing no take zones, closed areas & seasons, catch & size limits, prohibited species, etc.... (Where applicable – could be established without *Dina* or *fady* as well)</td><td>1.______________
2.______________
3.______________</td></tr>
<tr><td>13. ...the election of representatives for local area management committees? i.e. village council or FIMIHARA (e.g. answer: current members or local community members, etc....)</td><td>1.______________
2.______________
3.______________</td></tr>
<tr><td>14. ...day to day conservation practices? i.e. monitoring catches; enforcing area restrictions, etc....</td><td>1.______________
2.______________
3.______________</td></tr>
<tr><td>15. ...obtaining, & managing financial resources for the community conserved area?</td><td>1.______________
2.______________
3.______________</td></tr>
<tr><td>16. ...distribution of monetary and/ or non-monetary benefits (e.g. building of schools or clinics, capacity-building & skills development, etc....) to stakeholders? (Where applicable)</td><td>1.______________
2.______________
3.______________</td></tr>
<tr><td>17. ...changes in local level policy, governance or institutional structure? (i.e. rules of natural resource access & use and how management committees are formed & function)</td><td>1.______________
2.______________
3.______________</td></tr>
</table>

Appendix 8: Additional African-specific studies investigating CBC 'enablers' which informed the selection of the proposed 14 CBC enablers in Chapter 3, section 3.3.2.2..

Research Title/ Topic (Reference)	Country/ Region	Key Findings
Sharing invisible resources in the age of climate change: a transboundary groundwater sharing agreement in Sahel, Africa, analysed through Ostrom's design principles for collective action (Blanck, 2019)	Sahel	Ambiguous boundaries, a lack of congruence with local conditions and accountability of monitors, and insufficient attention to local governance constrain this "heavily state-centred" transboundary Sahelian water resource management project.
Sustainable Governance of Wildlife and Community-based Natural Resource Management (Child, 2019)	Southern Africa	Referring to the work of Ostrom (1990) and Murphree (2009), Child (2019) emphasizes the importance of 'emergent principles' notably long-tern tenure rights and the legitimacy of local institutions; clearly defined boundaries; resource value; and the ability of those affected by the rules to modify them face-to-face. Informed by these and other scholars Child (2019: p288) also proposes a Theory of Change for CBNRM emphasizing the need to evaluate the feasibility of CBNRM as an approach to the context; building a vison and coalition for change; building capacity for a collaborative adaptive management; implementation based upon creating economic value; governance characterized by informed participation and benefit-sharing; natural resource protection; and security and safety.
Inadequate community engagement hamstrings sustainable wildlife resource management in Zambia (Milupi et al., 2019)	Zambia	Communities are not involved in decision-making, Collective-choice decisions constrained by lack of legislation empowering communities. A lack of conflict-resolution mechanisms and inequitable benefit-sharing constrain success.
Design Principles, Common Land, and Collective Violence in Africa (Oyerinde, 2019)	Nigeria	Minimal recognition of rights, a lack of congruence with local conditions, and an inability to resolve conflict have caused violence related to these two Nigerian cases of common land practices.
Ostrom's Governance Principles and Sustainable Financing of Fish Reserves - Namibia (Wiederkehr et al., 2019)	Namibia	Fish Protection Area established in participatory and consultative process. Group cohesion is high as is congruence of rules with local conditions, declining fish stocks incentivizes protection. However, a lack of perceived monetary benefits and high cost of management inhibit progress.
The importance of Ostrom's Design Principles: Youth group performance in northern Ethiopia (Holden & Tilahun, 2018).	Ethiopia	Found strong correlation with the presence of Ostrom's design principles and success in Ethiopian youth groups managing CPRs. While internal conflict resolution emerged with a high level of satisfaction this was not true for outsiders. Sanctions were strictly enforced by not always graduated. Success was strongly linked to trust, political support and stability, and diversification of livelihoods.

Institutional multiplexity: social networks and community-based natural resource management (Schnegg, 2018)	**Namibia**	Examination focuses on fixed boundaries, proportional cost sharing, and formal sanctioning and finds these three DPs rarely emerge in practice in seven Namibian community water resource management initiatives. Reveals institutional multiplexity (i.e. inseparable sharing of resources and other domains of dependency) constrains successful implementation of DPs, which requires greater institutional flexibility.
An Analysis of the Global Applicability of Ostrom's Design Principles to Diagnose the Functionality of Common-Pool Resource Institutions (Gari et al., 2017)	**Case Studies from: Tanzania; Ethiopia; Kenya; Madagascar & South Africa**	Clear boundaries commonly present. Success dependent upon community participation in rule-making. Failures commonly due to lack of state support, graduated sanctions and conflict resolution mechanisms.
Property rights, institutional regime shifts and the provision of freshwater ecosystem services on the Pongola River floodplain, South Africa (Nkhata et al., 2017)	**South Africa**	Highlights the importance of explicitly defining the nature and context of resource-user rights for managing institutional regime shifts. In particular the importance of excludability to manage this water resource management initiative.
An assessment of community water governance on Mount Kenya based on Ostrom's Design Principles (Dell'Angelo et al., 2016)	**Kenya**	Emphasizes the importance of institution being adaptive to changing conditions, the negative impact of inequitable benefit-sharing, and increased effective collaboration for successful these water resource management initiatives.
Moral equality and success of common-pool water governance in Namibia (Schnegg et al., 2016)	**Namibia**	The three principles of delineation of boundaries, sharing of costs, and formal sanctions play a key role in the success of Namibian water resource management cases. Institutional approaches lacking consideration for social interactions are prone to failure.
Management of non-timber forestry products extraction: Local institutions, ecological knowledge and market structure in South-Eastern Zimbabwe (Mutenje et al., 2011)	**Zimbabwe**	Emphasizes the importance of consensus and enforcement of rules, graduated sanctions and shared social beliefs for local institutional management success. In turn, they illustrate how resource scarcity, market integration and infrastructure development exacerbate resource degradation. They also highlight the importance of enhanced ecological knowledge to long-term livelihood security.
Multilevel water, biodiversity and climate adaptation governance: evaluating adaptive management in Lesotho (Bisaro et al., 2010)	**Lesotho**	They emphasize that no ideal 'adaptive regime' exists in Lesotho. However, highlight the importance of decentralized decision-making, knowledge-sharing, and accounting for multiple user interests improve initiative outcomes.

		Emphasize strategies for successful CBNRM within the South African cases notably related to understanding social-ecological context; establishing and communicating a clear vision; building upon local institutions present; developing capacity; providing financial support; and creating lasting incentives. They also propose additional strategies including developing knowledge networks to benefit from experiences of a wide range of actors; establishing formalized multi-level decision-structures; ensuring acceptance of governance arrangements by community members; and clearly defining legitimate conflict resolution procedures.
Community-based natural resource management: governing the commons (Fabricius & Collins, 2007)	**South Africa**	
Design principles and common pool resource management: An institutional approach to evaluating community management in semi-arid Tanzania (Quinn et al., 2007)	**Tanzania**	Clear boundaries, conflict resolution and flexible institutions are emphasized as key to successful management in 12 semi-arid Tanzanian agricultural/ agro-pastoral villages. Moreover, local context emerged an important driver of institutions. Notably the ability to cope with change appeared weak in all management regimes.
A critique of and recommendations for a subsistence fishery, Lake St Lucia, South Africa (Crook & Mann, 2002)	**South Africa**	Conclude that a depleted resource would motivate compliance in this community gillnet fishery. Boundaries should have clear meaning to the fishers, rule should be developed by the fishers within the parameters of the local conservation agency, and monitoring is necessary to ensure compliance. Emphasizes the importance of the conservation authority for effective monitoring and enforcement. Key to compliance levels and thus effective management is the presence of a respected community authority.
Current Directions in Fisheries Management Policy: A Perspective on Co-Management and its application to South African Fisheries (Hutton & Pitcher, 1998)	**South Africa**	Conclude that the clear delineation of both resource and resource-user boundaries, local participation in rule design and mutually agreed upon allocation of cost and benefits are all key to success in South African fisheries co-management arrangements.

Case References:

. Biggs, D., Ban, N. C., Castilla, J. C., Gelcich, S., Mills, M., Gandiwa, E., ... & Possingham, H. P. (2019). Insights on fostering the emergence of robust conservation actions from Zimbabwe's CAMPFIRE program. *Global Ecology and Conservation*, *17*, e00538. DOI: 10.1016/j.gecco.2019.e00538

. Bisaro, A., Hinkel, J., & Kranz, N. (2010). Multilevel water, biodiversity and climate adaptation governance: evaluating adaptive management in Lesotho. *Environmental Science & Policy*, *13*(7), 637-647. DOI: 10.1016/j.envsci.2010.08.004

. Blanck, A. (2019). *Sharing invisible resources in the age of climate change: a transboundary groundwater sharing agreement in Sahel, Africa, analysed through Ostrom's design principles for collective action.* Retrieved from: http://www.diva-portal.org/smash/get/diva2:1302874/FULLTEXT01.pdf

. Child, B. (2019). *Application of Theories of micro-governance to CBNRM.* In: Child, B. (Ed.). *Sustainable Governance of Wildlife and Community-based Natural Resource Management: From Economic Principles to Practical Governance*, 268-286. New York: Routledge.

. Crook, B.J., & Mann, B.Q. (2002). A critique of and recommendations for a subsistence fishery, Lake St Lucia, South Africa. *Biodiversity and Conservation*, *11*, 1223–1235 (2002). DOI: 10.1023/A:1016074802295

6. Dell'Angelo, J., McCord, P. F., Gower, D., Carpenter, S., Caylor, K. K., & Evans, T. P. (2016). Community water governance on Mount Kenya: an assessment based on Ostrom's design principles of natural resource management. *Mountain Research and Development, 36*(1), 102-115. DOI: 10.1659/MRD-JOURNAL-D-15-00040.1

7. Fabricius, C., & Collins, S. (2007). Community-based natural resource management: governing the commons. *Water Policy, 9*(S2), 83-97. DOI: 10.2166/wp.2007.132

8. Gari, S. R., Newton, A., Icely, J. D., & Delgado-Serrano, M. M. (2017). An analysis of the global applicability of Ostrom's design principles to diagnose the functionality of Common-Pool Resource institutions. *Sustainability, 9*(7), 1287. DOI: 10.3390/su9071287

9. Holden, S. T., & Tilahun, M. (2018). The importance of Ostrom's Design Principles: Youth group performance in northern Ethiopia. *World Development, 104*, 10-30. DOI: https://doi.org/10.1016/j.worlddev.2017.11.010

10. Hutton, T., & Pitcher, T. J. (1998). Current directions in fisheries management policy: a perspective on co-management and its application to South African fisheries. *South African Journal of Marine Science, 19*. Retrieved from: https://www.ajol.info/index.php/ajms/article/view/66355

11. Milupi, I. D., Somers, M. J., & Ferguson, W. (2020). Inadequate community engagement hamstrings sustainable wildlife resource management in Zambia. *African Journal of Ecology, 58*(1), 112-122. DOI: https://doi.org/10.1111/aje.12685

12. Mutenje, M. J., Ortmann, G. F., & Ferrer, S. R. (2011). Management of non-timber forestry products extraction: Local institutions, ecological knowledge and market structure in South-Eastern Zimbabwe. *Ecological Economics, 70*(3), 454-461. DOI: https://doi.org/10.1016/j.ecolecon.2010.09.036

13. Nkhata, B. A., Breen, C., Hay, D., & Wilkinson, M. (2017). Property rights, institutional regime shifts and the provision of freshwater ecosystem services on the Pongola River floodplain, South Africa. *International Journal of the Commons, 11*(1), 97–118. DOI: 10.18352/ijc.615

14. Oyerinde, O. K. (2019). Design Principles, Common Land, and Collective Violence in Africa. *International Journal of the Commons, 13*(2), 993–1002. DOI: 10.5334/ijc.930

15. Quinn, C. H., Huby, M., Kiwasila, H., & Lovett, J. C. (2007). Design principles and common pool resource management: An institutional approach to evaluating community management in semi-arid Tanzania. *Journal of Environmental Management, 84*(1), p100-113. DOI: 10.1016/j.jenvman.2006.05.008

16. Schnegg, M., Bollig, M. & Linke, T. (2016). Moral equality and success of common-pool water governance in Namibia. *Ambio, 45*, 581–590. DOI: 10.1007/s13280-016-0766-9

17. Schnegg, M. (2018). Institutional multiplexity: social networks and community-based natural resource management. *Sustainability science, 13*(4), 1017-1030. DOI: 10.1007/s11625-018-0549-2

18. Wiederkehr, C., Berghöfer, A. & Otsuki, K. (2019). Ostrom's Governance Principles and Sustainable Financing of Fish Reserves. *Human Ecology, 47*, 13–25. DOI: 10.1007/s10745-019-0052-0

Appendix 9: South African CBC 'constrainers' and 'enablers' with case-study examples informing their selection.

South African CBC Constrainers/ Enablers		
Wildlife Case Studies (References)	**Forestry Case Studies (References)**	**Coastal-Marine Case Studies (References)**
Constrainers		
1. Slow land claims and CBC proclamation, planning & implementation processes • South African CBNRM overview (Fabricius & Collins, 2007) • Khomani San and Meir Contract Park – Kalahari Gemsbok National Park (Thondhlana et al., 2011, 2015) • Ndumo Game Reserve (Meer & Schnurr, 2013) • Masebe Nature Reserve (Boonzaaier, 2010, 2012)	• Eastern & southern Cape PFM overview (Brown, 2009) • Cwebe Forest (Grundy et al., 2002) • iGxalingenwa Forest (Robertson & Lawes, 2005) • Mngazana Mangrove Forests (Traynor & Hill, 2008) • Umzimvubu District Forestry (Obiri & Lawes, 2002)	• South African coastal & fisheries co-management overview (Hauck & Sowman, 2001) • Dwesa-Cwebe Reserve (Paterson & Mkhulisi, 2014; Sunde, 2014) • Mkambati Nature Reserve (Kepe et al., 2008) • Silaka Nature Reserve (Thondhlana et al. 2016)
2. High turnover & weak participation by under-capacitated local, national & provincial government • South African CBNRM overview (Fabricius & Collins, 2007) • Madikwe Game Reserve (Magome et al., 2000) • Mokala National Park (Kruger et al., 2016; Coetzee & Nell, 2019)	• South African PFM review (Watts, 2006; Holmes-Watts & Watts, 2008) • Mt Coke State Forest (Cundill, 2005) • Umzimvubu District Forestry (Obiri & Lawes, 2002)	• South African coastal & fisheries co-management overview (Hauck & Sowman, 2001) • Sokhulu Subsistence Mussel-Harvesting Project (Harris et al., 2003) • St. Lucia Gillnetting (Mann et al., 2003) • Kosi Bay Gillnet Joint-Management Project (Sunde, 2013) • Dwesa-Cwebe Reserve (Ntshona et al., 2010; Sunde, 2014; Paterson & Mkhulisi, 2014) • Silaka Nature Reserve (Thondhlana et al., 2016)

3. **Poor coordination amongst diverse actors due to poor communication and a lack of trust (and thus a lack of consideration for diverse interests)**	• South African CBNRM overview (Fabricius & Collins, 2007) • South African PA overview (Thondhlana & Cundill, 2017) • Khomani San and Meir Contract Park – Kalahari Gemsbok National Park (Mannetti et al., 2015) • Madikwe Game Reserve (Davies, 2000) • Masebe Nature Reserve (Boonzaaier, 2010, 2012) • Somkhanda Nature Reserve (WILDLANDS, 2016) • Mokala National Park (Kruger et al., 2016)	• South African PFM review (Holmes-Watts & Watts, 2008) • iGxalingenwa Forest (Robertson & Lawes, 2005) • Tsitsikamma Forest (Matose & Watts, 2010)	• South African coastal & fisheries co-management overview (Hauck & Sowman, 2001) • Dwesa-Cwebe Reserve (Sunde, 2014; Paterson & Mkhulisi, 2014) • Mkambati Nature Reserve (Kepe et al., 2008; Cundill et al., 2013) • St. Lucia Gillnetting (Mann et al., 2003)
4. **Lack of true devolution of decision-making power and community participation**	• Madikwe Game Reserve (Magome et al., 2000) • Makuleke Contract Park - Kruger National Park (Ramutsindela, 2003; Kepe et al., 2005) • Khomani San and Meir Contract Park – Kalahari Gemsbok National Park (Thondhlana et al., 2011, 2015) • Manyeleti Game Reserve (de Koning, 2010) • Masebe Nature Reserve (Boonzaaier, 2010, 2012)	• Tsitsikamma Forest (Holmes-Watts & Watts, 2008; Matose & Watts, 2010) • iGxalingenwa Forest (Robertson & Lawes, 2005)	• Dwesa-Cwebe Reserve (Cundill et al., 2013; Sunde, 2014) • Mkambati Nature Reserve (Kepe et al., 2008; Cundill et al., 2013) • Silaka Nature Reserve (Thondhlana et al., 2016)
5. **Lack of consideration for historical socio-political legacies leading to continued dependency or conflict**	• South African CBNRM overview (Fabricius & Collins, 2007) • South African CBNRM overview (Cundill & Fabricius, 2010) • South African PA overview (Thondhlana & Cundill, 2017) • Makuleke Contract Park - Kruger National Park (Ramutsindela, 2003; Kepe et al., 2005) • Khomani San and Meir Contract Park – Kalahari Gemsbok National Park (Thondhlana et al., 2011, 2015)	• Mngazana mangrove forests (Traynor & Hill, 2008) • Mt Coke State Forest (Cundill, 2005) • Ongoye Forest (Phadima, 2005) • Umzimvubu District Forestry (Obiri & Lawes, 2002)	• Mkambati Nature Reserve (Kepe et al., 2008; Cundill et al., 2013) • Dwesa-Cwebe Reserve (Sunde, 2014) • Silaka Nature Reserve (Thondhlana et al., 2016)

	• Ndumo Game Reserve (Meer & Schnurr, 2013)		
6. Lack of community rights to access, use & manage natural resources to derive both cultural and monetary benefits	• Ndumo Game Reserve (Meer & Schnurr, 2013) • Madikwe Game Reserve (Davies, 2000) • Masebe Nature Reserve (Boonzaaier, 2010, 2012) Khomani San and Meir Contract Park – Kalahari Gemsbok National Park (Thondhlana et al., 2011, 2015)	• Umzimvubu District Forestry (Obiri & Lawes, 2002) • Cwebe Forest (Grundy et al., 2002) • iGxalingenwa Forest (Robertson & Lawes, 2005) • Mngazana mangrove forests (Traynor & Hill, 2008) • Eastern & southern Cape PFM overview (Brown, 2009)	• Sokhulu Subsistence Mussel-Harvesting Project (Harris et al., 2003) • Dwesa-Cwebe Reserve (Ntshona et al., 2010; Sunde, 2014; Paterson & Mkhulisi, 2014) • Mkambati Nature Reserve (Kepe et al., 2008; Cundill et al., 2013) • Silaka Nature Reserve (Thondhlana et al., 2016)
7. Capture of, and conflict over benefits	• South African CBNRM overview (Fabricius & Collins, 2007) • South African PA overview (Thondhlana & Cundill, 2017) • Khomani San and Meir Contract Park – Kalahari Gemsbok National Park (Thondhlana et al., 2011, 2015) • Nama Contract Park - Richtersveld National Park (Fabricius & Collins, 2007) • Ndumo Game Reserve (Meer & Schnurr, 2013) • Makuleke Contract Park - Kruger National Park (Ramutsindela, 2003; Kepe et al., 2005) • Manyeleti Game Reserve (de Koning, 2010) • Masebe Nature Reserve (Boonzaaier, 2010, 2012) • Somkhanda Nature Reserve (McCann et al., 2015; WILDLANDS, 2016) • Mokala National Park (Kruger et al., 2016; Coetzee & Nell, 2019)	• Eastern & southern Cape PFM overview (Brown, 2009) • Tsitsikamma Forest (Matose & Watts, 2010) • Umzimvubu District Forestry (Obiri & Lawes, 2002)	• South African coastal & fisheries co-management overview (Hauck & Sowman, 2001) • Dwesa-Cwebe Reserve (Mitchell et al., 2012; Cundill et al., 2013; Sunde, 2014; Paterson & Mkhulisi, 2014) • Mkambati Nature Reserve (Kepe et al., 2008; Cundill et al., 2013) • Mkambati Ecotourism Project () • Kleinmond (Sutton & Rudd, 2014) • Silaka Nature Reserve (Thondhlana et al., 2016)

8. Weak & incapacitated local governance institutions	• South African CBNRM overview (Fabricius & Collins, 2007) • South African CBNRM overview (Cundill & Fabricius, 2010) • Nama Contract Park - Richtersveld National Park (Fabricius & Collins, 2007) • Ndumo Game Reserve (Meer & Schnurr, 2013) • Masebe Nature Reserve (Boonzaaier, 2010, 2012) • Somkhanda Nature Reserve (McCann et al., 2015; WILDLANDS, 2016) • Usuthu Gorge Nature Reserve (WILDLANDS, 2016) • Mokala National Park (Kruger et al., 2016; Coetzee & Nell, 2019)	• Eastern & southern Cape PFM overview (Brown, 2009) • iGxalingenwa Forest (Robertson & Lawes, 2005) • Mngazana mangrove forests (Traynor & Hill, 2008) • Mt Coke State Forest (Cundill, 2005) • Tsitsikamma Forest (Matose & Watts, 2010) • Umzimvubu District Forestry (Obiri & Lawes, 2002; Obiri et al., 2002)	• South African coastal & fisheries co-management overview (Hauck & Sowman, 2001) • St. Lucia Gillnetting (Mann et al., 2003) • Kleinmond (Sutton & Rudd, 2014)
9. Lack of alignment of State and customary institutions	• Masebe Nature Reserve (Boonzaaier, 2010, 2012; Boonzaaier & Wels, 2016) • Ndumo Game Reserve (Meer & Schnurr, 2013)	• Umzimvubu District Forestry (Obiri & Lawes, 2002; Obiri et al., 2002)	• South African small-scale fisheries management overview (Sunde et al., 2013) • Mkambati Nature Reserve (Kepe et al., 2008; Cundill et al., 2013) • Dwesa-Cwebe Reserve (Mitchell et al., 2012; Cundill et al., 2013; Sunde, 2014; Paterson & Mkhulisi, 2014) • Silaka Nature Reserve (Thondhlana et al., 2016)

1. External financial and technical support for the CBC initiative	• South African CBNRM overview (Cundill & Fabricius, 2010) • Khomani San and Meir Contract Park – Kalahari Gemsbok National Park (Mannetti et al., 2015) • Madikwe Game Reserve (Davies, 2000) • Makuleke Contract Park - Kruger National Park (Fabricius & Collins, 2007) • Manyeleti Game Reserve (de Koning, 2010) • Masebe Nature Reserve (Boonzaaier, 2010, 2012) • Somkhanda Nature Reserve (McCann et al., 2015; WILDLANDS, 2016) • Usuthu Gorge Nature Reserve (WILDLANDS, 2016) • Mokala National Park (Kruger et al., 2016)	• iGxalingenwa Forest (Robertson & Lawes, 2005) • Mngazana mangrove forests (Traynor & Hill, 2008) • Umzimvubu District Forestry (Obiri & Lawes, 2002)	• South African coastal & fisheries co-management overview (Hauck & Sowman, 2001) • Kosi Bay Gillnet Joint-Management Project (Kyle, 2003)
2. Relations of trust between actors for improved communication and coordination toward a clear, shared and formalised vision	• South African CBNRM overview (Fabricius & Collins, 2007) • South African CBNRM overview (Cundill & Fabricius, 2010) • South African PA overview (Thondhlana & Cundill, 2017) • Khomani San and Meir Contract Park – Kalahari Gemsbok National Park (Mannetti et al., 2015) • Nama Contract Park - Richtersveld National Park (Fabricius & Collins, 2007)	• South African PFM review (Holmes-Watts & Watts, 2008) • iGxalingenwa Forest (Robertson & Lawes, 2005) • Mngazana mangrove forests (Traynor & Hill, 2008)	• Kosi Bay Gillnet Joint-Management Project (Kyle, 2003) • Kleinmond (Sutton & Rudd, 2014)

Criterion			
3. Formal decision-making structures jointly creating and enforcing rules for natural resource access and use in collaboration with the community	• South African PA overview (Thondhlana & Cundill, 2017) • Khomani San and Meir Contract Park – Kalahari Gemsbok National Park (Mannetti et al., 2015) • Nama Contract Park - Richtersveld National Park (Fabricius & Collins, 2007)	• South African PFM review (Holmes-Watts & Watts, 2008) • iGxalingenwa Forest (Robertson & Lawes, 2005) • Mngazana mangrove forests (Traynor & Hill, 2008) • Umzimvubu District Forestry (Obiri & Lawes, 2002; Obiri et al., 2002)	• Sokhulu Subsistence Mussel-Harvesting Project (Harris et al., 2003) • Silaka Nature Reserve (Thondhlana et al., 2016)
4. Presence of 'champions' to motivate actors and drive CBC implementation and governance processes	• South African CBNRM overview (Fabricius & Collins, 2007) • South African UNESCO MAB programme (Pool-Stanvliet et al., 2018) • South African People & Parks Programme (DEA, 2019b)		• South African coastal & fisheries co-management overview (Hauck & Sowman, 2001) • Macubeni (Fabricius & Cundill, 2010)
5. Understanding of the social-ecological context (including recognition of livelihood and culturally significant practices)	• South African CBNRM overview (Fabricius & Collins, 2007) • South African PA overview (Thondhlana & Cundill, 2017) • Khomani San and Meir Contract Park – Kalahari Gemsbok National Park (Thondhlana & Shackleton, 2015)	• South African PFM review (Holmes-Watts & Watts, 2008) • Mngazana mangrove forests (Traynor & Hill, 2008) • Mt Coke State Forest (Cundill, 2005)	• Kleinmond (Sutton & Rudd, 2014)
6. Sustainable and tangible incentives for continued community participation and commitment	• South African CBNRM overview (Fabricius & Collins, 2007) • Madikwe Game Reserve (Davies, 2000) • Mokala National Park (Kruger et al., 2016; Coetzee & Nell, 2019)	• southern Cape PFM (Muchapondwa et al., 2009) • iGxalingenwa Forest (Robertson & Lawes, 2005) • Umzimvubu District Forestry (Obiri & Lawes, 2002)	• South African coastal & fisheries co-management overview (Hauck & Sowman, 2001)
7. Clearly identify and legitimise conflict resolution strategies	• South African CBNRM overview (Fabricius & Collins, 2007) • South African CBNRM overview (Cundill & Fabricius, 2010) • South African PA overview (Thondhlana & Cundill, 2017) • Mokala National Park (Kruger et al., 2016)	• South African PFM review (Holmes-Watts & Watts, 2008) • iGxalingenwa Forest (Robertson & Lawes, 2005) • Mngazana mangrove forests (Traynor & Hill, 2008)	• South African coastal & fisheries co-management overview (Hauck & Sowman, 2001) • Dwesa-Cwebe Nature Reserve (Ntshona et al., 2010)

8. **Collaboratively developed knowledge & management capacity of community and partners**	• South African CBNRM overview (Fabricius & Collins, 2007) • South African CBNRM overview (Cundill & Fabricius, 2010) • Nama Contract Park - Richtersveld National Park (Fabricius & Collins, 2007) • Makuleke Contract Park - Kruger National Park (Fabricius & Collins, 2007) • Khomani San and Meir Contract Park – Kalahari Gemsbok National Park (Mannetti et al., 2015).	• South African PFM review (Watts, 2006; Holmes-Watts & Watts, 2008) • Mngazana mangrove forests (Traynor & Hill, 2008) • Umzimvubu District Forestry (Obiri & Lawes, 2002)	• Kosi Bay Gillnet Joint-Management Project (Kyle, 2003) • Sokhulu Subsistence Mussel-Harvesting Project (Harris et al., 2003)
9. **Continuous monitoring, learning and adapting of CBC initiatives through an iterative and community inclusive process**	• South African CBNRM overview (Fabricius & Collins, 2007) • South African CBNRM overview (Cundill & Fabricius, 2010)	• South African PFM review (Holmes-Watts & Watts, 2008) • Mngazana mangrove forests (Traynor & Hill, 2008)	• South African coastal & fisheries co-management overview (Hauck & Sowman, 2001) • Sokhulu Subsistence Mussel-Harvesting Project (Harris et al., 2003)

www.ingramcontent.com/pod-product-compliance
Lightning Source LLC
LaVergne TN
LVHW042342190726
843493LV00005B/899